PHYSICS RESEARCH AND TECHNOLOGY

SPACE-TIME GEOMETRY AND QUANTUM EVENTS

PHYSICS RESEARCH AND TECHNOLOGY

Additional books in this series can be found on Nova's website
under the Series tab.

Additional e-books in this series can be found on Nova's website
under the e-books tab.

PHYSICS RESEARCH AND TECHNOLOGY

SPACE-TIME GEOMETRY AND QUANTUM EVENTS

IGNAZIO LICATA
EDITOR

nova publishers
New York

For permission to use material from this book please contact us:
Telephone 631-231-7269; Fax 631-231-8175
Web Site: http://www.novapublishers.com

Additional color graphics may be available in the e-book version of this book.

Library of Congress Cataloging-in-Publication Data
Available upon request.

ISBN: 978-1-63117-455-1

Published by Nova Science Publishers, Inc. † New York

CONTENTS

PREFACE

IN SEARCH OF THE UNDERLYING STRUCTURES. THE VIEWPOINT OF A BOHMIST

It is well-known that the fundamental problem in contemporary theoretical physics is the "pacific coexistence" between General Relativity and Quantum Mechanics. What it fails to be pointed out is that the proliferating of "everything theories" – whether they are inspired to the philosophy of General Relativity or to the particles physics one – does not solve the problem, sometimes that makes it amplified. Moreover, it is forgotten the difficult conciliation of them both with QFT, which is actually – despite its flaws – the most powerful physics theory we have and the only real "everything theory". It seems to require new ideas suggesting a conceptual and mathematical structure able to overcome "coexistence" and join the two souls of contemporary physics into a single one.

Only then will it be possible to start an authentic unification program of Physis able to connect the manifestations with the history of matter.

As usual in history of Physics, it is a matter which finds its formulation in a well specific mathematical request: conciliating the GR continuous and local language with the discrete and mom-commutative aspects of QM. A crucial point among the others – in our opinion the one which maybe contains the most interesting indications – is related to extending the GR constitutive principle, the equivalence principle, to the quantum world (D. W. Sciama & P. Candelas, 1983). The other big question deals with the "internal" relationships - anything but clarified – between QM and QFT, which is to say between a theory built on a wave/corpuscle dualism, and one where there are neither waves nor particles, but intertwined "field modes", which thing defines the still less explored area of relationships between non-locality and discrete networks of events.

The scenarios of the explorable relationships between classical space-time and quantum land are various: the geometrodynamic one (by a proper extension of geometry [Francaviglia & Macedo, 1991]), the stochastic fractal one (defining a middle land mediated by QFT-like hypotheses), the emergent one (from a physical viewpoint, by the collective behaviours of discrete entities, which mathematically means that the geometry derives from an algebraic structure of events). The writer has a typical Bohmist point of view (Bohmist rather that Bohmian, in the same sense as the word Marxian is opposed to Marxist): we do not believe that there exists a single level of description, but a possible plurality in relation to the range under consideration. Nevertheless, we are convinced that it will be necessary to put the continuous aside and the theory able to pacify QM, QFT and GR will be theory of the relationships between explicate/implicate order where the terms of implication and the general constraints conditioning the space time descriptions will be expressed in the "purely algebraic" language Einstein spoke of in his last year (1955) [Bohm, 2002; see also: Majid, 2008]. It will maybe imply a redefinition of vacuum over and above the Heisenberg fields, as an informational, atemporal and "predynamic" matrix "at the World bottom" [Silberstein et al., 2013, Chiatti, 2008; Majid, 1991], a question that has been put shyly since the first formulations of the holographic principle [Susskind, 1995; Bousso, 2002; see for a review: Davies, 2010].

Without any pretension of exhaustiveness, and away from the most beaten tracks, this anthology includes some of the most significant voices on the problem of the possible relations between the space-time dynamics and the quantum networks of events. As the Editor I'm grateful to all the authors who enthusiastically joined the volume, sometimes undertaking a *tour de force* against time.

I have the pleasure and the honour, devoted to my Bohmist vision of the description levels, to collaborate with most of them.

It is a demanding challenge and we do not know which strategy will be prove to be fecund and where it will lead us, but the old wisdom reminds us that ""It is not your responsibility to finish the work, but you are not free to desist from it either" (Rabbi Tarphon).

Ignazio Licata
Full Prof. of Theor. Phys.
ISEM, Institute for Scientific Methodology, Palermo
and
School of Advanced International Studies on Theoretical and non Linear Methodologies of Physics, Bari, Italy

2013-12-27

References

Bohm, D. *Wholeness and the Implicate Order*, Routledge, Reissue edition, 2002.

Bousso, Raphael (2002), The holographic principle, *Reviews of Modern Physics* 74 (3): 825–874.

Chiatti, L. (2008), Archetypes, Causal Description and Creativity in Natural World, in I. Licata & A. Sakaji, *Physics of Emergence and Organization*, World Scientific, 2008.

Davies, P. (2010), Universe from Bit, in P. Davies & N. H. Gregersen (eds), *Information and the Nature of Reality*, Cambridge Univ. Press, 2010.

Francaviglia, M., Macedo, P., The Evolution of the Concept of Ether and its Underlying Geometry, *Phys. Essays*, 4(3):384-388.

Majid, S (1991), Principle of Representation- Theoretic Self-Duality, *Phys. Essays*, 4(3): 395-405.

Majid, S. (ed), *On Space and Time*, Cambridge Univ. Press, 2008.

Sciama, D.W., Candelas, P. (1983) Is there a quantum equivalence principle? *Phys. Rev. D* 27, 1715–1721.

Silberstein, M., Stuckey W. M., McDevitt, T (2013), Being, Becoming and the Undivided Universe: A Dialogue Between Relational Blockworld and the Implicate Order Concerning the Unification of Relativity and Quantum Theory, *Found. Phys.* 43(4): 502-532.

Susskind, L. (1995) The World as a Hologram, *Journal of Mathematical Physics* 36 (11): 6377–6396.

In: Space-Time Geometry and Quantum Events
Editor: Ignazio Licata, pp. 1-10
ISBN: 978-1-63117-455-1

Chapter 1

SPACE-TIME GEOMETRY AND QUANTUM EVENTS

*Ignazio Licata**
ISEM, PA, Italy
and
Advanced International Studies on Theoretical and non Linear Methodologies of Physics, Bari, Italy

Abstract

Over a span of a few years, Theoretical Physics has widened its horizons. It has outlined new perspectives on classical and quantum systems, discussed the meaning of matter at Planck's wall and the role of Quantum Information. Nevertheless, it seems that such rapid colonization of new territories runs the risk of forgetting some foundational problems which the inner unity of Physics knowledge depend on. It is here that it is taken into consideration the difficult relationship between geometry and dynamics established by the current epistemological arrangement of Quantum Mechanics (QM) and General Relativity (GR). It is also necessary to find classes of languages able to conciliate, on different scales and ranges, the space-time descriptions with the quantum ones, starting from the consolidated paths of the Journey.

1. Classic and Quantum, Global and Local

General Relativity (GR) may be considered the highest achievement of classical Physics, a formidable synthesis of the notions of space and time as a theatre of coordinates, of the description of the field of gravity, and of the equations of motion. Moreover, it has furnished the modern conceptual approach to studying the large-scale structure of space-time and the processes that take place in hyperdense matter (Hawking & Ellis, 1975). The essential form of the tensorial equations of Einsteinian gravity can be written as:

$$G_{\alpha,\beta} = 8\pi T_{\alpha,\beta} \tag{1}$$

* E-mail address: Ignazio.licata@ejtp.info

with G being the Einstein tensor and T the stress-energy tensor.

The peculiarity of eqs (1) leads to an effect of non-linearity – gravity gravitates! – which makes it very difficult to apply the traditional "local" processes of quantization to Einstein's space-time curve. Moreover, a much more radical effect of non-locality is at the center of Quantum Mechanics (QM), the theory that threw many of the certainties of the classical image about the World into crisis and that provided the tools for the investigation of Particle Physics and condensed matter (see for ex. Hughes, 1992).In the EPR-Bell phenomena, in fact, objects separated in space and time show space-like correlations that are inexplicable in classical terms, and that characterize the quantum statistics from which the most interesting properties of the structure of matter derive. Since the non-local correlations do not transport energy, they do not violate Relativity but nonetheless remain outside the bounds of the classical picture of the world. This situation goes under the name of "pacific coexistence" between Relativity and Quantum Physics (Shimony). The problem of reconciling the classical image of Einstein's space-time and quantum non-locality is that of constructing a dynamic model of space-time in which the quantum processes also find a place, or that of developing a quantum geometrodynamics. This line clearly takes its inspiration from the philosophical and geometrical approach of General Relativity.

2. Quantum Potential and Active Information

The most recent attempts at a geometric approach to quantum processes are owed to Wheeler (Wheeler, 1990) based on Weyl's geometry, and then to Wood and Papini (Wood & Papini, 1995) using a modification of the Weyl-Dirac theory. These theories have suggested that QM could be incorporated as a corresponding "deformation" of space-time. More recently Sidharth (Sidharth, 2002) has proposed a geometric interpretation ofQM on the basis of non-commutative and non-integrable geometries. Nonetheless, all of these attempts seem to elude the epistemological core of QM Standard Interpretation. As Heisenberg observed at the dawning of quantum theory, Copenhagen's interpretation is radically a-causal, and quantum processes cannot therefore be retraced to a space-time vision. This profoundly limits every attempt to understand Quantum Physics within the traditional Einsteinianspace-time arena, however "extended" it may be.

In recent years interest in D. Bohm's interpretation of QM has grown. Here let's recall that Bohm's interpretation reproduces all the results of QM without any ambiguity regarding the role of the observer and allows an easy extension to the Field Theory formalism (Bohm & Hiley, 1995; Durr et al. 2004; Nikolic, 2007, Licata & Fiscaletti, 2014). In this interpretation non-locality is not an "unexpected visitor," as in the standard interpretation, but derives directly from quantum potential Q, a necessary term for the conservation of energy in Schrödinger's equation:

$$Q = -\frac{\hbar^2}{2m}\frac{\nabla^2 R}{R} \tag{2}$$

The form of potential (2) reveals some interesting properties: it depends on the amplitude R of the wave function and its action is like-space, exactly that called for in the EPR-Bell processes. Quantum potential contains global information of physical processes, defined as "active information" by Bohm, or the contextual information of the system under observation and its environment, information that is not "external" to space-time but should rather be considered a type of geometric information "woven" into space-time itself. It is, further, a dynamic entity. Paraphrasing J.A. Wheeler's famous saying about GR, we can say that the evolution of the state of a quantum system changes active global information, and this in turn influences the state of the quantum system, redesigning the non-local geometry of the universe.

3. Geometries of Non-Locality

The geometry subtending quantum potential has been explored by various authors (see Carroll, 2006). One very interesting result is that of Shojai & Shojai (Shojai & Shojai, 2004), who studied the behavior of particles at spin 0 in a space-time curve, demonstrating that quantum potential contributes to the curvature that is added to the classic one and that reveals profound and unexpected connections between gravity and quantum phenomena. All of this is expressed by a metric conformal such as:

$$\tilde{g}_{\mu\nu} = \frac{M^2}{m^2} g_{\mu\nu} \tag{3}$$

where the expression for mass is: $M^2 = m^2 \exp Q$, with Q as the quantum potential.

This is a perfect image of quantum geometrodynamics that combines the gravitational and quantum aspects of matter, at least in terms of the level of macroscopic description of physical processes. In reality, once again, things are not so simple. Non-locality remains a phenomenon that rests uncomfortably with a "mechanical" vision of the universe, and it is not by chance that Bohm referred to QM and its interpretation as *quantum non-mechanics*, to reiterate that they could not be in any way understood as a return to the classical, but rather as the partial recovery of a "fuzzy realism." (Bosca, 2013 Hiley et al., 2000). Taking up the ideas of Heisenberg, G. Chew and D. Finkelstein (Finkelstein, 1987; 1996; 2013), the Birbeck group (Hiley& Monk, 1998; Brown, 2002) demonstrated that the entire symplectic non-commutative geometry identified by quantum potential can be derived from Weyl's discrete algebra. The recent works by Hiley on Wigner-Moyal distributions are particularly interesting, they create a solid bridge with the non-commutative structure of dissipative QFT (Hiley, 2010).

In more directly physical terms, this means that there are two epistemological interventions to do towards Eddington's quantum sheet of geometrodynamics:

a) it is taken as primary and non-local, and therefore it is necessary to introduce additional hypotheses about its deep structure, or
b) the space-time manifold must be considered an emergence of the deepest processes situated at the level of quantum gravity.

We have to remember here, at least, the original proposal of Sacharov of deducing the gravity as "metric elasticity" of quantum vacuum (Sacharov, 1968; Visser, 2002) and more recent one by Consoli on ultra-weak excitations in a condensed as a model for the gravity and Higgs mechanism (Consoli, 2009).An interesting synthesis on the “discrete condensate” and one of the most convincing scenario to describe the emergence of classical space-time from the collective behaviour of discrete, pre-geometric atoms of quantum space can be found in Oriti (Oriti, 2013; 2011)

Using the now famous image of complementarity in D. Bohm’s version, we can say that the entire connected and local structure of both space-time and the Shannon-Turing information we use to compute the events in it is the explicit order of a hidden, implicit order, which acts as a “fabric of reality” at a sub-quantum level, fundamentally discrete and non-commutative (Licata, 2008).

4. The Quantum Foam of Implicit Order

The idea of a structure of relations subtending the observable forms of matter, energy and space-time was defined by J. A. Wheeler as “quantum foam,” with the precise intent of evoking the erosion of traditional notions toward the Planck scale typical of quantum gravity. Despite the ongoing lack of a strong unifying principle, the various versions of String theory have had a certain success in overcoming some of the impasses of Particle Physics, and it has been suggested that space-time manifold is the result of the interaction between p-branes, and that the acquisition of the masses in Higg’s Ocean finds its natural explanation in the mechanisms of uncurling and compactification (see Riotto, 2000; Sundrum& Randall, 1999). In reality, the majority of the versions of strings works, just like Quantum Field Theory, which is its closest relative, with a flat Minkowski space-time , while a correct, authentically relativistic (in the sense of GR) theory should be independent from the background, or not presuppose any metric signature. Various theories have these requirements. One is Penrose’s Twistor theory (Huggett & Todd, 1994). To use Penrose’s own words, “a twist or is an object similar to the two-faced Janus, unitary but with one face turned toward QM and the other toward GR.” More precisely a twistor is and object without mass and charge and with spin, invariant for the conformal group, so as to find again the light-cone of Minkowski’s space-time. The famous representation by Robinson is based on a stereographic projection of Clifford’s algebra that defines the structure of twistors and allows the essential characteristics of the dynamic non-local “fragments” of space-time to be intuitively taken from it. Another very elegant theory that has the right relativistic requirements is Rovelli and Smolin’s “Loop quantum gravity” (Rovelli, 2007). The loops are closed field lines that do not depend on the coordinate system and therefore provide the basis for a relational description of space-time in the spirit of Leibniz. The theory presupposes a very particular space-time structure at the Planck scale: the operators associated with area and volume in fact have a discrete spectrum, giving birth to a complex and fascinating graph structure and thus furnishing a discrete combinatorial view of Physics.

We have to mention also some interesting attempts to model a quantized space-time as a crystal lattice (Licata, 1991; Kleinert et al. 2010) and the Preparata Plank Lattice where the quantum foam structure itself acts as the Higgs mechanism and allows the emerging of a spectrum of masses selected by the lattice (Preparata & Xue, 1994).The idea of new constraints between micro and macro leads to the Double Special Relativity (see for. ex. Nassif, 2012).

5. A Radical Criticism to "EverythingTheories"

Holger Nielsen has directed a radical criticism at Everything Theories in his Random Dynamics (Nielsen, 1989; Gaeta, 1993). The key idea is quite simple, in his own words: "Could the fundamental physical "laws" be enormously complicated, but our well-know laws come out in a limit?". Nielsen observed that any directly verifiable statement about physical world – from experimental view-point - is structurally connected to Yang-Mills Theories and Gauge Symmetries. Consequently, what we can say about the "fundamental constituents" of the world at an actually inaccessible range is they are compatible with some very general mathematical structures. The same essential lesson seems to come out of Garret Lisi's "Exceptionally Simple" Theory of Everything (Garret Lisi, 2007). In the frame of a radical emergentist approach, Robert Laughlin starts from the instability of Yang-Mills equations to criticize any fundamentalist nomological attitude: *if a strategy to solve such equations is adopted, it should be better not to speak of a Theory of Everything, but just of patenting a technology) to calculate them.*

A reasonable and provisional conclusion deriving from such reflections is that we should not to think the world structure in terms of fundamental objects, but rather as informational patterns acting as matrices generating the physical processes.

6. Who Needs QM Interpretations?

We could now ask whether it is not the case to examine the QM foundational problems from another viewpoint. It is not from the past that a modern conductor reaches Mozart – however good as a philologist he may be -, but from the historical understanding of his legacy. Analogously we should maybe analyze QM from its ripest fruit: the Quantum Field Theory (QFT) that is indeed considered the nucleus of the early "Theory of Everything" (Srednicki, 2007). Most of the interpretative debate is still centered on 1927 Schrödinger equation that is surely a very useful formulation but loaded with a classical burden responsible for the so-called wave-particle dualism. Recently, M. Cini (Cini, 2003) proposed to come back to the P. Jordan original approach, so deriving all the QM characteristics from Planck's field quantization and the consequent uncertainty principle. The traditional statistical properties thus derive from the Wigner-Feynman pseudo-probabilities without any reference to "first quantization" and its tough conceptual problems (see Feynman, 1987).

G. Preparata (1942 – 2000) has followed the same line by his "realistic interpretation" completely based on QFT (Preparata, 2002).It is shown that QM is a discrete approximation of QFT for dilute systems, but has extremely significant effects on the instability of quantum vacuum for the emerging of condensed states of coherence.

The above-mentioned two simple cases show not only that QM gets rid of any "Alice-in-wonderland" features if we look at it from the superior viewpoint of QFT, but also that it has a greatly explicative potential. The old wave/particle dualism simply becomes the consequence of continuity/discontinuity aspects between the field modes, non-locally "intertwined" and obviously subjected to superposition and interference phenomena. The detection of a quantum object ("collapse") is nothing but the "local" click of a quantum within an apparatus according to Planck. Let's remember that it is due to the heuristic images of Einstein and Thompson if the idea to consider a quantum as a localized particle became was established as tradition.

A breath of fresh air is represented by the new QM transactional interpretations (Chiatti, 1994; 1995; Kastner, 2013; Licata; 2013). In this theory, a legacy from the semi-classical approach by Cramer –Wheeler-Feynman, the space-time of classical physics, the "traditional" QM and the huge richness of QFT and its manifestations, emerge from a *network of non-local transactions* which so becomes the *load bearing structure* of the physical world. These are not exotic events, the transactions are already at the origin of Quantum Theory, as the famous *quantum jumps* in Bohr atom. The theory has innovative and cogent implications both for Quantum Gravity and Cosmology and includes QM and QFT in a natural way and harmonizes them in a unitary structure. In particular, it is possible to derive the traditional machinery: Schrödinger equation, fluctuations, path integrals and Feynman diagrams. The novelty, and the price to pay, is the introduction of an *atemporal element* in apparatus, and in general in the description of the physical world. The transactional network constitutes the *logic fabric of the physic world*: dynamics and locality, just like the space-time where they are defined, become *emergent* characteristics (Silberstein et al., 2013).

In this way QM, far from being a baffling puzzle, is the unsteady historical and conceptual passage unifying the early Quantum Theory to the strongly powerful QFT. So the fundamental questions shift to cosmological level, on the origin, the boundary conditions and the evolution of the quantum informational fabric of the Universe.

7. Quantum Information and Cosmology

Any theory of interactions is not complete without a general scenario where to set it. The importance of cosmology for the physics of elementary particles has become evident with the developing of Gauge theories, where the unification project strictly depends on the ranges of the temperatures of the Universe history.

Recently, one of the fundamental goals for theoretical physicists has been to merge Einstein cosmology and Quantum Physics into a single frame (Hartle& Hawking, 1983; Vilekin, 1984). Different attempts *ad hoc* have shed a new light on the De Sitter model and the cosmological constant role (Einstein biggest mistake!). Although a general agreement has not been reached yet, the old Big Bang conception as a "thermodynamic balloon" seems to be irreparably compromised by now. We just quote here the Author and L. Chiatti work on Archaic Universe where the starting point is the quantum improvement of Fantappiè-Arcidiacono group approach based on DeSitter Universe. The elimination ofthe initial singularity and the adoption of DeSitter 5 hyper-sphere as the quantum vacuum's geometrical shape make possible a very concise description of the boundary conditions necessary for the evolution of the observed physical universe. Such "pre-space" we define as "archaic" has not to be considered as antecedent to "Big-Bang", but rather as a spatial and a-temporal substrate of the usual space-time metric containing *in nuce* all the evolutionary possibilities that the General Projective Relativity (GPR) equations indicate. After eliminating any geometrical singularity with Euclidean substrate, the description of the Universe evolution can be seen as an extended nucleation from a coherent state with very high non-local information to an observable mix of matter-energy. The passage from the archaic to the evolutionary state is defined by a sort of "holomovement" (Bohm, 1995) due to a Wick rotation which characterizes the appearence of the dynamics and time arrow starting from the general constraints on the pre-dynamic, archaic condition. It is remarkable that the structure itself of the theory simplifies any speculation about dark matter and inflation, and gives a purely geometrical description to the cosmological constant (Licata, 2006; Licata &Chiatti, 2009, 2010).

So, Archaic Quantum Information fixes the broadest "matrix of reality" compatible with experimental observations, requires a derivation of the usual field theory from an algebraic and topological theory and preludes to an ambitious project of 5-dimensonal unification between space and matter.

Such scenario is not incompatible with a recent suggestion by A. Valentini (2002; 2009). The quantum phase we observe now, characterized by the Born rule $\rho = |\psi|^2$, could be the fossil of a previous very high correlation phase where non-locality could has allowed the hypercomputational processes and played a decisive role in the formation of "frozen" structures nearly to the threshold of the physics of living systems. Besides, it is patent by now that as soon as we remain within the "Turing Cage" the morphogenic possibilities of quantum information will not be fully comprehended (Licata, 2008; 2010; Blume-Kohout & Zurek, 2006; Davies et al., 2009).

8. Beyond: Heraclitean and Parmenedian Aspects in Contemporary Physics

A rapid review of the relationships between the explicate order of space-time manifold and the theories that investigate the fine structure of quantum foam invites interesting reflection, both epistemological and cognitive. The entire history of Physics may be considered a progressive refinement of the models of space-time , from Newton's absolute one to non-Euclidean and conformal geometries of the various classical and quantum geometrodynamics. All of these theories are characterized by the notion of "process," understood as the evolution of a set of observables in space and in time. Quantum Physics has created an irreversible leak in the self-cohesion of such kind of view of the world, and the exploration of quantum gravity seems to propose the introduction of new geometric and algebraic structures that identify the weaving of relationships in the implicit order from which the very concepts of space, time, and evolution emerge. D. Finkelstein wrote to the author of these notes, some times ago: "*Weizsäcker put Parmenides into the quantum theory instead of relativity where he belongs. I only realized rather recently how Parmenides still limits us today (...) I have been wondering who to pair against Parmenides. Someone who puts trial-and-error before axioms, or doxis before Logos, as source of the little knowledge that is possible. Usually I use Cusano, C S Peirce, Buddha and Heisenberg. Maybe it should be Heraclitus*"

A unitary vision of the relationship between GR and QM will require, therefore, new conceptual terms to describe the deep complementarity between the Heraclitean and Parmenedian aspects of the physical world.

References

Arcidiacono, G. (1976), A New Projective Relativity based on the De Sitter Universe, *Gen. Rel.Grav.*,7, 885-889.

Blume-Kohout, R. &Zurek, W.H. (2006), Quantum Darwinism: Entanglement, Branches, and the Emergent Classicality of Redundantly Stored Quantum Information, *Phys. Rev.* A 73, 062310.

Bohm, D, Hiley, B. (1995), The Undivided Universe, Routledge.

Bosca' M. C. (2013) Some Observations upon "Realistic" Trajectories in Bohmian Quantum Mechanics, *Theoria* 76: 45-60.

Brown, M. R. (2002), *The quantum potential: the breakdown of classical symplectic symmetry and the energy of localisation and dispersion,* http://arxiv.org/abs/quant-ph/9703007v3.

Carroll, R. (2006), *Fluctuation, Information, Gravity and the Quantum Potential,* Springer.

Chiatti, L.(1994) Wave Function Structure and Transactional Interpretation, in *Waves and Particles in Light and Matter*, A. van der Meerwe, A. Garuccio Eds, Springer,181-187.

Chiatti, L. (1995) Path integral and transactional interpretation, *Found. Phys.,* 25 (3):481-490.

Consoli, M. (2009), Ultraweak Excitations of the Quantum Vacuum as Physical Models of Gravity, *Class & Quantum Gravity,* vol. 26, 225008.

Davies,P.C.W., Abbott, D., Pati, A.K. (Eds), *Quantum Aspects of Life*, Imperial College Press, 2009.

Durr,D., Goldstein, S, Tumulka, R., Zanghì, N. (2004), Bohmian Mechanics and Quantum Field Theory, *Phys. Rev. Letters,* 93,090402.

Feynman, R. (1987), Negative Probabilities, in *Quantum Implications. Essays in Honour of David Bohm* (B. J. Hiley& F. D. Peat Eds), Rutledge & Kegan Paul, London.

Finkelstein, D. (1987), All is Flux, in *Quantum Implications: Essays in Honour of David Bohm* (B. J. Hiley& F. D. Peat Eds) Routledge & Kegan Paul.

Finkelstein, D., *Quantum Relativity*, Springer, 1996.

Finkelstein, D. (2013), Palev Statistics and the Chronon, in Lie Theory and Its Applications in Physics (V. Dobreved), *Springer Proceedings in Mathematics & Statistics* Vol. 36:25-38.

Gaeta, G. (1993), Breaking of Permutation Symmetry and Diagonal Group Action: Nielsen Model and the Standard Model as Low-Energy Limit, *Int. Journ. of Theor. Phys.,* 32, 5.

Garrett-Lisi, A(2007), *An Exceptionally Simple Theory of Everything,* http://arxiv.org/pdf/0711.0770.

Hartle, J., Hawking, S. (1983), Wave Function of the Universe, *Physical Review* D 28 2960.

Hawking, S. Ellis, G. (1975), *The Large Scale Structure of Space-time* , Cambridge Univ. Press.

Hiley, B., Monk, N. A. M. (1998), A Unified Algebraic Approach to Quantum Theory, in *Found. Phys. Lett.,* 11, 4, 371-377.

Hiley, B., J., Callaghan, R.E., Maroney, O. (2000) *Quantum trajectories, real, surreal or an approximation to a deeper process?* arXiv:quant-ph/0010020v2.

Hiley, B., Callaghan R.E. (2010), *The Clifford Algebra approach to Quantum Mechanics A: The Schroedinger and Pauli Particles*; http://arxiv.org/abs/1011.4031 The Clifford Algebra Approach to Quantum Mechanics B: The Dirac Particle and its relation to the Bohm Approach, http://arxiv.org/abs/1011.4033.

Hiley, B. (2010), On the Relationship Between the Wigner-Moyal and Bohm Approaches to Quantum Mechanics: A Step to a More General Theory?, *Found. Phys.* 40 (4): 356-367.

Hughes, R. (1992), *The Structure and Interpretation of Quantum Mechanics*, Harvard Univ. Press.

Huggett, S. A., Todd, K. P. (1994) *An Introduction to Twistor Theory*, Cambridge Univ. Press.

Kastner, R. (2013), *The Transactional Interpretation of Quantum Mechanics*, Cambridge Univ. Press.

Kleinert, H., Jizba, P., Scardigli, F. (2010), Uncertainty Relation on a World Crystal and its Applications to Micro Black Holes, *Phys. Rev.*D81, 084030.

Laughlin, R. (2005) *A Different Universe: Reinventing Physics from the Bottom Down*, Basic Books.

Licata, I. (1991), Minkowski Space- Time and Dirac Vacuum as Ultrareferential Fundamental Frame, *Hadr. J.,* 14: 225-250.

Licata, I (2006), Universe Without Singularities. A Group Approach to De Sitter Cosmology, *Electronic Journal of Theoretical Physics,* Vol.3, No10, 211-224, also in "Majorana Legacy in Contemporary Physics", Ignazio Licata Ed., EJTP/Di Renzo, Roma, 2006.

Licata, I (2008), Emergence and Computation at the Edge of Classical and Quantum Systems, in Licata, I. &Sakaji, A. (eds), *Physics of Emergence and Organization*, World Scientific, 2008.

Licata, I, Chiatti, L. (2009),The Archaic Universe: Big Bang, Cosmological Term and the Quantum Origin of Time in Projective Cosmology, *Int. Journ. of Theor. Physics,* 48, 4, 1003-1.

Licata, I., Chiatti, L. (2010), Archaic Universe and Cosmological Model: "Big-Bang" as Nucleation by Vacuum, *Inter. Jour. of Theor. Phys.*, 49, 10, 2379-2402.

Licata, I. (2013) Transaction and Non Locality in Quantum Field Theory, *Europ. Jour. of Phys.,* 60, *in press.*

Licata, I., Fiscaletti, D. *Quantum Potential. Physics, Geometry, Algebra*, Springer, 2014.

Linde, A. (1979), Phase Transitions in Gauge Theories and Cosmology, *Rep. Prog. Phys.* 42, 389.

Nassif, C. (2012), Double Special Relativity with a minimum speed and the Uncertainty Principle, *Int. J. Mod. Phys.* D, 21, 1250010.

Nielsen, H. (1989),Random Dynamics and Relations between the Number of Fermion Generation and the Fine Structure Constant, *Acta Physica Polonica,*B20, 5.

Nikolic, H.(2007), Bohmian Mechanics in Relativistic Quantum Mechanics, Quantum Field Theory and String Theory, *J. Phys.: Conf. Ser.* 67 012035.

Oriti, D (2013), *Disappearance and emergence of space and time in quantum gravity,* arXiv:1302.2849 [physics.hist-ph].

Oriti, D. (2011), *On the depth of quantum space,*arXiv:1107.4534 [physics.pop-ph].

Preparata, G, Xue, She Sheng (1994), *Quantum gravity, the Planck lattice and the Standard* Model, http://arxiv.org/abs/hep-th/9503102.

Preparata, G. (2002), *An Introduction to a Realistic Quantum Physics,* World Scientific, Singapore.

Randall, L., Sundrum, R. (1999), Large Mass Hierarchy from a Small Extra Dimension, in *Physical Review Letters,* 83, 17, 3370–3373.

Riotto, A (2000), *D*-branes, string cosmology, and large extra dimensions, in *Phys. Rev.* D 61, 123506.

Rovelli, C. (2007), *Quantum Gravity*, Cambridge Univ. Press.

Sacharov, A (1968) Vacuum Quantum Fluctuations in Curved Space and The Theory of Gravitation, *Sov. Phys. Dokl.*12 1040.

Shojai, A., Shojai, F. (2004), Constraint algebra and equations of motion in the Bohmian interpretation of quantum gravity, in *Class. Quantum Grav.*21 1-9.

Sidharth, B. G. (2002), *Geometry and Quantum Mechanics*, http://arxiv.org/abs/physics/0211012.

Silberstein, M., Stuckey W. M., McDevitt, T (2013),Being, Becoming and the Undivided Universe: A Dialogue Between Relational Blockworld and the Implicate Order Concerning the Unification of Relativity and Quantum Theory, *Found. Phys.* 43(4): 502-532.

Srednicki, M. (2007) *Quantum Field Theory*, Cambridge University Press.

Wheeler, T. (1990), Quantum measurement and geometry, *Phys. Rev.* D 41, 431 – 441.

Wood, W.R., Papini, G. (1995), A geometric approach to the quantum mechanics of de-Broglie–Bohm and Vigier", in The present status of quantum theory of light, *Proc. of Symposium in honor of J.P. Vigier,* York University.

Valentini, A. (2002), Subquantum Information and Computation, *Pramana J. Physics,* 59(2), 269–277.

Valentini, A. (2009), Beyond the Quantum, Physics World, November 2009, 32—37.

Vilenkin, A. (1984),Quantum Creation of Universes, *Phys. Rev.* D 30, 509–511.

Visser, M., (2002), Sacharov's Induced Gravity: A Modern Perspective, *Mod. Phys. Lett.* A17, 977-992.

Vitiello, G. (2005), Classical Trajectories and Quantum Field Theory, *Brazilian Journal of Physics,* vol. 35. no. 2A.

In: Space-Time Geometry and Quantum Events
Editor: Ignazio Licata, pp. 11-43
ISBN: 978-1-63117-455-1

Chapter 2

The Transaction as a Quantum Concept

Leonardo Chiatti*
AUSL VT Medical Physics Laboratory, Viterbo (Italy)

Abstract

This essay presents a novel approach to the concept of "transaction" in quantum physics. The central ideas of this approach were outlined by this author in two essays in the 1990s [1, 2], while a more detailed treatment was published in a volume in 2005 [3]. Breaking with Cramer's original theory, the transaction is not connected to the simultaneously retarded and advanced spacetime propagation of classical fields, as in the spirit of Wheeler-Feynman electrodynamics. Instead, the transaction is seen as an archetypal structure intrinsic within the quantum formalism. The present approach is advantageous in that, while preserving the essential point of Cramer's theory, it is also fully consistent with the standard quantum formalism. In particular, it has the advantage of avoiding the introduction of elements which are completely extraneous to quantum formalism (such as the propagation of real physical waves in four-dimensional spacetime, the phase difference between offer and confirmation waves which is necessary for the elimination of "tails", the echoing mechanism, etc.) and which have led to misunderstandings, such as Maudlin's objection. Furthermore, this approach elucidates the relationship between transactional mechanism and block universe, implicate order, and acausality.

1. Introduction

Transaction is normally associated with a particular interpretation of quantum formalism, the so called Transactional Interpretation (TI) introduced by Professor J.G. Cramer in the 1980s [4,5,6]. All interpretations of quantum formalism introduce elements extraneous to the formalism itself. These elements (which can be broadly associated with Bell's "beables") are introduced with the aim of making quantum formalism understandable by placing it in the context of a given ontology.

* E-mail address: fisica1.san@asl.vt.it

In TI the basic quantum process - which includes the preparation of the initial quantum state, the detection of the final quantum state and the connection between these two events - is described through the introduction of the following beables:

- A retarded offer wave propagating from the preparation event towards the possible detection events;
- A source of the aforementioned wave, essentially consisting of the preparation event;
- An advanced confirmation wave propagating from the detection event to the various possible preparation events;
- A source of the aforementioned wave, consisting of the actualized detection event.

Transaction is the feedback (or handshake) procedure that closes the loop between the effectively realized preparation and detection events. Such a procedure is clearly non-local, albeit in principle relativistically invariant. It can take place in one step or become completed in a finite number of echoes in a sort of pseudotime. Apart from this latter detail, this scheme resembles that of Wheeler-Feynman's electrodynamics [7,8], which is a pre-quantum or classical theory. Cramer's aim therefore is to derive the basic quantum process by using classical pre-quantum concepts, especially in those aspects that deviate from classical theory the most: the quantum localization and the randomness of the final (initial) state with an assigned initial (final) state.

It must be noted that the basic quantum process is, according to TI, a dynamic process taking place in spacetime. The offer and confirmation waves propagate in spacetime and the preparation and detection events are localized in spacetime. In the prevalent terminology during the 1980's, the TI can be defined as a non-local realist interpretation.

In order to achieve his aim, Cramer attributes to his beables certain patterns of behavior. Beside the above mentioned echoing mechanism, Cramer assumes that the confirmation wave is out of phase with respect to the offer wave, so that the tails (those regions of the offer wave that follow the detection, and those regions of the confirmation wave that precede the preparation) cancel each other via destructive interference. Cramer also assumes that the two waves can be represented as a superposition of the eigenfunctions of the pertinent observables, with coefficients that are exactly those same necessary to ensure compliance with Born's rule. This latter requirement is not altogether obvious and leads to difficulties that have been clearly evidentiated by Maudlin [9] and discussed by Berkovitz [10] and Kastner [11].

This essay intends to show that once Cramer's time-symmetrical point of view is adopted, the formal structure of the transactional loop is already expressed by the current quantum formalism, making the introduction of the beables unnecessary. As a matter of fact, transaction is already present in formalism and is fully consistent with it, so that special interpretations such as TI are unnecessary. Given this, Cramer's solution should be regarded not so much as the proposal of a new interpretation, but rather as the discovery of a formal structure underlying the quantum theory that has previously gone unrecognized or at least underestimated.

The necessary step seems to be rather the adoption of an ontology of the basic quantum process strictly consistent with formalism and its current metatheoretical interpretation. Therefore, only the events of preparation of the initial state (creation) and of detection of the final state (annihilation or destruction) are assumed to be real physical events. The connection between these two events is ensured by their common origin in an atemporal and aspatial background outside spacetime. The process amplitudes are a mathematical fiction suited to calculate the statistics of the connection itself. Accordingly, the virtual processes obtained by expanding these amplitudes into partial amplitudes are also mathematical fiction. The acausality of the basic quantum porcess is intrinsic to its emergence from a background which is outside the spacetime. The dynamics of the basic quantum process is in reality *a statics*: the forward and backward transition amplitudes (which substitute here the Cramer's offer and confirmation waves respectively) are not associated with any propagation into spacetime.

Strictly speaking, only the events of creation/annihilation associated with an actual localization in the spacetime can be connected to the pointevents of the relativistic chronotope; generally, however, such events cannot be represented in spacetime. Transaction is therefore an archetypal structure, functioning as a bridge between two levels of physical reality: one is synchronic (a sort of *arché*[1]) and the other diachronic. This bridge, of course, cannot be exclusively represented on the diachronic level by means of purely diachronic mechanisms, albeit time-symmetrical, such as those envisioned in traditional TI.

Research along these lines should focus on the definition of a theory postulating the emergence of physical reality from the *arché*. A complete derivation of quantum formalism, of quantum dynamics and of transactional mechanism should be a particular aspect of such general theory. This is actually the program that Bohm had already sketched out when dealing with clarifying the relationship between *implicate* and *explicate order* [12]. In what follows, I will confine myself to illustrating some aspects of the relationship between the transactional mechanism and quantum statistics while I remind to ref. [3] for an extensive presentation of a possible example of such a derivation.

The outline of the present essays is as follows: Section 2 contains an overview of transaction as a structure built in the quantum formalism. Section 3 deals in details with some logical elements of the problem. Section 4 shows the derivation of quantum statistics. Owing to the limits of this essay dynamics laws are not worked out; the reader is referred to another work [3] for this subject. Section 5 contains a simple justification of the well-known correspondence between thermodynamics and quantum mechanics, which is here taken to be a rule of the *unfolding* of the transactional loop.

2. Transactions: An Overview

We assume that the only truly existent "thing" in the physical world are the events of creation and destruction (or, if one prefers, physical manifestation and demanifestation) of certain qualities. In the language of QFT these events are the "interaction vertices", while the different sets of manifested/demanifested qualities in the same vertex are the "quanta".

As an example, in a certain vertex a photon (E, p, s) can be created, where E is the (created) energy of the photon, p is the (created) impulse of the photon and s is the (created) spin of the photon. In a subsequent event this photon can be absorbed and this corresponds to a packet of properties (-E, -p, -s) where –E is the (absorbed) energy of the photon, - p is the (absorbed) impulse of the photon and – s the (absorbed) spin of the photon; (-E, -p, -s) is the absorbed, i.e. destroyed, photon. It is assumed that the first event chronologically precedes the

[1] Greek for "origin" or "first cause".

second, and that E is positive. A process is therefore being described in which a positive quantity of energy is created and then destroyed. In an absolutely symmetrical manner, it can be said that we are describing a process which proceeds backwards in time and during which a share of negative energy –E is firstly created (in the second event) and then destroyed (in the first event). The two descriptions are absolutely equivalent.

There are two reasons for which, in the process described, one cannot have $E < 0$, i.e. the propagation of a positive energy photon towards the past. A photon with $E > 0$ which retropropagates towards the past, yielding energy to the atoms of the medium through which it travels (such as, say, a X photon which ionizes the matter through which it travels), is seen by an observer proceeding forward in time as a photon with $E < 0$ which *absorbs* energy from the medium (13). This photon would be spontaneously "created" by subtracting energy from the medium, through a spontaneous coordination of uncorrelated atomic movements which is statistically implausible, and has never been observed experimentally.

From a theoretical point of view, the probability of the occurrence of a creation/destruction event for a quantum Q in a point event x is linked to the probability amplitude $\Psi_Q(x)$, which can be a spinor of any degree. Each component $\Psi_{Q,i}(x)$ of this spinor satisfies the Klein-Gordon quantum relativistic equation $(-\hbar^2 \partial^\mu \partial_\mu + m^2c^2)\,\Psi_{Q,i}(x) = 0$, where m is the mass of the quantum[2]. At the non-relativistic limit, this equation becomes a pair of Schrödinger equations (14):

$$-\frac{\hbar^2}{2m}\,\Delta\Psi_{Q,i}(x) = i\hbar\,\partial_t\,\Psi_{Q,i}(x) \tag{1}$$

$$-\frac{\hbar^2}{2m}\,\Delta\Psi^*_{Q,i}(x) = -\,i\hbar\,\partial_t\,\Psi^*_{Q,i}(x) \tag{2}$$

The first equation has only retarded solutions, which classically correspond to a material point with impulse p and kinetic energy $E = p\cdot p/2m > 0$. The second equation has only advanced solutions, which correspond to a material point with kinetic energy $E = -\,p\cdot p/2m < 0$. Thus there are no true causal propagations from the future.

One may wonder whether eq. (2) can lead to hidden advanced effects. The answer to this question is affirmative. To understand this topic, one ought to reconsider the photon example seen above. The creation of the $E > 0$ energy followed by its subsequent absorption and, conversely, the creation of a $-E < 0$ energy preceded by its destruction are clearly two different descriptions of the same process. This however is true so long as the interaction vertices are considered i.e., according to the here adopted ontology, the true substance of the physical world[3].

From the point of view of the dynamical laws for the probability amplitudes of these events, matters are quite different, however. The creation of quality Q is associated with the

---

[2] It is necessarily positive, because it is a measure of the energy which must be released in order to create the quantum; a physical entity can certainly not be created in the vacuum by subtracting energy from it.

[3] As we will see in the following, the term "interaction vertex" has to be intended as referred to an elemental event of interaction between real particles, or more generally to an elemental event involving an objective "reduction" of the quantum state.

initial condition for $\Psi_{Q,i}(x)$ in eq. (1); the destruction of quality Q is associated with the "initial", actually the final, condition for $\Psi^*_{Q,i}(x)$ in eq. (2). In general, however, the two conditions are different and therefore generate different solutions for the two equations, which are not necessarily mutual complex conjugates.

It is a fundamental fact that the destruction event is not described by eq. (1) and that the creation event is not described by eq. (2); this remains true even if the Hamiltonians of interaction with the remaining matter are introduced into the two equations. Thus, at the dynamic laws level, the process of the creation and destruction of Q is completely described solely by the loop:

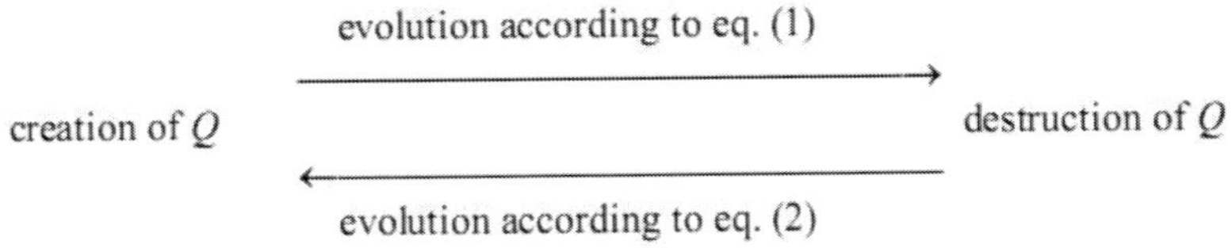

and not only by the upper or lower half-loop.

More generally speaking, we shall have at $t = t_1$ the event of the creation-destruction of a quality Q ($|Q><Q|$) and at $t = t_2$ the event of the creation-destruction of a quality R ($|R><R|$). These two processes will be linked by a time evolution operator S according to the ring:

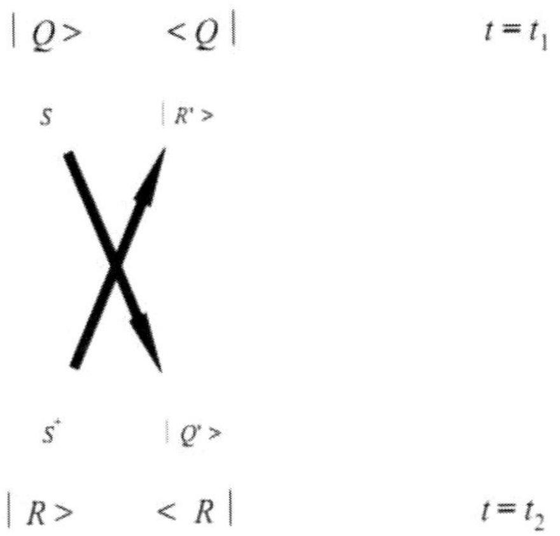

$$
\begin{array}{ccc}
|Q> & <Q| & t=t_1 \\
S & |R'> & \\
& \times & \quad (3)\\
S^+ & |Q'> & \\
|R> & <R| & t=t_2
\end{array}
$$

In other words, $|Q>$ is transported from S into $|Q'>$ and projected onto $<R|$, $|R>$ is transported by S^+ into $|R'>$ and projected onto $<Q|$. The amplitudes product:

$$<R|S|Q><Q|S^+|R> = |<R|S|Q>|^2$$

is immediately obtained, which is the probability of the entire process. If quality Q is constituted by a complete set of constants of motion then $R = Q$ and this is the type of process which can describe the propagation of a photon-type quantum[4], otherwise it is the generic process of the creation of a quality Q causally linked (by means of S) to the destruction of a quality R. Moving to the representation of the coordinates, by substituting bras and kets with wavefunctions, we once again obtain as a particular case the result already seen with the non-relativistic expressions (1), (2). The ring process described above will be called *transaction*; within quantum formalism, it plays the role that in Cramer's TI is reserved to the response of confirmation waves to an offer wave. The transaction exists if the propagation S^{+} is as "real" as the S; to ascertain this, one must see whether experimental situations exist in which the initial condition $|Q\rangle$ and the final one $\langle R|$ are connected in a nonlocal way. As is well known, the answer is affirmative: the phenomena EPR and GHZ [15] are highly valid examples of quantum-mechanical predictions which violate locality. In the case of EPR phenomena there is now confirmed solid experimental evidence [16-19].

Some remarks.

1) Quantum nonlocal effects do not follow from the forward propagation S only. However, the existence of such effects is well established at present, which implies that another factor is involved. The existence of advanced propagation S^{+} is a logical inference since it is reflected in the Born probability and in the probability amplitudes.
2) In the same hypothesis, i.e., considering only the forward propagation, one cannot explain the destruction of the "quantum state" as a phenomenon that takes place at a defined instant t_2. Well known paradoxes, such as that of Schrödinger's cat [20], derive from this.
3) From an algebraic point of view, the transactional ring is a sort of identity operator, because $SS^{+} = S^{+}S = 1$ and the qualities Q, R are simultaneously created and destroyed. One has the impression that every quantum process (therefore all matter) and time itself are emitted from an invariant substratum and re-absorbed within it. This substratum cannot be observed because it is invariant and outside of space and time, which originate from it: a sort of "motionless motor".

The transactional ring may be described by circular inference rules which establish a self-consistency rather than, as in traditional logic, by linear inference rules which establish a deduction. A circular logic, not a linear one, is applied to a ring; it appears therefore that individual quantum processes are self-generated in a non-causal manner through an extra-spacetime mechanism.

Self-generation implies acausality. At the same time, however, the creation/destruction events that occur at $t = t_1$, $t = t_2$ take place because of interactions with other rings. This implies the existence of rules on how the rings connect and therefore the acausality does not turn into complete arbitrariness. This appears to be a natural and entirely convincing explanation of the simultaneous presence of causality and acausality in quantum processes.

4) The energy is propagated only in one time direction, and the causal effects thus proceed from the past towards the future.

[4] We can also describe this process as the propagation of a single quality. In conventional TI this would be a single transaction with probability of unity

To sum up, the two extreme events of a transaction correspond to two reductions of the two state vectors which describe the evolution of the quantum process in the two directions of time. They constitute the "R processes" (R stands for *reduction*) of the Penrose terminology [21-23] and, from our perspective, are the only real physical processes. They are constituted of interaction vertices in which real elementary particles are created or destroyed; these interactions are not necessarily acts of preparation or detection of a quantum state in a measurement process.

The evolution of probability amplitudes in the two directions of time constitutes, in Penrose terminology, a U process (where U stands for *unitary*). From our ontological viewpoint, U processes are not real processes: both the amplitudes and the time evolution operators which act on them are mathematical inventions whose sole purpose is to describe the causal connection between the extreme events of the transaction, i.e. between R processes. This connection is possible because the two events derive from the transformation of the same aspatial and atemporal "substratum". As a specific consequence of this assumption, all virtual processes contained in the expansion of the time evolution operator are deprived of physical reality.

According to this approach, therefore, the history of the Universe, considered at the basic level, is given neither by the application of forward causal laws at initial conditions nor by the application of backward causal laws at final conditions. Instead, it is assigned as a whole as a complete network of past, present and future R processes. Causal laws are only rules of coherence which must be verified by the network and are *per se* indifferent to the direction of time; this is the so called *block universe* in Putnam and Rietdijk's conception [24, 25].

A transaction that begins with the creation of quality q and ends with the destruction of quality r can be represented simply by the form q^+r^-. For clarity's sake, however, we will use here also another symbolism, the one introduced by Costa de Beauregard [26-29]. According to it, the creation q^+ of the property q is indicated with $|q)$, while the destruction r^- of property r is indicated with $(r|$. The transaction q^+r^- is then denoted as $(r|\,q)$. We remark that a symbol as, e.g., $|q)$ represents an operator belonging to a suitable algebraic structure and should be not confused with the associated ket $|q>$ which is instead an amplitude.

As an example of a transactional network, let us consider a process well known in QFT, constituted by the decay of a microsystem, prepared in the initial state 1, into two microsystems 2, 3 which are subsequently detected. The preparation consists of the destruction of quality 1 [which we shall indicate with $(1 |$] which closes the transaction which precedes it, and of the creation of quality 1 [which we shall indicate with $| 1)$] which opens a new transaction. It will be represented by the form $| 1)(1 |$. The decay consists of the destructions of qualities 2, 3 which close the transaction that began with the preparation, and of the creations of qualities 2, 3 which open a new transaction which will be closed with the detection of microsystems 2, 3. It will be represented by the form $[| 2)| 3)]\ [(2 |(3 |]$.

The detection of microsystems 2, 3 will be constituted by the destructions of qualities 2, 3 which close the transaction that began with the decay, and by the creations of qualities 2, 3 which open subsequent transactions. It will be represented by the interaction events $| 2)(2 |$, $| 3)(3 |$. The double transaction described here corresponds to the process usually associated with the probability amplitude $< 2, 3\, | S |\, 1 >$.

Another example is Young's classical double slit experiment. The preparation of the initial state of the particle can be represented by the form $| 1)\ (1 |$, following the same reasoning as in the previous case. Instead of the decay, here we have the encounter with slits 2 and 3, i.e. the interaction between a particle and a double slit screen represented by [| 2)| 3)] [(2 |(3 |]. Instead of the detection of the two particles created in the decay, here we have the sole event of the detection of the particle on the second screen at a certain position 4, i.e.: | 4)(4 |. Two transactions are involved: the first starts with the preparation of the particle and ends with its interaction with the first screen; the second begins with this second interaction and ends with the interaction of the particle with the second screen. The latter interaction then constitutes the beginning of the following transaction. The process is that which corresponds to the probability amplitude $< 4 \mid S \mid 1 >$.

We note that the forward time evolution of the amplitudes, represented with $S| \; 1 >$, contains both the kets $\mid 2 >$, $\mid 3 >$; nevertheless, processes relating to the passage through the individual slit a (where $a = 2, 3$) do not exist. Such processes would require an intermediate event represented by| a) (a |, which effectively does not take place. It is in this sense that processes that can be associated with compound probability amplitudes $< 4 \mid S \mid a >< a \mid S \mid 1 >$ are "virtual" and not real. The process of the crossing of one of the two slits becomes real when the other slit is closed.

3. The Physical Universe as Network of R Microevents

Let us focus on some of the concepts summarily reviewed in the previous Section. A single event separating two transactions is denoted as a product of destructions (which close the preceding transaction) and creations (which open the subsequent transaction). For example, the event $A = |s)(u|$ consists of the destruction of u taking place in time $t_A - \varepsilon$, and the creation of s taking place in time $t_A + \varepsilon$, where t_A is the instant in which the event A occurs and $\varepsilon \rightarrow 0$.

Also, the symbol $A = |s)|u)$ denotes the event A consisting of the creation of u simultaneous with the creation of s, and so forth.

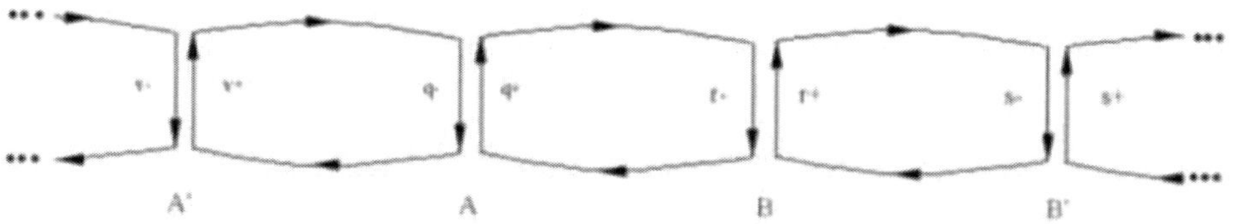

Figure 1.

Two or more transactional rings can be hooked to each other in a succession or sequence as shown in Figure 1. As one can see, the generic event $A = |q)(q|$ closes, with its component q^-, the preceding transaction (whose initial event can possibly be placed to the past infinity $t = -\infty$); with its component q^+ it opens up the subsequent transaction (whose final event can possibly be placed to the future infinity $t = +\infty$). Therefore, the event A joins the transactions v^+q^- and q^+r^-. In a similar way, the event $B = |r)(r|$ joins the transactions q^+r^- and r^+s^-, and so on. Two or more such sequences can cross each other, intersecting in a certain event. In Figure 2 the event $O = (w^+, w^-, z^+, z^-)$ belongs to the sequence C1 through operations w^-, w^+

which respectively close a transaction in C1 and open the following one in C1; the same event also belongs to the sequence C2 through operations z^-, z^+ which respectively close a transaction in C2 and open the following one in C2. Several sequences, by intersecting each other, will create a net provided with edges and vertices: a net of interconnected physical events. The physical world consists of the set of such nets.

The sequences are real (not virtual) physical processes. Each event (that is, each vertex of the net) represents an *interaction.* If it is placed at the intersection of two or more sequences then it represents the real (not virtual) interaction between the real physical processes embodied by these sequences, as the decay $1 \rightarrow 2 + 3$ considered above; otherwise the interaction is really a null interaction, as in quantum measurement with a negative result. For example, let us consider a Young's interferometer; when the particle's passage through the first slit is not detected, this means that the particle went through the second slit. With q = "position of the second slit", the event q^+q^- is then generated; this event is a null interaction.

We remark that, if an algebra exists which reflects the structure of the interaction in A and $|q) = |r)|s)$ is a well formed formula of this algebra, then the event of interaction $A = |q)(q|$ can actually take the form $A = |r)|s)(q|$, as in the examples discussed in the previous Section.

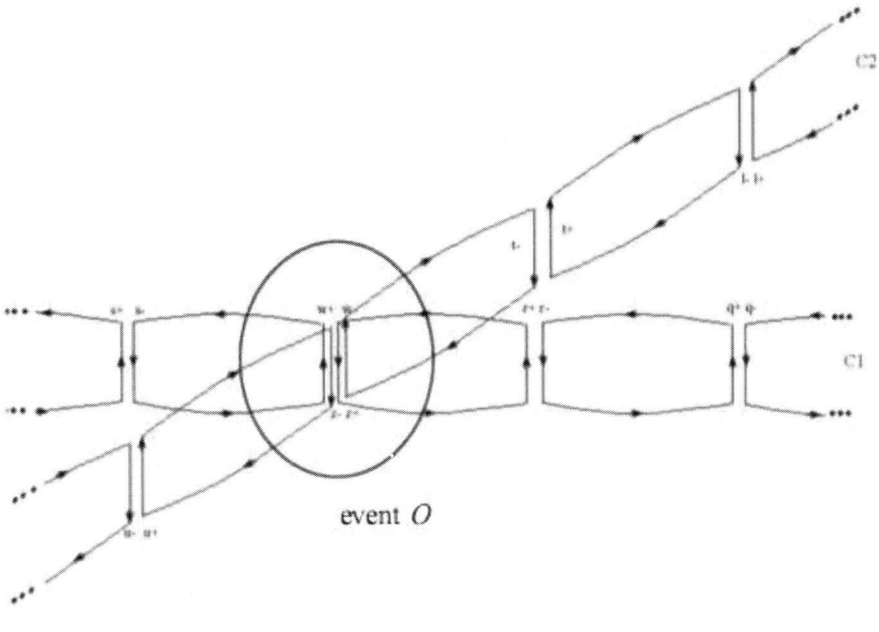

Figure 2.

4. The Statistics of Transactions

Each transaction is a self-connection of the aspatial and timeless physical vacuum. This process can be described as the concurrence of two distinct transformations. In the first transformation, the vacuum divides into two pairs of opposites: q^+, q^- and r^+, r^- ; the first pair forms "event A", the second pair forms "event B" :

$$A = |q)(q| \, , B = |r)(r| \, .$$

The expression $|q)(q|$ can be considered as representative of an infinite set of loops as that depicted in Figure 3a (oriented self-connections of the event A). Analogously, the expression $|r)(r|$ can be considered as representative of an infinite set of loops as that depicted in Figure 3b (oriented self-connections of the event B)[5].

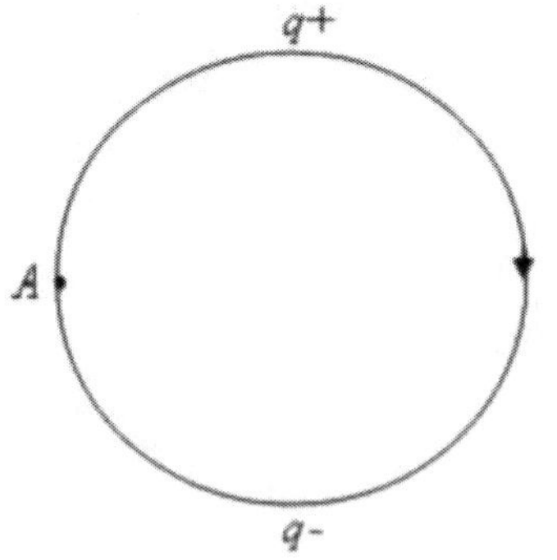

Figure 3a.

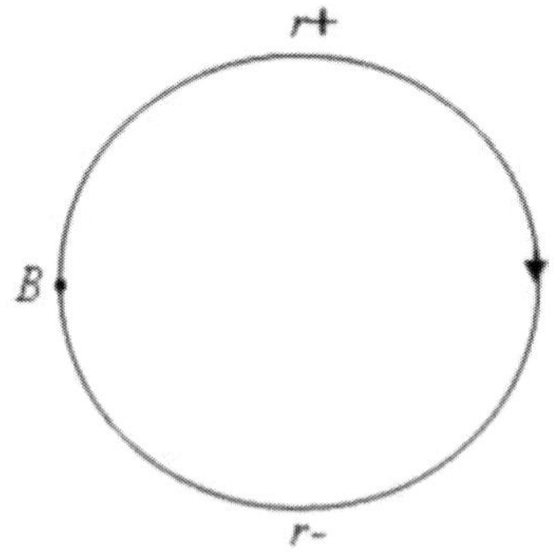

Figure 3b.

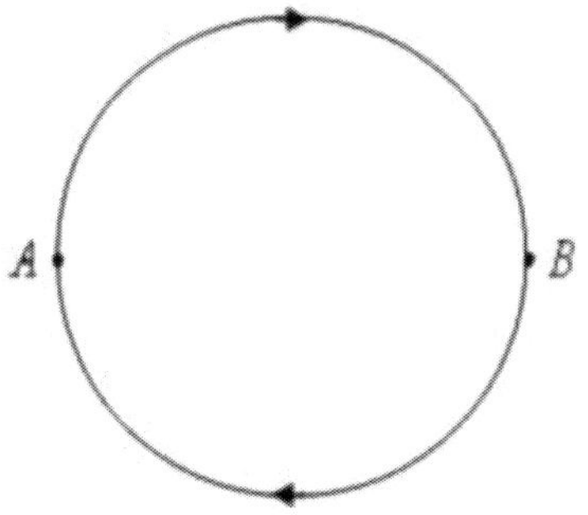

Figure 3c.

[5] Graphs 3a, 3b resemble Feynman graphs of vacuum-vacuum coupling; however, these last present at least two vertices. What we are really describing here is the emergence of the quality q from an aspatial and timeless vacuum. Let us remark that q can be (or can include) also a spacetime position. Therefore, the vacuum we refer to is actually the source of spacetime, rather than the state of minimum energy of a field defined on the spacetime. It should not be confused with the vacuum as intended in Quantum Field Theory (which underlies to Feynman graphs calculation). For an elucidation of the relation between these two notions of physical "vacuum" we remind to ref. (3).

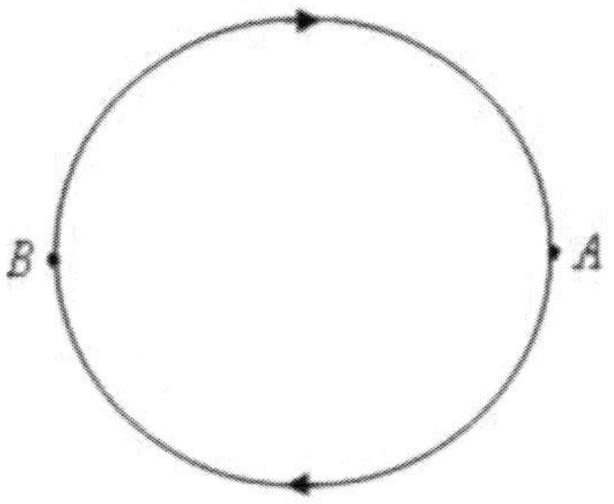

Figure 3d.

The second transformation is the real generation of the transaction having the events A and B as its extremities. It consists in the breaking of N loops of the first group and M loops of the second group. For what concerns the first group, the free ends of the broken lines are connected to the event B (Figure 3c); analogously, the free ends of the broken lines of the second group are connected to the event A (Figure 3d). Finally, A and B events are connected by means of $M+N$ lines oriented from A towards B, and $M+N$ lines having opposite verse.

We postulate that each line of the first (second) group entering into the points A, B joints with every line of the same group exiting from the same point, so generating a closed loop. In this way N^2 closed loops are formed with lines belonging to the first group, and M^2 closed loops are formed with lines belonging to the second group. If one assumes the *a priori* equiprobability of all these self-connections of vacuum, the global statistical weight of the process is expressed by $N^2 + M^2$.

These self-connections are acausal. Let us now consider a set of transactions having the same event at one extremity (we say, A) and the other extremity (we say, B) freely varying. Therefore the created quality q is exactly determined, while the destroyed quality r can assume a value selected in the set $\{r_1, r_2, ..., r_L\}$.

In other terms:

$$A = |q)(q| \, , \, B = B_i = |r_i)(r_i| i = 1, 2, ... L \, .$$

We thus have L possible distinct $q^+ r_i^-$ transactions ($i = 1, 2, ... L$), each of these involving $N(i)$ loops of the first group and $M(i)$ loops of the second group. For the reasons detailed above, the statistical weight of the i-th transaction is $[N(i)]^2 + [M(i)]^2$. Thus, the probability that the effectively bootstrapped transaction coincides with the k-th process amounts to:

$$P(q^+ r_k^-) = \{ [N(k)]^2 + [M(k)]^2 \} / \Sigma_i \{ [N(i)]^2 + [M(i)]^2 \}$$

$$i \, , k = 1, 2, ..., L \, .$$

Let us introduce, therefore, the “probability amplitude” of the k-th transaction:

$$\langle r_k | q \rangle = Z^{-1/2} \{ [N(k) + M(k)] + i \, [N(k) - M(k)] \} \, , \tag{4}$$

where $Z = 2 \, \Sigma_j \{ [N(j)]^2 + [M(j)]^2 \}$.

One immediately obtains that $P(q^+ r_k^-) = |\langle r_k | q \rangle|^2$.

The transaction in which the quality r_k is created rather than destroyed, and the quality q is destroyed rather than created, differs from the transaction in question solely by the exchange of events *A* and *B*. This is equivalent to exchanging *N*(*i*) and *M*(*i*). Thus, the amplitude of this second transaction is:

$$\langle q | r_k \rangle = Z^{1/2} \{ [M(k) + N(k)] + i\,[M(k) - N(k)] \}, \qquad (5)$$

where *Z* maintains the same value. Comparing the expressions of the two amplitudes one thus has that:

$$\langle q | r_k \rangle = \langle r_k | q \rangle^*.$$

The probability of this inverse transaction is therefore $P(r_k^+ q^-) = |\langle q | r_k \rangle|^2$. It is equal to that of the direct transaction: $P(r_k^+ q^-) = P(q^+ r_k^-)$.

We can observe that absorbing the factor $Z^{1/2}$ in *M*(*k*), *N*(*k*), relations (4), (5) can be inverted so obtaining:

$$M(k) = \mathrm{Re}[\langle q | r_k \rangle (1 - i)]/2 \qquad (4')$$

$$N(k) = \mathrm{Re}[\langle r_k | q \rangle (1 - i)]/2 \qquad (5')$$

Eqs. (4’), (5’) are extremely important, because they connect some transempirical properties of the vacuum (as *M*(*i*) and *N*(*i*) are) with the usual quantum formalism.

Let us return to the direct transaction (*r* |*q*) we have considered at the beginning of this section, associated with a probability amplitude $\langle r | q \rangle$. Let us suppose that, in the graph in Figure 3a, the set of self-connections (loops) *A*→*A* is subdivided into a certain number of subsets, say, two for the sake of example. Therefore, the transaction involves N_1 loops belonging to the first subset, and N_2 loops belonging to the second subset. Obviously $N = N_1 + N_2$.

We assume the same partition holds for the loops $B \rightarrow B$ of the second group (Figure 3b). Therefore, we have M_1 loops belonging to the first subset and M_2 loops belonging to the second subset. Again, it is $M = M_1 + M_2$.

One can therefore write:

$$\langle r | q \rangle =$$

$$Z^{1/2} \{ [N + M] + i\,[N - M] \} =$$

$$Z^{1/2} \{ [N_1 + N_2 + M_1 + M_2] + i\,[N_1 + N_2 - M_1 - M_2] \} =$$

$$Z^{1/2}\{ [N_1 + M_1] + i\,[N_1 - M_1] \} +$$

$$Z^{1/2}\{ [N_2 + M_2] + i\,[N_2 - M_2] \} =$$

$$= \langle r|q\rangle_1 + \langle r|q\rangle_2 \,.$$

Consequently:

$$P(\,q^+ r_k^-\,) = |\langle r|q\rangle|^2 =$$

$$|\langle r|q\rangle_1|^2 + |\langle r|q\rangle_2|^2 +$$

$$\langle r\,|q\rangle_1 \langle r|q\rangle_2^* + \langle r|q\rangle_2 \langle r|q\rangle_1^* =$$

$$Z^{1}\{ [N_1^2 + M_1^2] + [N_2^2 + M_2^2] + 2[N_1N_2 + M_1M_2] \}. \tag{6}$$

The appearance of the last term constitutes the well-known phenomenon of self-interference. With reference to the popular example of the double slit, the index 1 (2) represents “particle paths” crossing the slit 1 (2), while q and r represent respectively the initial and the final state of the particle.

The origin of this phenomenon is far from mysterious. We remember that A and B events are connected by means of M+N lines oriented from A towards B, and M+N lines having opposite verse. Clearly, is well possible that a line of the first group belonging to the subset 1 joints with a line of the first group belonging to the subset 2, etc.

There are four possibilities:

1) Through A and B a half-loop of the subset 1 is connected to an inverse half-loop of the subset 1. First term of Eq. (6).
2) Through A and B a half-loop of the subset 2 is connected to an inverse half-loop of the subset 2. Second term of Eq. (6).
3) Through A and B a half-loop of the subset 1 is connected to an inverse half-loop of the subset 2. Last term of Eq. (6).
4) Through A and B a half-loop of the subset 2 is connected to an inverse half-loop of the subset 1. Last term of Eq. (6).

Naturally, if only one subset of connections is realized in the actual transaction, self-interference vanishes because the sum of the amplitudes is then confined to the sole term actually present (1 or 2). This situation occurs in the case, for example, of the double slit experiment when only one slit is left open and the other is closed.

An important case is that in which the subdivision into subclasses takes place on the basis of the results which one would have had by subdividing the transaction into two distinct *successive* transactions. To understand this topic, one must firstly observe that the probability of the process consisting of the two successive transactions $(r_k|q)$ and $(s|r_k)$ is obviously given by the product of the probabilities of the two transactions, i.e.:

$$P(\, q^+ r_k^- \,) \times P(\, r_k^+ \, s \,) =$$

$$|\langle r_k|q\rangle|^2 \times |\langle s|r_k\rangle|^2 =$$

$$|\langle r_k|q\rangle\langle s|r_k\rangle|^2.$$

The rule whereby the probability of the process is given by the modulus squared of the amplitude (Born rule) therefore remains valid in this case, too, provided that the amplitude of the process $(s|r_k)(r_k|q)$ is defined in the following manner:

$$\langle s|r_k\rangle\langle r_k|q\rangle \,.$$

Let us now consider a transaction $(s|q)$ during which the quality q is created and the quality s is destroyed. It can be imagined as the simultaneous actuation of all the processes:

$$(s|r_k)(r_k|q) \; ; k = 1, 2, ..., L \,.$$

In other words, each transactional half-loop connecting the events $|q)(q|$ and $|s)(s|$ can be seen as the succession of two edges: one connecting $|q)(q|$ to $|r_k)(r_k|$, the other connecting $|r_k)(r_k|$ to $|s)(s|$. The transactional half-loops can therefore be subdivided into subsets corresponding to the various r_k ($k = 1, 2, ..., L$) values of r.

As we have seen, the total transaction amplitude can thus be expressed as the sum of the partial amplitudes relating to the various subsets:

$$\langle s|q\rangle = \sum\nolimits_k \langle s|r_k\rangle\langle r_k|q\rangle \,.$$

This result allows the *ket* to be defined:

$$|q\rangle = \sum\nolimits_k |r_k\rangle\langle r_k|q\rangle \,.$$

The transaction probability amplitude is thus obtained by "left-multiplying" this expression by $\langle s|$. Alternatively, one can define the *bra*:

$$\langle s| = \sum\nolimits_k \langle s|r_k\rangle\langle r_k| \,.$$

The transaction probability amplitude is thus obtained by "right-multiplying" this expression by $|q\rangle$. It obviously follows from the relations:

$$\langle r_k|q\rangle = \sum_i \langle r_k|r_i\rangle\langle r_i|q\rangle$$

since the qualities r and q are arbitrary, that $\langle r_k|r_i\rangle = \delta_{ki}$. Bra and ket are therefore "vectors" defined with respect to a "complete" orthonormalized basis.

Some remarks.

1) We have assumed for simplicity that $N(i)$, $M(i)$ are integer positive numbers. Actually, this condition may be removed. The quality r, for example, can be a physical quantity which assumes continuous values on a real segment (which can possibly be extended, with a passage to the limit, to the real straight line or to a half-straight line). This segment can thus be subdivided into L identical contiguous segments and the index i can be applied to distinguish these segments. In other words, one can let $r = r_i$ if the value of r falls within the i-th segment. One can then thicken the subdivision of the domain of r by making L tend to infinity ($L\rightarrow\infty$). In this passage to the limit the numbers $N(i)$, $M(i)$ remain defined if they are re-interpreted as the *fraction* of transactions for which r falls within the i-th interval. With this re-definition (implicit in the introduction of the normalization factor Z), they become real for every finite L and remain real (infinitesimals) in the passage to the limit $L \rightarrow\infty$. The actual amplitude is the limit value of the amplitude for $L\rightarrow\infty$, provided that this limit exists.

Regarding the sign of $N(i)$, $M(i)$ it must be borne in mind that at the end of the procedure these numbers appear in the definition of the transaction amplitude. The probability of the transaction is given by the modulus squared of the amplitude and is therefore not affected by the multiplication of the amplitude by a phase factor $exp(i\delta)$. In particular, if $\delta = \pi$ this factor is equal to -1. Thus, by multiplying the amplitude by -1, the probability of the transaction remains unchanged. But to multiply the amplitude by -1 means to multiply the numbers $N(i)$, $M(i)$ which appear in its definition by -1, i.e. to redefine them as negative. Thus the condition that these quantities be positive or null can also be removed.

With this removal, the meaning of the numbers N, M changes slightly. A positive value of N signifies a *creation* of connections of the first group (Figure 3a), while a negative value of N signifies the *destruction* of an equivalent number of such connections. Similarly, a positive value of M signifies a *creation* of connections of the second group (Figure 3b), while a negative value of M signifies the *destruction* of an equivalent number of such connections. The sum and product of these numbers becomes the sum and product of operations that create and destroy a certain type of connection.

2) The concept of probability implies that of randomness. There is nothing mysterious about quantum randomness. The transaction $(r|q)$ *is constituted* by its extremes which are the events $|q)(q|$ and $|r)(r|$, and is therefore completely defined once these two

extremes are given. There is therefore no effective indetermination - once the initial and final conditions have been assigned the transaction is completely determined, because it actually only consists of the pair of these two conditions.

The problem arises when only one of the two conditions is given and one wants to know which the other one can be. The other condition is then determined by the physical reality bootstrap process which, as we have seen, comprises acausal elements; to predict this condition with certainty, therefore, is generally not possible.

One must therefore reason in statistical terms, introducing a probability of the transaction taking place. In the process in question, the vacuum is coupled with itself through a certain number of closed rings or loops. Each ring is broken into two half-rings whose extremes, A and B, are equal for all the rings in question; the events $A = |q)(q|$ and $B = |r)(r|$ are associated with these extremes. Two oppositely-oriented half-rings can reconnect to these extremes, giving rise to a new closed ring which comes to constitute the transaction. This reconnection can be done in n^2 ways, where n is the number of half-rings having a given orientation and belonging to a given group (that is, $n = N$ or $n = M$). All these ways are identical with respect to the bootstrap process and are therefore *a priori* equiprobable. The probabilistic reasoning described above now follows.

3) The transaction $(s|q)$ has been associated with the amplitude $\langle s|q\rangle$ with:

$$|q\rangle = \Sigma_k |r_k\rangle\langle r_k|q\rangle.$$

The quality q also includes, in its definition, the instant $t(q)$ at which the event $|q)(q|$ occurs, so that the ket $|q\rangle$ depends on $t(q)$. The preceding equation thus implies that also the amplitudes $\langle r_k|q\rangle$ and kets $|r_k\rangle$ are dependent upon $t(q)$. Yet $\langle r_k|q\rangle$ is the amplitude of the $(r_k|q)$ transaction, in which two distinct time instants appear: that of the event $|q)(q|$ and that of the event $|r_k)(r_k|$; that is, $t(q)$ and $t(r)$, respectively. The dependence of the right-hand member of the equation on $t(q)$, $t(r)$ is compatible with the dependence of the left-hand member on $t(q)$ only if it is assumed that $t(q) = t(r)$. Therefore the kets $|q\rangle$, $|r_k\rangle$ and the bras $\langle r_k|$ must be considered at the same time instant.

The same reasoning applies to the expression:

$$\langle s| = \Sigma_i \langle s|r_i\rangle\langle r_i| ,$$

for which it therefore turns out that $t(s) = t(r)$. On the other hand, the amplitude $\langle s|q\rangle$ can be constructed by left-multiplying the ket $|q\rangle$ by the bra $\langle s|$. This amplitude therefore makes sense only if the conditions $t(q) = t(r)$, $t(s) = t(r)$ are simultaneously satisfied; by the transitive property of equality, one thus obtains $t(s) = t(q)$. If $t(s)$ and $t(q)$ are different, then the amplitude $\langle s|q\rangle$ makes sense only if a transport rule exists for the ket $|q\rangle$ from $t = t(q)$ to $t = t(s)$, or a transport rule for the bra $\langle s|$ from $t = t(s)$ to $t = t(q)$. In the first case the ket $|q\rangle$ is transported to $t = t(s)$ and then is left-multiplied by $\langle s|$; in the second case the bra $\langle s|$ is transported to $t = t(q)$ and is then right-multiplied by $|q\rangle$. The transport rule is defined by the time evolution operator S, but the quantum dynamics will not be discussed here for reasons of space; interested readers are referred to other works (3). Strictly speaking, therefore, the portion of formalism developed here is only valid for the qualities q, r, s associated with kets or bras which do not vary over time (constants of motion), or which vary to a negligible degree within the interval $(t(q), t(s))$.

4.1. Quantum Formalism

Essentially, there is nothing mysterious about quantum formalism, provided that it is gradually and properly introduced. We have seen that a generic vector (ket or bra) can be decomposed into a complete basis of orthonormalized vectors. For example:

$$|\psi\rangle = \Sigma_i c_i |\varphi_i\rangle$$

$$\langle\varphi_k|\varphi_i\rangle = \delta_{ki} ,$$

where c_i are complex numbers. The probability of the $(\varphi_k|\psi)$ transaction, therefore, is expressed by Born's rule, bearing in mind the caveats on the time variation of the quantities already discussed in the previous section:

$$P(\varphi_k|\psi) = |\langle\varphi_k|\psi\rangle|^2 / \langle\psi|\psi\rangle = c_k^* c_k / \Sigma_i c_i^* c_i .$$

Let us now suppose that a certain physical quantity O, in the realization of the $(\varphi_k|\psi)$ transaction, assumes the value $o(k)$. Let us assume that this value is a real number; this assumption does not constitute a loss of generality because it is practically always possible to satisfy it by adopting a suitable system of definitions.

Let us suppose that the quality ψ is fixed, while the quality φ can freely vary on its support set $\{\varphi_k\}$. Therefore the *a priori* probability that the outcome is quality φ_k is $P(\varphi_k|\psi)$; the expectation value of O is given by:

$$\langle O\rangle_\psi = \Sigma_k P(\varphi_k|\psi) o(k) =$$

$$\Sigma_k o(k)\, c_k^* c_k / \Sigma_i c_i^* c_i .$$

This formula is entirely general. A very important special case is that in which an appropriate self-adjoint linear operator Ω exists on the rigged space of kets and bras, such that:

$$\Omega|\varphi_i\rangle = o(i)\, |\varphi_i\rangle \text{for every value of } i.$$

From the self-adjointness of Ω follows that:

$$\langle\varphi_i|\Omega = \langle\varphi_i|o(i) \text{ for every value of } i.$$

It immediately follows from the linearity of Ω that:

$$\Omega|\psi\rangle = \Sigma_i c_i \Omega|\varphi_i\rangle = \Sigma_i c_i o(i)\,|\varphi_i\rangle\,;$$

$$\langle\psi|\Omega|\psi\rangle = (\,\Sigma_k c_k^{\,*}\langle\varphi_k|\,)\,(\,\Sigma_i c_i o(i)\,|\varphi_i\rangle\,) = \Sigma_k o(k)\, c_k^{\,*} c_k\,.$$

and since:

$$\langle\psi|\psi\rangle = (\,\Sigma_k c_k^{\,*}\langle\varphi_k|\,)\,(\,\Sigma_i c_i|\varphi_i\rangle\,) = \Sigma_i c_i^{\,*} c_i\,,$$

one obtains:

$$\langle O\rangle_\psi = \langle\psi|\Omega|\psi\rangle\,/\,\langle\psi|\psi\rangle\,.$$

This formula is less general but frequently used in quantum physics. The operator Ω is thus said to be "associated" with the quantity O. The $o(k)$ values in this case form the "spectrum" of O (or of Ω) . The Hermitianity of Ω ensures on the one hand that the $o(k)$ values are real and on the other that the procedure is symmetrical. Indeed, we would have obtained the same result by developing $\langle\psi|\Omega$ and then right-multiplying by $|\psi\rangle$. One ought to bear in mind that, given a physical quantity O, a self-adjoint linear operator associated with it does not always exist; a well-known example is the time t. The "postulate" set out in many text books, whereby to every O there corresponds an Ω, is not correct. In many important cases, though, Ω exists and the aforementioned formula allows to draw advantage from this.

4.2. Structure of Transactions

It seems appropriate to elucidate a question which could easily be a source of misunderstanding. Let us consider a generic transaction $(\varphi|\psi)$ during which the quality ψ is created and the quality φ is destroyed. It is constituted by the two events $A = |\psi)(\psi|$ and $B = |\varphi)(\varphi|$, which take place at the instants t_A and t_B, respectively. These two events are the only manifest reality in the physical world. The "facts" are constituted solely by these two events.

These two events, on the other hand, are correlated. The correlation between A and B is represented, as we have seen, by the "amplitude" of the transaction, $\langle\varphi|S|\psi\rangle$. This amplitude can generally be developed as the sum of the partial amplitudes corresponding to different paths joining A and B. Each of these partial amplitudes can then be represented as the product of amplitudes of free propagations joining interaction vertices.

For example, let us consider the well-known double-slit experiment. In such a case, $|\psi\rangle$ is the ket associated with the "preparation" of the "particle" in the source and $\langle\varphi|$ is the bra associated with the "detection" of the "particle" behind the screen containing the slits.

Under the action of S , $|\psi\rangle$ evolves freely up to the interaction with the screen. If the screen is completely absorbent, then, indicating as x the position of a generic point on its rear wall, one has $\langle x|\psi\rangle = 0$, unless x corresponds to one of the two slits. The following time evolution of $|\psi\rangle$ is still free. Therefore, indicating the free time evolution operator of the particle as S and the two slits as f_1, f_2, one can write:

$$\langle\varphi|S|\psi\rangle = \langle\varphi|S|f_1\rangle\langle f_1|S|\psi\rangle + \langle\varphi|S|f_2\rangle\langle f_2|S|\psi\rangle \,.$$

As can be seen, one has the sum of two partial amplitudes, one of which corresponds to a "passage through slit f_1", the other to a "passage through slit f_2" . These two amplitudes interfere, for the reasons seen above. Each of the two amplitudes is represented, in turn, as the product of two free propagation amplitudes of the "particle": one corresponding to the propagation from the source to the slit, the other from the slit to the detector.

Hence: either the "particle" interacts with the atoms which compose the screen, and is absorbed, or it does not interact with the screen at all. Only in this second case can it reach the detector, passing through f_1 or through f_2. The terms f_1, f_2 which appear in the expression of the amplitude are therefore associated with interactions with the atoms of the screen [to be precise, "negative" interactions, i.e. non-interactions or absence of actual interactions]. The correlation between events A and B is therefore conditioned by the atoms of the screen.

And what are the atoms of the screen? Nothing other, in fact, than events of the $|\Lambda)(\Lambda|$ type, where Λ is the "atomic quality". Therefore, the correlation between events A and B is affected by the other events of the Universe, in this specific case by the events $|\Lambda)(\Lambda|$ which constitute the screen. The structure of $\langle\varphi|S|\psi\rangle$ includes these influences.

Now, the fundamental thing to understand is that *the interaction events which appear in* $\langle\varphi|\, S\,|\psi\rangle$ *are not real, in fact they are not events.* For example, the (non) interaction events f_1, f_2 in the example we have just described are not real events. By this we mean to say that these events are not pairs of emission/re-absorption of qualities by the physical vacuum, whereas the extreme events A, B are.

One must be careful to consider, however, that the events $|\Lambda)(\Lambda|$ are real, too, and that the transaction amplitude actually expresses the influence of the other real events of the Universe, such as these very events $|\Lambda)(\Lambda|$, on the correlation between the real events A and B.

This fact that the virtual events appearing in the partial amplitudes as mere artifices of calculation do not have a physical reality of their own should not lead to believe, however, that they bear no relation with physical reality. Indeed, as has been seen, they represent actual aspects of the transaction bootstrap from the vacuum: the partial amplitudes are associated with the half-loops. Furthermore, virtual events can become real. For example, a "particle" counter can be placed behind f_1 or f_2. In this case, though, one is no longer dealing with the original transaction, rather with other entirely different transactions, for example $(f_1|\psi)$ [6].

[6] The event $|f_1\ )\ (f_1|$ is real, whether the "particle" is detected behind f_1, or is not detected behind f_2, if third possibilities are excluded. In this second case, one has a null interaction vertex.

The free propagations $\langle f_1|S|\psi\rangle$, $\langle\varphi|S|f_1\rangle$, $\langle\varphi|S|f_2\rangle$, $\langle f_2|S|\psi\rangle$ can be represented, as is well known, in the form of the propagator of a particle of energy E, impulse p and spin s (and given internal quantum numbers): $\langle E, p, s\ |S|E\ , p\ , s\rangle$. The "particle" therefore is actually nothing other than a packet of constants of motion which are propagated, and does not exist as substantial reality.

In this respect, an issue which has given rise to quite a number of misunderstandings is that which concerns the sign of energy E. Energy E appears from the vacuum at the time of creation, and vanishes into the vacuum at the time of destruction, as do, in fact, the other quantities which make up the quantum.

From a merely numeric standpoint, one can view creation as the release by the vacuum of an E amount of energy to the manifest physical world. Destruction will thus be the release by the manifest physical world of an amount of energy E to the vacuum. Now, it is known that a system's exchanges of energy with the "outside" must be taken with a different sign according to their direction. We may therefore adopt two conventions.

First convention. We may assume that the energy released by the vacuum is positive and that absorbed by the vacuum is negative. In this case, we shall have at creation the exchange of an energy $E> 0$, and at destruction the exchange of an energy $-E< 0$.

Second convention. We may assume that the energy released by the vacuum is negative and that absorbed by the vacuum is positive. In this case, we shall have the exchange of an energy $E< 0$ at creation, and the exchange of an energy $-E> 0$ at destruction.

Now, any transaction falls within a chain of transactions unlimited in the two directions of time and to which a total order relation applies, which is chronological ordering. The direction of such an ordering must be the same for all the interacting chains which make up the physical world, if one is to have a global time coordinate that is definable. Once the direction of the time ordering has been chosen, creation precedes destruction, based on the convention that creation is called the extreme of the transaction which occurs first. Thus, with the first convention we have the creation of an energy $E> 0$ at t_0 and the absorption of an energy $E< 0$ at $t_1>t_0$. This corresponds to the forward propagation in time of a quantum of energy $E> 0$, or to the propagation backwards in time of a quantum of energy $E< 0$. Indeed the creation of a quantum of energy $E< 0$ at t_1 is equivalent to the destruction of a quantum $E> 0$ and the destruction of a quantum $E< 0$ at t_0 is equivalent to the creation of a quantum of energy $E> 0$.

With the second convention we have the creation of an energy $E< 0$ at t_0 and the absorption of an energy $E> 0$ at $t_1>t_0$. This corresponds to the forward propagation in time of a quantum of energy $E< 0$, or to the propagation backwards in time of a quantum of energy $E> 0$. Indeed, the creation of a quantum of energy $E> 0$ at t_1 is equivalent to the destruction of a quantum $E< 0$ and the destruction of a quantum $E> 0$ at t_0 is equivalent to the creation of a quantum of energy $E< 0$. Clearly the two conventions, though equally possible, are incompatible: either one or the other must be chosen. If the first is chosen, as is usual, retropropagations in time of quanta with positive energy cannot exist[7]. One thus has a "time arrow" of entirely microphysical origin and which is absolutely elementary. This concept can also be applied to the solutions of any wave equation, for example the D'Alembert wave equation with a four-current term in the right-hand. This entails the non-existence of advanced potentials; indeed, they do constitute valid solutions, but under the opposite

[7] Should this occur, one is actually considering a positive energy propagation, forward in time, of the conjugated of the quantum in question (CPT theorem).

convention. The existence of a time arrow also in a mechanism which is of itself totally symmetrical in time, such as a transaction, is a fact of capital importance. It implies the causality of quantum phenomena, and daily proof that a cause always precedes its effect. Time, therefore, flows in one direction only, even if the transaction - the elemental process of the physical world - is symmetrical in time. In discussing the double slit experiment we considered a situation in which two events A and B are connected by an operator S which can be represented as the sum of partial propagation amplitudes of *individual* quanta. This situation is typical of microphysics experiments, but not of everyday reality. In everyday life one deals with transaction aggregates, rather than with individual transactions. A piece of "common matter" is actually an aggregate of transactions which take place in enormous numbers per second and per cubic centimeter. It is absolutely impossible, in a typical situation, to follow the alternation of individual transactions, i.e. to do that which in technical jargon is known as "following the time evolution of the system microstate". One therefore confines oneself to taking into consideration certain *global* qualities which in the course of time evolution are preserved or show a sufficiently "continuous" trend.

Thus the idea is born of a persistent "object", which constitutes the basis of classical physics and which instead has no part to play at the individual transaction level.

4.3. Wavefunctions

The general quantum formalism introduced in the preceding sections, and which basically constitutes the Dirac formalism (30) admits, as a special case, of a less general formalism known as the "wavefunction" formalism.

Let us consider the transaction $(q|\psi)$ during which the quality ψ is created at the time t_1 and the quality q is destroyed at the time $t_2>t_1$. We assume that the quality q can be represented by a set of N values of an equal number of real variables q_i ($i = 1, 2, ..., N$), called "positions" or "generalized coordinates". The amplitude associated with this transaction is $\langle q|S(t_2, t_1)\,|\psi\rangle$.

In the limit $t_2 \to t_1$ one has $S(t_2, t_1) \to 1$, and the amplitude tends to $\langle q|\psi\rangle = \psi(q)$. In general this limit amplitude depends on $t = t_1 = t_2$, and it is indicated, therefore, with $\psi(q, t)$. This is the "wavefunction". Wavefunctions are therefore a special class of amplitudes and their formalism can be obtained from the Dirac general formalism by particularization. It is possible, however, to introduce the quantum formalism by reasoning directly on the wavefunctions. This approach is less general than the one seen above, but is of considerable theoretical interest, so we propose to describe it here.

The wavefunction concept is useful when there exists a real function $A(q, t)$, called "action", of the "Lagrange coordinates" q and of the time t such that the "generalized pulses" $p_i = \partial A(q, t)/\partial q_i$ are constants of motion in the free evolution.

The function $\psi(q, t)$ is assumed to be decomposable, by means of a Fourier transform, into the sum of harmonic components (letting $\hbar = h/2\pi = 1$):

$$\phi(p)\, exp(i\, p\bullet q - i\, Et)\,.$$

On thus obtains:

$$-i(\partial/\partial q_i)\,[\phi(p)\,exp(i\,p\bullet q - i\,Et)] = p_i\,[\phi(p)\,exp(i\,p\bullet q - i\,Et)]\,.$$

It follows from this relation that: a) the operator associated with p_i is $-i(\partial/\partial q_i)$; b) the harmonic components are eigenfunctions of this operator. Therefore p assumes a definite value on these components (that is, its fluctuations are null). In addition, one obtains:

$$i(\partial/\partial t)\,[\phi(p)\,exp(i\,p\bullet q - i\,Et)] = E\,[\phi(p)\,exp(i\,p\bullet q - i\,Et)]\,.$$

That is: a) the operator associated with E is $i(\partial/\partial t)$; b) the harmonic components are eigenfunctions of this operator. Therefore E assumes a definite value on these components (that is, the fluctuations of E are null). That E is the energy, i.e. an eigenvalue of the Hamiltonian H, is easily seen by using the very definition of Hamiltonian as an exponent in the evolution operator. If one takes $\phi(p)\,exp(i\,p\bullet q)$ as the initial wavefunction, one must have:

$$\phi(p)\,exp(i\,p\bullet q - i\,Et) = exp(-i\,Ht)\,\phi(p)\,exp(i\,p\bullet q)\,.$$

This relation implies $H = E\cdot$, i.e. the fact that the harmonic component is an eigenfunction of H with an eigenvalue E. On the other hand, the operator associated with E is $i(\partial/\partial t)$, so that in conclusion $H = i(\partial/\partial t)$.

Classically, the Hamiltonian function depends on q and on p. A common, in many cases useful, empirical method for constructing the Hamiltonian operator is to substitute p_i in the classical expression with $-\,i(\partial/\partial q_i)$. One can thus write in symbols $H = H\,(q, -i\partial_q)$. Assuming that:

$$\psi(q\,,t) = \Sigma_p\phi(p)\,exp(i\,p\bullet q - i\,Et)$$

One has, given the linearity of the operators H and $i(\partial/\partial t)$:

$$[H - i(\partial/\partial t)]\;\psi(q\,,t) = \Sigma_p\phi(p)\,[H - i(\partial/\partial t)]\,exp(i\,p\bullet q - i\,Et) =$$

$$= \Sigma_p\phi(p)\,[H - E]\,exp(i\,p\bullet q - i\,Et) = 0\,.$$

i.e. the equation of motion:

$$H(q, -i\partial_q)\;\psi(q\,,t) = i(\partial/\partial t)\;\psi(q\,,t)\,.$$

This equation defines the time evolution of the ψ. It is fundamental to understand that the q, on which the ψ depends, are not spatial coordinates but generalized coordinates introduced on the assumption that an action exists for the process being studied. As is well known, the Lagrange coordinates coincide with the spatial coordinates in one case only: that of the classical "material point". Therefore the function ψ does not represent an actual physical field on spacetime.

To understand the actual physical meaning of the wavefunction, one must consider its role in a defined transaction process.

Let us firstly consider a transaction (E' , p' $|E$, p), in which at the time t_1 a harmonic component is created which has a moment p and energy E, and at the time $t_2 \geq t_1$ a harmonic component is destroyed which has a moment p' and energy E'. In the limit $t_2 \rightarrow t_1$, the amplitude of this transaction is

$$\langle p' | p \rangle \langle E' | E \rangle = \delta(p - p') \, \delta(E - E') .$$

Therefore, in the instantaneous limit, the amplitude of the transaction is equal to 1 if $p = p'$ and $E = E'$, otherwise it is 0. In other words, in this limit the interaction effect disappears, so that p and E are preserved. The meaning of this formalism is therefore the following (spin is disregarded and it is assumed that $E > 0$) :

a) A quantum (E, p) is emitted at $t_1 = t$.
b) A quantum (E' , p') is absorbed at $t_2 = t + \varepsilon$. $\varepsilon \rightarrow 0$, $\varepsilon > 0$.
c) The emission and absorption of the quantum are possible only if $p = p'$ and $E = E'$.
d) In this case, the number of quanta emitted and absorbed (i.e. 1) is given by the amplitude (which is indeed equal to 1).
e) In the same transaction, a quantum ($-E'$, $-p'$) is emitted at $t_2 = t + \varepsilon$.
f) In the same transaction, a quantum ($-E$, $-p$) is absorbed at $t_1 = t$.
g) The emission and absorption of the quantum are possible only if $p = p'$, $E = E'$.
h) In this case, the number of quanta emitted and absorbed (i.e. 1) is given by the complex conjugate amplitude of the preceding one (which is indeed equal to $1^* = 1$).

The quanta propagating forward in time are the half-loops, pointing towards the future, which connect the two transaction events. The quanta propagating back in time are the half-loops, pointing towards the past, which connect the two transaction events. The number of complete loops obtained by the rejoining of the two half-loops, one pointing to the future and the other pointing towards the past, is therefore equal to the number of quanta propagated forward in time by the number of quanta propagated backwards in time. In this specific case, the number of closed loops is 1 x 1 = 1. Generally speaking, this number is proportional to the transaction probability.

Let us now turn to the general case of a transaction ($\phi|\psi$) in the course of which the quality ψ is created at the time t_1, and the quality ϕ is destroyed at the time $t_2 \geq t_1$. The amplitude of the process forward in time is $\langle \phi | S (t_2, t_1) | \psi \rangle$, and the amplitude of the process backwards in time is the complex conjugate of the latter. Letting:

$$\langle q | S (t_2, t_1) | \psi \rangle = \frac{1}{(2\pi)^{N/2}} \int dp \, e^{i \, p \cdot q} \, \alpha(p)$$

$$\langle \phi | q \rangle = \frac{1}{(2\pi)^{N/2}} \int dp' \, e^{-i \, p' \cdot q} \, \beta^*(p') \ ,$$

the forward amplitude is expressed by:

$$\frac{1}{(2\pi)^N} \int dp\,dp'\alpha(p)\beta^*(p')\int dq\,e^{i\,p\bullet q}\;e^{-i\,p'\bullet q}$$

$$= \int dp\,dp'\alpha(p)\beta^*(p')\delta(p-p') = \int dp\,\alpha(p)\beta^*(p)$$

which is equal to a certain complex number $\Gamma = \Gamma_0 exp(i\delta)$, where Γ_0 is real and positive. The amplitude backwards in time is the complex conjugate $\Gamma^* = \Gamma_0 exp(-i\delta)$ and the number of closed loops reconstituted by the joining of the half-loops is $\Gamma^*\,\Gamma$, real and positive. This number is proportional to the transaction probability.

One could object that Γ, being a generic complex number, cannot be the number of half-loops - which obviously must be either integer and positive or real and positive. On must note, however, that by multiplying the two wavefunctions α and β under the integral sign by $exp(i\eta)$, $exp(-i\varphi)$, respectively, with $\eta - \varphi = -\delta$, the phase factor $exp(i\delta)$ in Γ is cancelled out and only Γ_0 survives, real and positive. The number of loops is therefore given by $\Gamma_0^*\Gamma_0$, which is still equal to $\Gamma^*\,\Gamma$. It follows from the Schrödinger equation, on the other hand, that multiplying the wavefunction by a constant phase factor does not entail any change in the dynamics. Therefore, multiplying the wavefunctions by the phase factors $exp(i\eta)$, $exp(-i\varphi)$ does not have any physical effects.

The fact that the transaction probability turns out to be proportional to $|\Gamma|^2$ constitutes Born's rule. In the expression shown, the integral in q corresponds to the count of a single emitted-absorbed quantum. The double integral in p and p' corresponds to the sum of these individual counts, i.e. to the half-loop count.

The construction of the quantum formalism proceeds in an entirely parallel fashion to that seen with bras and kets. If it is possible to associate with a physical quantity O, having a spectrum $\{o(i)\,;\, i = 1, 2, ...\}$, a self-adjoint linear operator Ω such that:

$$\Omega\phi_i(q) = o(i)\;\phi_i(q)\;\; i = 1, 2, ...$$

then it is also possible, by the well-known spectral theorem on self-adjoint linear operators, to orthonormalize the basis:

$$\int dq\,\phi_i(q)\phi_k^*(q) = \delta_{ik}\,.$$

Assuming therefore that:

$$\psi(q) = \Sigma_n \alpha_n \phi_n(q)\,,$$

the amplitude of the $(\phi_k|\psi)$ transaction in the instantaneous limit is expressed by:

$$\frac{\int dq\,\psi(q)\phi_k^*(q)}{\left[\int dq\,\psi(q)\psi^*(q)\right]^{1/2}} = \alpha_k / (\Sigma_n|\alpha_n|^2)^{1/2}.$$

And by normalizing the ψ to 1, one thus has, for the transaction probability, the value $|\alpha_k|^2$. Therefore if ψ is the fixed extreme of the transaction and ϕ_k is the free extreme, the value $o(k)$ of the quantity O will be realized with a probability of $|\alpha_k|^2$. The average value of the quantity O on the ensemble of all the $(\phi_n|\psi)$ transactions, where n is variable, is therefore:

$$\Sigma_n o(n)|\alpha_n|^2 = \int dq\,\psi^*(q)\Omega\psi(q)\,,$$

as can be verified immediately- This is the rule for calculating $<\Omega>_\psi$.

5. Transactional Meaning of the Wick Rotation

The formalism of wavefunctions is especially appropriate for approaching, from a transactional point of view, one of the most time-honoured mysteries of theoretical physics: the connection between time and thermodynamics mediated by the Wick rotation [31].

5.1. Quantum Leap in the Energy Representation

Let us consider the transactional loop:

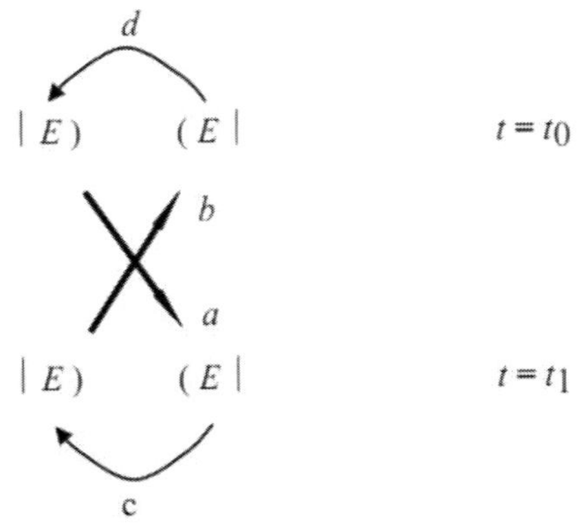

The vacuum is connected to itself through the *acbd* cycle or through the inverse cycle *dbca*. The two events $|E)(E|$ which occur at the instants t_0 and $t_1>t_0$ represent the emission and the absorption, respectively, of a quantum (electron, photon, muon, etc.) of exactly defined energy equal to *E*.

The hypothesis is that the entire process - say, *acbd* - takes place as a virtual self-coupling process of the archaic vacuum. If, at the time instants t_0 and t_1, the coupling is possible with other transactional loops - that is, if the events $|E)(E|$ *actually exist* - the virtual process becomes real and one has the unfolding of a transaction over spacetime. This is how the network of events over spacetime which forms the observed material Universe emerges. Since the coupling satisfies the principles of conservation, macroscopic causality is complied with.

In the absence of the events $|E)(E|$, the virtual process *acbd* is an indivisible whole; the entire cycle must be covered in order to return to the point of departure, i.e. to the vacuum. From the point of view of the vacuum, therefore, this cycle is an elementary event. We can hypothesize that an occurrence probability is associated with this elementary event:

$$P = exp(-2E/kT) ,$$

with $T > 0$. We can divide the cycle into the two processes *da*, *cb* which are not true events (they cannot occur by themselves) and factorize *P* according to the symmetrical expression:

$$\mathrm{P} = \mathrm{P_{da}} \times \mathrm{P_{cb}}$$

$$\mathrm{P_{da}} = \mathrm{P_{cb}} = \exp(-E/kT) .$$

Obviously, the factors P_{da} , P_{cb} are not probabilities because, as we have said, the processes *da*, *cb* are not events.

When the spacetime unfolding of the transaction occurs, the half-processes *da, cb* become true distinct processes on spacetime, instead. Process *da* consists of event $|E)(E|$ which occurs at $t = t_0$ and of the forward connection with the event $|E)(E|$ which occurs at $t = t_1$. Process *cb* consists of the event $|E)(E|$ which occurs at $t = t_1$ and of the backward connection with event $|E)(E|$ which occurs at $t = t_0$. We are dealing therefore with the two sub-processes which constitute the transaction.

The spacetime unfolding corresponds to a Wick rotation. For the first process the Wick rotation takes the form:

$$1/kT \rightarrow -i(t_1 - t_0)/\hbar ,$$

while for the second process it takes the form:

$$1/kT \rightarrow -i(t_0 - t_1)/\hbar .$$

In these equations, the variable *t* indicates the proper time of the quantum of energy *E* exchanged between the two events. Consequently, the factors P_{da} , P_{cb} become:

$$P_{da} \rightarrow \Pi_{da} = exp[iE(t_1 - t_0)/\hbar]$$

$$P_{cb} \to \Pi_{cb} = exp[-iE(t_1 - t_0)/\hbar] \,.$$

Thus, the product of these factors is the transform of P according to the Wick rotation; this product is equal to 1. It is still a probability, but its meaning now is entirely different. This is the probability that, given the emission (absorption) of a quantum of energy E, the energy of the successively absorbed (previously emitted) quantum is E. Since the propagation of the quantum is free (we are considering a single transaction without intermediate vertices) this probability is certainly 1. Alternatively, one can say that 1 is the number of quanta of energy E propagated between the emission and absorption events.

More generally, let us suppose that an action variable S exists such that:

$$E = -\,\partial_t S.$$

If E is a constant of motion, as in the example being considered, then:

$$E = -\,\Delta S/\Delta t$$

and therefore:

$$P_{da} = exp(-E/kT) = exp[-\Delta S/(\Delta t kT)] \to exp[-(\Delta S/\Delta t)\,(-i\Delta t/\hbar)] = exp(i\Delta S/\hbar) \,;$$

$$P_{cb} = exp(-E/kT) = exp[-\Delta S/(\Delta t kT)] \to exp[-(\Delta S/\Delta t)\,(i\Delta t/\hbar)] = exp(-i\Delta S/\hbar) \,;$$

where $\Delta t = t_1 - t_0$. One must note that the last equalities obtained remain valid even if the quantum mass is null. They are also relativistically invariant, and are therefore valid in a general frame of reference in which $\Delta S = E\Delta t - p\Delta x$, where the vector x represents the generalized coordinates of the quantum and the vector p represents its conjugate pulses.

In conclusion, the Wick rotation connects the probability of a virtual process to the probability of the same process once it has become real.

5.2. Quantum Statistics in the Energy Representation

Let us now consider the case in which the event at $t = t_1$ actually were $|E')(E'|$ with $E' \neq E$. In this case, the coupling would not have been possible, as this would have violated the principle of the conservation of energy. The probability of this transaction would thus have been zero. Since, in this case,

$$P_{cb} \to exp[-iE'(t_1 - t_0)/\hbar],$$

the two cases are combined into one, assuming that the probability of the transaction is expressed by:

$$(1/2\pi\hbar)\int exp[iE(t_1 - t_0)/\hbar]\ exp[-iE'(t_1 - t_0)/\hbar]\ d(t_1 - t_0) = \delta(E - E').$$

This expression can be written:

$$\int \psi\varphi^*\ d(t_1 - t_0) = \delta(E - E'),$$

letting:

$$\psi = (1/2\pi\hbar)^{-1/2}\Pi_{da}, \varphi = (1/2\pi\hbar)^{-1/2}(\Pi_{cb})^*.$$

Let us now consider a more general transaction connecting the event $|\psi)(\psi|$ which occurs at $t = t_0$ with the event $|\psi)(\psi|$ which occurs at $t = t_1$. Let us suppose that the ψ is represented, in this case, by the linear superposition:

$$\psi = (1/2\pi\hbar)^{-1/2}\sum_E a_E exp[iE(t_1 - t_0)/\hbar].$$

Then $\varphi = \psi$ and the probability of the transaction is expressed by the integral:

$$(1/2\pi\hbar)\int\sum_{E,E'} a_E a_{E'}^*\ exp[iE(t_1 - t_0)/\hbar]\ exp[-iE'(t_1 - t_0)/\hbar]\ d(t_1 - t_0) =$$

$$= \sum_{E,E'} a_E a_{E'}^*\ \delta(E - E') = \sum_E a_E a_E^*.$$

As can be seen, the number of quanta of energy E exchanged in the course of this transaction is $a_E a_E^*$, and they represent a fraction:

$$a_E a_E^* / \sum_E a_E a_E^*$$

of the total number of quanta exchanged. In other words, the probability of emitting/absorbing a quantum of energy E is expressed by the fraction:

$$a_E a_E^* / \sum_E a_E a_E^* = \left|\int \psi\xi^*\ d(t_1 - t_0)\right|^2 / \int \psi\psi^*\ d(t_1 - t_0)$$

where:

$$\xi = exp[iE(t_1 - t_0)/\hbar].$$

In the transaction considerated above, the initial "state" ψ (emission) and the final "state" ψ (absorption) coincide. Let us now suppose that instead of the event $|\psi)(\psi|$ which occurs at $t = t_1$ there is an event $|\varphi)(\varphi|$; let us suppose that the φ is represented, in this case, by the linear superposition:

$$\varphi = (1/2\pi\hbar)^{-1/2}\sum_E b_E exp[iE(t_1 - t_0)/\hbar] .$$

For simplicity, let us also suppose that ψ and φ are normalized to 1. The previous formula for the transaction probability is thus generalized:

$$|\int \psi\varphi^* \, d(t_1 - t_0)\,|^2 .$$

This result constitutes Born's rule and its validity extends beyond the energy representation (E-representation), as has been shown in the previous sections.

5.3. Wick Rotation and Uncertainty Principle

Let us now return to the expression of the Wick rotation; to fix the concept in our minds, we shall consider the "forward" rotation:

$$1/kT \rightarrow -i(t_1 - t_0)/\hbar ,$$

but the line of reasoning is also applicable to the "backward" rotation obtained from this one by exchanging t_1 and t_0. By comparing the expressions:

$$P_{da} = exp(-E/kT) \text{ and } \Pi_{da} = exp(i\Delta S/\hbar)$$

it can be seen the uncertainty principle $\Delta S \geq \hbar$ takes the "non-rotated" form $E \geq kT$. The meaning is obvious: energy exchanges of smaller amount than the thermal energy kT cannot be distinguished from the "substratum" of thermal fluctuations within the thermostat and are therefore "virtual".

5.4. Action as Information

By inverting the relation:

$$P_{da} = P_{cb} = exp(-E/kT)$$

one obtains:

$$E = -kT \ln(\rho)$$

where $\rho = P_{da}$, P_{cb} . We can introduce a "number of cases" variable equal to $W = 1/\rho$ and an entropy:

$$\Lambda = klnW = kln(1/\rho) ,$$

so that the obtained relation becomes $E = \Lambda T$. Therefore, according to the Wick rotation the quantity:

$$E/kT = \Lambda/k$$

is transformed into the quantity $\pm i\Delta S/\hbar$. In other words, the action is the "rotated" expression of entropy. This equivalence between action and entropy had already been evidenced by De Broglie [32]. We also note that an action value of $\Delta S = \hbar ln(2)$ corresponds to a binary choice ($\rho = \frac{1}{2}$). One can therefore measure the action adimensionally, as an amount of information (in bits):

$$\Sigma = \Delta S/\hbar ln(2) .$$

It is clear from the above that quantum statistics in E-representation is *de facto* equivalent to the statistics of a canonical ensemble with a thermal bath at the absolute temperature T. It is this fact which led De Broglie to conjecture the presence of a "hidden thermostat" [32].

It must be noted, however, that the representation of the thermal bath in the form of an aether of particles having chaotic motion, located on spacetime and with which real particles supposedly interact [32], is absolutely naive. Such an approach views real particles as persistent objects which describe a continuous - though non-differentiable - trajectory in spacetime. Actually, the "thermal" representation outlined here considers the elementary quantum process of emission-absorption (transaction) as a whole and it is therefore nonlocal. Time, in particular, emerges from the archaic physical vacuumas a consequence of the unfolding described by means of the Wick rotation, so that the "thermostat" does not live in time but rather belongs to a timeless dimension which unfolds through the rotation. Any attempt to re-obtain locality through a completion of this representation with hidden variables is therefore doomed to failure.

Conclusion

In our opinion, the reformulation of the transaction concept presented in this work maintains the essential premises of Cramer's approach. Firstly, the extremes of a transaction are events of creation and destruction of certain physical qualities. In particular circumstances, these qualities can be mapped on the values assumed by a maximal complete set of compatible observables which are constants of motion, and in this case one is dealing with the propagation of a *quantum*. This is a consequential generalization of Cramer's idea, which indeed identifies such extremes with emission/absorption events in the context of a classical (time-symmetric) theory of fields. In addition, the reformulation presented is fully consistent with the current formalism, if R processes are identified with *actual* - and not purely virtual - interaction processes (*objective* reduction postulate)[8].

[8] For what concerns the quantum reduction within a time-simmetric context we remind here to Elio Conte previous works (33,34,35).

What for Cramer is the complex consisting of offer and confirmation waves here becomes, more simply and consistently with the standard formalism, the set of the two "forward" and "backward" amplitudes of the elementary quantum process, associated with the evolution operators S and S^+, respectively.

In this formulation, none of the paradoxes raised against TI appears. For example, Maudlin's argument [9] is unfounded because the change in the experimental setup brought about by moving detector B, though triggered by a signal coming from detector A, leads to a completely different transaction. In the first case, the microevent B^+B^- is never produced, in the second case it is certainly produced. Since one of the two extreme microevents of the transaction changes, the transaction itself changes and therefore, in accordance with the arguments set out above, the time evolution operator of this process changes. We note, on the other hand, that the null signal from A, i.e. the event (not $A)^+$(not $A)^-$ is itself a transaction termination (null interaction). The second transaction assumes as its input state the output state of the first.

For the same basic reasons, no problem can arise with Afshar's experiment [36]. We remark, *inter alia*, that a correct QED description of this particular experiment involves the transaction connecting the initial amplitude $|\,1>|\,0>+|\,0>|\,1>$ (mutually exclusive firing of detector D1 *or* D2) and the final amplitude $|\,1>|\,0>$ (D1 fires) or $|\,0>|\,1>$ (D2 fires). This transaction not happens in spacetime, so no "photon trajectory" is really tracked; in addition, also the not firing detector is involved in the process.

If the proposed reformulation appears to be an improvement for the purpose of understanding the elementary quantum process, no less interesting are the possibilities for its application to sectors of physics research still being debated. For example, the Wick rotation becomes, in a physically clear way, a rule for the unfolding of the transactional loop from the *implicate order*; this agrees with the approach suggested by Bohm. We are dealing here with a connection between synchronic and diachronic realms of the physical world, which is also an emergence of real phenomena from a background of virtual possibilities, and an emergence of *time*. An application to the problem of initial conditions in cosmology has recently been proposed [37].

From a more speculative perspective, it is possible that the network of transactions, defined consistently with the proposal summarized here, constitutes the causal aspect of a more general modality for connections between microevents, which we may call *archetypal*. For a definition of archetypes and their possible role in natural history, see ref. [38]; in ref. [39] the scheme is described of a class of experiments addressing their possible detection.

It must be noted, finally, that the spirit of the proposal, especially as regards the reference to a background outside of space and time, can also be found in the so-called "possibilist TI" formulated by R.E. Kastner in her most recent works [40], [41].

Acknowledgments

This work has been written at the kind invitation of Ruth Elinor Kastner, whom I thank wholeheartedly for her encouragement and the confidence she has placed in me. The author is indebted with Prof. Elio Conte for his warm support.

References

[1] Chiatti L.; "Wave function structure and transactional interpretation" in "*Waves and particles in light and matter*", A. Van der Merwe and A. Garuccio eds., pp. 181-187. (Proceedings of Trani workshop 24 – 30 Sept. 1992).

[2] Chiatti L.; *Found. Phys.* 25 (3), 481 (1995).

[3] Chiatti L.; *The Archetypal Structures of the Physical World*; Di Renzo, Rome, 2005 (*in Italian*).

[4] Cramer J.G.; *Phys. Rev. D* 22, 362 (1980).

[5] Cramer J.G.; *Rev. Mod.Phys.* 58 (3), 647 (1986).

[6] Cramer J.G.; *Int. Journ. Theor. Phys.* 27 (2), 227 (1988).

[7] Wheeler J.A., Feynman R.P.; *Rev. Mod. Phys.* 17, 157- 181 (1945).

[8] Wheeler J.A., Feynman R.P.; *Rev. Mod. Phys.* 21, 425–433 (1949).

[9] Maudlin T.; *Quantum Non-Locality and Relativity* (2nd ed.); Blackwell, Oxford 2001.

[10] Berkovitz, J.; "On Causal Loops in the Quantum Realm," in "*Non-locality and Modality*", Placek T. and Butterfield J. eds, pp. 233-255 (*Proceedings of the NATO Advanced Research Workshop on Modality, Probability and Bells Theorems*); Kluwer, Dordrecht 2002.

[11] Kastner R.E.; *Synthese* 150, 1–14 (2006); preprint arXiv:quant-ph/0408109v1.

[12] Bohm D.; *Wholeness and Implicate Order*; Routledge & Kegan Paul, London, 1980.

[13] Selleri F., Vigier J.P.; *Nuovo Cim. Lett.* 29, 7 (1980).

[14] Bjorken J.D., Drell S.D.; *Relativistic Quantum Mechanics*; Mc Graw-Hill, New York, 1965.

[15] Greenberger D.M., Horne M., Zeilinger A.; "Going Beyond Bell's Theorem", in "*Bell's Theorem, Quantum Theory and Conceptions of the Universe*", M. Kafatos ed., Kluwer, Academic, Dordrecht, 1989.

[16] Aspect A. et al.; *Phys. Rev. Lett.* 49, 91 (1982).

[17] Aspect A. et al.; *Phys. Rev. Lett.* 49, 1804 (1982).

[18] Tittel C. et. al.; *Phys. Rev. Lett.* 81, 3563 (1998).

[19] Ou Z.Y.; *Phys. Rev. Lett.* 68, 3663 (1992).

[20] Schrödinger E.; *Naturwiss.* 49, 53 (1935).

[21] Hawking S.W., Penrose R.; *The Nature of Space and Time*; Princeton University Press, 1996.

[22] Penrose R.; The Emperor's New Mind; Oxford University Press, 1989.

[23] Penrose R.; *The Road to Reality*; Oxford, 2004.

[24] Putnam H.; *Journal of Philosophy* 64, 240-247 (1967).

[25] Rietdijk C.W.; *Philosophy of Science* 33, 341-344 (1966).

[26] Costa de Beauregard O.; "Relativity and probability, classical or quantal"; in "*Bell's Theorem, Quantum Theory and Conceptions of the Universe*", M. Kafatos ed., pp. 117-126; Kluwer Academic, Dordrecht, 1989.

[27] Costa de Beauregard O.; Time, *The Physical Magnitude*; Reidel, Dordrecht, 1987.

[28] Costa de Beauregard O.; *Found. Of Phys. Lett.* 5 (3), 291-294 (1992).

[29] Costa de Beauregard O.; "Causality as identified with conditional probability and the quantal nonseparability", in "*Microphysical Reality and Conceptions of the Universe*", A. Van der Merwe et al. eds., pp. 219-232; Kluwer, Dordrecht 1988.

[30] Dirac P.A.M.; *Principles of quantum mechanics*, 4nd edition; Clarendon, Oxford, 1958.
[31] Sakurai J.J.; *Modern Quantum Mechanics*, Chapt. 2.5; Benjamin, 1985.
[32] De Broglie L.; *Thermodinamique de la particule isolée*; Gauthiers-Villars, Paris, 1964.
[33] Conte E.; *Epistemological Letters* 41, 1-7 (1981).
[34] Conte E.; *Lett. Nuovo Cimento* 31, 380-382 (1981).
[35] Conte E.; *Lett. Nuovo Cimento* 32, 286-288 (1981).
[36] Afshar S.S.; *Proc. SPIE* 5866, 229-244 (2005).
[37] Licata I., Chiatti L.; *Int. Jour. Theor. Phys.* 49 (10), 2379-2402 (2010); preprint arXiv:gen-ph/1004.1544 (2010).
[38] Chiatti L.; "Archetypes, Causal Description and Creativity in Natural World"; in "*Physics of Emergence and Organization*", World Scientific; I. Licata, A. Sakaji eds, 2008; preprint arXiv:physics/0607082.
[39] Chiatti L.; *Quantum Biosystems*, 3, 194-201 (2008); preprint arXiv:physics/0807.3253.
[40] Kastner R.E.; preprint arXiv:quant-ph/1006.4902.
[41] Kastner R.E.; *Studies in History and Philosophy of Modern Physics* 42, 86-92 (2010).

In: Space-Time Geometry and Quantum Events
Editor: Ignazio Licata, pp. 45-154
ISBN: 978-1-63117-455-1

Chapter 3

STOCHASTIC FOUNDATION OF QUANTUM MECHANICS AND THE ORIGIN OF PARTICLE SPIN

***L. Fritsche*[1,*] *and M. Haugk*[2]**
[1]Institut für Theoretische Physik der Technischen Universität Clausthal, Clausthal-Zellerfeld, Germany
[2]Kaffeeberg, Herrenberg, Germany

Abstract

The present contribution is aimed at removing most of the obstacles in understanding the quantum mechanics of massive particles. We advance the opinion that the probabilistic character of quantum mechanics does not originate from uncertainties caused by the process of measurement or observation, but rather reflects the presence of objectively existing vacuum fluctuations whose action on massive particles is calibrated by Planck's constant and effects an additional irregular motion. As in the theory of diffusion the behavior of a single particle will be described by an ensemble of identically prepared but statistically independent one-particle systems. Energy conservation despite the occurrence of a Brownian-type additional motion is achieved by subdividing the ensemble into two equally large sub-ensembles for each of which one obtains an equation of motion that has the form of a Navier-Stokes- or "anti"-Navier-Stokes-type equation, respectively, the latter involving an anti-Brownian motion enhancing process. By averaging over the total ensemble one obtains a new equation of motion which can be converted into the time-dependent Schrödinger equation. We clarify the problem of the uniqueness of the wave function and the quantization of orbital momentum. The concept allows the inclusion of electromagnetic fields and can be extended to interacting N-particle systems. We analyze the problem of how an experimental setup can consistently be decomposed into the quantum system under study and the residual quantum system "apparatus". The irregular extra motion of the particle under study allows a decomposition of the associated ensemble into two subensembles the members of which perform, respectively, a right-handed or left-handed irregular circular motion about a given axis which becomes physically relevant in the presence of a magnetic field. We demonstrate that this orientation-decomposed "Zitterbewegung" behaves - in accordance with Schrödinger's original idea - as a spin-type angular momentum which appears in addition to a possible orbital angular moment of the particle.

*E-mail address: lfritsche@t-online.de (Corresponding author)

We derive the non-relativistic time-dependent Pauli equation and propose a theory of the Stern-Gerlach experiment. The Dirac equation proves to be derivable by drawing on similar arguments used in obtaining the Pauli equation. We, further, attempt to put Bell's theorem and the Kochen-Specker theorem into perspective.

Pacs: 02.50.Fz; 02.50.Cw; 02.50.Ey; 03.65.Ta; 05.30.Ch; 05.40.Jc; 05.40.-a

Keywords: Stochastic mechanics, fundamentals of quantum mechanics

"... to skeptics, heretics and naïve realists everywhere.
Keep doubting; let others keep the faith."

David Wick in: *The Infamous Boundary* [1]

"I have never been able to discover any well-founded reasons as to why there exists so high a degree of confidence in the .....current form of quantum theory."

David Bohm in: *Wholeness and the Implicate Order* [2]

1. Introduction

There exists a rich literature on attempts that have been made to derive non-relativistic quantum mechanics from a concept of dissipationless stochastic point mechanics. A precursor of the idea of correlating the probabilistic character of quantum mechanics with the action of a stochastic background field may be seen in the paper by Bohm and Vigier [3]. The present contribution draws on later work on this subject but avoids certain implications that have often been criticized during the past 20 years. (S. e.g. W. Weizel [4], E. Nelson [5], [6], Guerra and Morato, [7], M. Baublitz [8], L. de la Peña and A. H. Cetto [9], Petroni and Morato [10], T. C. Wallstrom [11] and numerous references therein. For a rather complete review see R.F. Streater [12]. An earlier review covering work up to 1986 is given in a book by Namsrai [13].) A more recent contribution is due to Fritsche and Haugk [14].

The derivation of the time-dependent Schrödinger equation constitutes the focus of the following considerations. A very interesting alternative to our approach that also relates to vacuum fluctuations, but draws on non-equilibrium thermodynamics has recently been put forward by Grössing [15],[16]. Based on the concept of our derivation one is led to conclude that every conceivable situation of a physical system is exhaustively described by the respective solution to the Schrödinger equation, and that there can be no independent measurement problem. As for this point we side with J. Bell [17] who argues that the attempt to base the interpretation of quantum mechanics on some notion of "measurement" has raised more problems than it has solved. As we shall outline in Section 23 "measurements" relate outcomes, e. g. detector readings, to characteristic properties of a quantum system by using solutions to the Schrödinger (or Pauli) equation as primordeal information. Without this equation and its solutions "measurements", i. e. in general, detector or "pointer" readings, constitute a set of worthless data.

2. Origin of Quantum Mechanical Randomness

We interpret the fact that microscopic particles move and behave differently from macroscopic objects as reflecting the active role of the vacuum providing the space for energy fluctuations. The latter will henceforth be referred to as vacuum fluctuations. The consequences of their existence have already been discussed quite some time ago, s. e. g. Bess [18], Puthoff [19], [20], Boyer [21], Calogero [22], Carati and Calgani [23]. Present day quantum mechanics is strongly shaped by historical contingencies in its development, and it has become almost impossible to tell fiction from facts. Statements on "the measurement of positions at different times" and "there is no momentum of a particle in advance of its measurement" are typical of this school of thought (s. e. g. Streater [12]), yet they are definitely void of meaning. What kind of experimental setup should allow a perfectly accurate position measurement at a perfectly accurate time point? And how does the setup look like that allows the measurement of a particle's momentum in the spirit of orthodox quantum mechanics; i. e. with zero variance? It is totally impossible to perform non-fictional measurements on quantities that would conform to their quantum mechanical definition, e. g. measuring commuting observables like energy and angular momentum at the same time. The "observables" around which a substantial portion of quantum mechanical literature revolves are in reality non-observables. Further, there is simply no evidence of a causal interrelation between the probabilistic character of quantum mechanics and indeterminacies introduced by "the observer".

By contrast, there is every reason to believe that vacuum fluctuations are real and constitute an objective property of nature. Zero point motion of particles constitutes the most obvious evidence of their existence. It is this zero point motion which, for example, keeps liquid ^{4}He "molten" down to the very lowest temperatures and explains this extraordinary material property.

One could view vacuum fluctuations as caused by an exchange of energy between the mechanical system in question and the embedding vacuum that serves as an energy reservoir in terms of virtual particles: if that reservoir reduces its content of virtual particles, the energy of the system under study increases so that the energy of the entire system comprising this "vacuum reservoir" is conserved. Considerations of Calogero [22] point in a similar direction. In that sense quantum mechanical systems may be viewed as open systems like classical point mass systems in contact with a heat bath. This analogy will become particularly visible in our treatment. We shall use the terms "point mass" and "point charge" with the reservation that the actual size of the particles in question might well be finite of the order 10^{-13} cm, but very small compared to atomic diameters of the order 10^{-8} cm. Occasionally "point mass" will stand for the centroid of an atom or some composite system.

In the following we shall focus on the description of the subsystem "point mass in real-space" which is open toward the active vacuum and whose energy is therefore conserved only on average.

An implication of this concept is that charged point masses, despite their irregular motion, do not emit or absorb radiation on the average. In stationary situations a charged point mass will exchange photons with the vacuum in a way that does not change its average energy and momentum.

Radiation only occurs when the particle's probability density of being at its various positions in space, or the associated current density becomes time-dependent.

This is analogous to a system kept by non-heat conducting fibers in a vacuum chamber whose walls serve as a heat bath. In a stationary state situation the system exchanges constantly photons with the heat bath without changing its average energy. However, if its temperature is, for example, higher than that of the wall, the system starts radiating, that is, there is now a net flow of photons leaving the system.

If one disregards the details of the energy transfer between the two systems, vacuum fluctuations appear as an irregular temporary departure of the particle in question from its energy conserving trajectory in that it changes its energy by an average amount ΔE for an average time interval Δt so that $\Delta E\,\Delta\,t = f\,\hbar$ where $h = 2\,\pi\hbar$ denotes Planck's constant and the factor f is about unity. It is this departure from classical energy conservation which explains, as already alluded to, why a harmonic oscillator in its state of lowest energy is irregularly driven out of the position where it would be classically at rest. Furthermore, it explains the stability of a hydrogen atom in its ground-state (which applies quite generally to all atoms and their compounds), the zero-point motion of atoms in molecules and solids and the "tunneling" of particles through a potential wall which actually amounts to overcoming that wall.

Zero-point motion is commonly associated with the uncertainty relation which, however, merely shifts the problem of understanding a non-classical phenomenon to understanding the origin of a non-classical relation. Moreover, it amounts to keeping a blind eye on the fact that one is dealing here with a ground-state phenomenon which is certainly not observer-induced. Only if the contrary would apply, one would be justified in referring to the uncertainty relation.

A Boltzmann distribution over the energy levels of some system is completely independent of the details and the kind of the energy exchange between the heat bath and the system. The distribution contains only one universal parameter, viz. Boltzmann's constant. Similarly, a system's stationary zero-temperature states that emerge from exchanging energy with the vacuum do not depend on the details of this exchange and on the kind of particles involved, but only depend on another universal constant, viz. Planck's constant.

The envisaged derivation implies that particle trajectories persist under the influence of the stochastic vacuum forces. Their existence becomes particularly obvious with tracks of α-particles in a track chamber, but also with the trajectories of electrons in a field electron microscope. Their property of forming straight lines from the field-emission tip (assumed semi-spherical) to the monitoring screen is actually presupposed in calculating the magnification of the microscope. Conversely, purely quantum mechanical behavior occurs at lowest energies when the trajectories do no longer possess a classical reference in the limit $\hbar \to 0$. Trajectories still persist in that case, but the respective particle now performs a purely irregular motion.

The existence of particle trajectories is denied by the Copenhagen school of thought because "things that cannot be observed do not exist". Supporter of this view have to live with the conflict that a complex-valued wavefunction or its associated state vector, which constitutes the center of quantum mechanics, cannot be observed as well. By contrast, we believe that the validity of assumptions can only be scrutinized by checking the consistency of the resulting theory against experimentally accessible quantities and laws. We are here in complete accord with Ballentine who states in his seminal article [24]:

"...quantum theory is not inconsistent with the supposition that a particle has at any instant both a definite position and a definite momentum, although there is a widespread folklore to the contrary."

In summarizing one can state that the notion of an empty, inert vacuum is at variance with the occurrence of the zero point motion of particles, the phenomen of spreading wave packets, the tunneling effect and the temporary appearance of massive particles - like pions - which mediate the strong forces between nucleons. An active vacuum should therefore be regarded as the primordial origin of the non-classical behavior of particles. One might compare the properties of this "active vacuum" with a realistic model, viz. superfluid ^{4}He which is known to consist of ^{4}He-atoms that are not completely close-packed and move irregularly due to zero-point motion. Hence, if one would inject a particle into the fluid, it would hit various atoms on its way and be scattered off its trajectory along which it would move if the ^{4}He-atoms were absent. However, if its average velocity remains below the critical velocity it cannot lose linear momentum to the fluid on average. This defines superfluidity. If one were to place a two-slit diaphragm across the direction of its motion it can, of course, continue its motion only by moving through one of the slits, but its trajectory evolves differently depending on whether or not the other slit is open.

In Section 3 we briefly discuss the construction of ensemble averages of quantities that appear in the Navier-Stokes equation given in Section 4. We regard this equation as a mathematical object that derives entirely from classical conccpts. Details of its derivation, which goes essentially back to Gebelein [25], will be relegated to the Appendix, Section 33. We discuss the construction of a "Brownian" and an "anti-Brownian" sub-ensemble. The motional behavior of the latter is governed by an "anti-Navier-Stokes" equation. We explain why a system of statistically independent particles moves according to the arithmetic mean of these two equations when their motion is governed by classical mechanics plus "conservative" stochastic forces. On forming this arithmetic mean we arrive at an equation that can be converted into the time-dependent Schrödinger equation. We demonstrate that Wallstrom's objection [11] against the legitimacy of this conversion and his arguments in favor of the standard approach to the quantization of orbital momentum are based on a misunderstanding and ignore fundamental considerations of Pauli [26] and Born and Jordan [27] in the early days of "conventional" quantum mechanics. In Section 14 we show how the derivation of the time-dependent Schrödinger equation can be extended by including electromagnetic fields. The derivation can be extended further to interacting many-particle systems.

3. Defining Ensembles and Averages

As in the theory of diffusion we start with considering a point-like particle that is driven by an external conservative force $\boldsymbol{F}(\boldsymbol{r})$ and moves in an environment where it is exposed to additional stochastic forces. To gain access to quantities that are commonly discussed within this framework we construct a sufficiently large set of N identical systems (an ensemble of systems) under the supposition that there is no correlation between the stochastic forces of different systems. As a fundamental consequence, one is led then, as in the theory of diffusion, to a form of quantum mechanics that merely describes ensemble behavior. But this, again, is in accord with Ballentine's view [24]: *"..in general, quantum theory predicts nothing which is relevant to a single measurement (excluding strict conservation laws like those of charge, energy or momentum)."*

The relative freqency with which the particle appears at the time t in an elementary volume $\Delta^3\boldsymbol{r}$ around the point $\boldsymbol{r}$ is given by

$$\frac{n(\boldsymbol{r},t)}{N} = \rho(\boldsymbol{r},t)\,\Delta^3\boldsymbol{r} \tag{1}$$

where $n(\boldsymbol{r},t)$ is the number of particles in $\Delta^3\boldsymbol{r}$, and $\rho(\boldsymbol{r},t)$ denotes the probability density. We, furthermore, introduce $N_{\boldsymbol{r}}$ for the number of elementary volumes into which the total volume $\mathcal{V}$ is thought to be subdivided. Since the sum over all elementary cells yields N particles we have

$$\sum_{\boldsymbol{r}}^{N_{\boldsymbol{r}}} \frac{n(\boldsymbol{r},t)}{N} = 1 \quad \text{that is} \quad \int_{\mathcal{V}} \rho(\boldsymbol{r},t)\,d^3\boldsymbol{r} = 1\,. \tag{2}$$

We refrain here from discussing the proper limiting case $N \to \infty$ and relating relative frequencies to probabilities, as this matter has extensively been analyzed elsewhere (s. e. g. Streater [12]). We assume that there will always be a smooth function $\rho(\boldsymbol{r},t)$ for any finite N that provides a least mean square fit to the actually histogram-type function $\frac{n(\boldsymbol{r}_j,t)}{N}$ in real-space where j numbers the cubes into which the normalization volume $\mathcal{V}$ is thought to be subdivided, and $\boldsymbol{r}_j$ denotes the centroid of the particle positions in the respective cube.

The relative frequency $\frac{n(\boldsymbol{r},t)}{N}$ which we shall below express as the mod squared of some wave function $\psi(\boldsymbol{r},t)$, refers - when multiplied by $\Delta^3 r$ - to the subset of identically prepared systems where the particle appears at $\boldsymbol{r}$ and nowhere else simultaneously, otherwise the term "particle" would be meaningless. We think that the commonly used phraseology "probability of **finding** the particle at $\boldsymbol{r}$" is inappropriate because it suggests that one would have placed a detector at $\boldsymbol{r}$ monitoring the occurrence of that particle. However, a detector would - apart from causing various uncontrollable perturbations - terminate the motion of the particle on impact, and hence there would be a shadow area behind the detector where $\frac{n(\boldsymbol{r},t)}{N} \approx 0$, different from the original unperturbed situation. Wherever in the following the quantity $\frac{n(\boldsymbol{r},t)}{N}\,\Delta^3 r$ or $\rho(\boldsymbol{r},t)\,\Delta^3 r$ will appear it is clearly to be understood as the probability of the particle **being** in $\Delta^3 r$ around $\boldsymbol{r}$.

We temporarily number the particles in $\Delta^3\boldsymbol{r}$ at time t by an index i, ($i = 1, 2 \ldots n(\boldsymbol{r},t)$). The particles move, in general, at different velocities $\boldsymbol{v}_i(t)$. We define the ensemble average of the latter as

$$\boldsymbol{v}(\boldsymbol{r},t) = \frac{1}{n(\boldsymbol{r},t)} \sum_{i=1}^{n(\boldsymbol{r},t)} \boldsymbol{v}_i(t)\,. \tag{3}$$

As is familiar from the theory of diffusion, the individual velocities $\boldsymbol{v}_i(t)$ in $\Delta^3\boldsymbol{r}$ will in general be quite different from $\boldsymbol{v}(\boldsymbol{r},t)$ which we shall come back to later in Section 10. By contrast, in Bohm's version of quantum mechanics [28] the true particle trajectories are, for no obvious reason, identified with the flowlines of the velocity field $\boldsymbol{v}(\boldsymbol{r},t)$. This is one of the points where our approach differs fundamentally from Bohm's and reflects a concomitant feature of our definition of $\boldsymbol{v}(\boldsymbol{r},t)$:

In performing the average according to Eq.(3) one sums over velocities $\boldsymbol{v}_i(t)$ of different trajectories that run through sometimes very different regions of the available space of the one-particle system. Hence, they are influenced by the classical field $\boldsymbol{F}(\boldsymbol{r})$ in those regions. This carries over to the ensemble average $\boldsymbol{v}(\boldsymbol{r},t)$. That means: if one places a diaphragm somewhere away from $\boldsymbol{r}$ so that a continuous subset of trajectories is blocked out, $\boldsymbol{v}(\boldsymbol{r},t)$ changes. That kind of non-local sensitivity explains why the flowlines of the field $\boldsymbol{v}(\boldsymbol{r},t)$ are affected by portions of the space which may be far away. The unfamiliar feature of non-locality will be illustrated by a particularly surprising example in Section 9.

As a general property of the stochastic forces that act on the respective particle in each system, we require them to ensure ergodicity in the following sense:
If the system is not explicitly time-dependent, that is, when it is in a bound stationary state and if one would follow the particle on its trajectory within the range it is bound to, one would see it successively occur in all the cubes $\Delta^3 r$ over which - in the ensemble average - all particles of the ensemble are distributed at a certain instant t. Thus, instead of forming the ensemble average according to Eq.(1) it can for a single particle just as well be defined as

$$\lim_{T\to\infty} \frac{\overline{\Delta t}(\boldsymbol{r})}{T} = \rho(\boldsymbol{r})\,\Delta^3 r \tag{4}$$

where $\overline{\Delta t}(\boldsymbol{r})$ denotes the overall time which the particle has spent occurring repeatedly in $\Delta^3 r$ around $\boldsymbol{r}$ within the total time span T.

The velocity $\boldsymbol{v}(\boldsymbol{r})$ can be defined analogously

$$\boldsymbol{v}(\boldsymbol{r}) = \frac{1}{\hat{n}(\boldsymbol{r})} \sum_{i=1}^{\hat{n}(\boldsymbol{r})} \boldsymbol{v}(t_i) \tag{5}$$

where $\hat{n}(\boldsymbol{r}) = \rho(\boldsymbol{r})\,\Delta^3 r$ is the number of times the particle has occurred in $\Delta^3 r$ around $\boldsymbol{r}$, and t_i denotes some point within the time span the particle has spent there the ith time. In realistic cases in which the system under study undergoes transitions between quasi-stationary states, one has to allow T to be finite, and quasi-stationarity can only be ensured if the changes are sufficiently slow on a time scale of unit length T. Practical experience shows, that this applies to the majority of cases. However, in Section 22 we shall give an example where T must be expected to be far too long to justify a classification of the states in a photo emission transition as quasi-stationary.

Yet, the bulk of this article will deal with ensemble averages.

4. Navier-Stokes Equations

If a particle of mass m_0 moves in an environment of kinematic viscosity ν the resulting ensemble average of its velocity $\boldsymbol{v}(\boldsymbol{r},t)$ is just the sum of the so-called "convective velocity"

$\boldsymbol{v}_c(\boldsymbol{r},t)$ and a "diffusive velocity" $\boldsymbol{u}(\boldsymbol{r},t)$ driven by the stochastic forces of the embedding medium:

$$\boldsymbol{v}(\boldsymbol{r},t) = \boldsymbol{v}_c(\boldsymbol{r},t) + \boldsymbol{u}(\boldsymbol{r},t) \tag{6}$$

Employing the Smoluchowski equation (s. Section 33) for the probability density $\rho(\boldsymbol{r},t)$, similarly for the probability current density $\boldsymbol{j}_c(\boldsymbol{r},t) = \rho(\boldsymbol{r},t)\,\boldsymbol{v}_c(\boldsymbol{r},t)$ and invoking Einstein's law [29] for the mean square displacement we obtain a Navier-Stokes-type equation of the form

$$\frac{\partial}{\partial t}(\boldsymbol{v}-\boldsymbol{u}) + [(\boldsymbol{v}+\boldsymbol{u})\cdot\nabla(\boldsymbol{v}-\boldsymbol{u})] - \nu\,\Delta(\boldsymbol{v}-\boldsymbol{u}) = \frac{1}{m_0}\boldsymbol{F}(\boldsymbol{r})\,. \tag{7}$$

with $\boldsymbol{F}(\boldsymbol{r}) = -\nabla V(\boldsymbol{r})$ denoting the external conservative force acting on the particle. The "osmotic" or "diffusive" velocity $\boldsymbol{u}(\boldsymbol{r},t)$ is defined by

$$\boldsymbol{u}(\boldsymbol{r},t) = -\nu\,\frac{\nabla\rho(\boldsymbol{r},t)}{\rho(\boldsymbol{r},t)}\,, \tag{8}$$

or equivalently in terms of the diffusion current density $\boldsymbol{j}_D$

$$\boldsymbol{j}_D(\boldsymbol{r},t) = -\nu\,\nabla\rho(\boldsymbol{r},t) \quad \text{“Fick's law”} \tag{9}$$

where

$$\boldsymbol{j}_D(\boldsymbol{r},t) = \rho(\boldsymbol{r},t)\,\boldsymbol{u}(\boldsymbol{r},t)\,. \tag{10}$$

In the special case when $\boldsymbol{v}_c \equiv 0$ the equation of continuity reduces to

$$\frac{\partial}{\partial t}\rho + \nabla\cdot\rho\,\boldsymbol{u} = 0\,, \tag{11}$$

which on insertion of $\boldsymbol{u}(\boldsymbol{r},t)$ from Eq.(8) attains the form of the diffusion equation

$$\frac{\partial}{\partial t}\rho = \nu\,\Delta\rho\,. \tag{12}$$

On the other hand, when $\nu = 0$ one has $\boldsymbol{u}(\boldsymbol{r},t) \equiv 0$, and hence all particles move now along smooth trajectories $\boldsymbol{r}(t)$ so that the various velocities $\boldsymbol{v}_i(t)$ under the sum in Eq.(3) become equal: $\boldsymbol{v}_i(t) = \boldsymbol{v}(\boldsymbol{r}(t))$. Thus

$$\frac{\partial}{\partial x_k}\boldsymbol{v}(\boldsymbol{r}(t)) \equiv 0 \quad (k=1,2,3) \quad \rightarrow \boldsymbol{v}\cdot\nabla\,\boldsymbol{v} \equiv 0\,,$$

and consequently Eq.(7) reduces to Newton's second law.

The set of equations (7) to (9) will be derived in Section 33.

Eq.(8) may be rewritten

$$\boldsymbol{u}(\boldsymbol{r},t) = -\nu\frac{\nabla\rho(\boldsymbol{r},t)}{\rho(\boldsymbol{r},t)} = -\nu\,\nabla\,\ln[\rho(\boldsymbol{r},t)/\rho_0] \tag{13}$$

where ρ_0 denotes a constant density that has merely been inserted for dimensional reasons. As $\boldsymbol{u}(\boldsymbol{r},t)$ can be expressed as a gradient of a function, we have

$$\nabla \times \boldsymbol{u}(\boldsymbol{r},t) = 0\,, \tag{14}$$

and hence

$$(\boldsymbol{u}\cdot\nabla)\boldsymbol{u} = \nabla\,\frac{\boldsymbol{u}^2}{2}\,. \tag{15}$$

If we, further, make use of the identity

$$\nabla \times (\nabla \times \boldsymbol{a}) = \nabla(\nabla\cdot\boldsymbol{a}) - \Delta\boldsymbol{a} \tag{16}$$

and observe Eq.(14) we obtain $\Delta\boldsymbol{u}(\boldsymbol{r},t) = \nabla(\nabla\cdot\boldsymbol{u}(\boldsymbol{r},t))$. Thus, Eq.(7) in conjunction with Eq.(8) may be cast as

$$m_0\frac{d}{dt}\boldsymbol{v}(\boldsymbol{r},t) = \boldsymbol{F}(\boldsymbol{r}) - \nabla V_{stoch}(\boldsymbol{r},t) + \vec{\Omega}(\boldsymbol{r},t)\,, \tag{17}$$

where $V_{stoch}(\boldsymbol{r},t)$ and $\vec{\Omega}(\boldsymbol{r},t)$ are abbreviations which stand for

$$V_{stoch} = \nu^2\left[\frac{1}{2}\left(\frac{\nabla\rho}{\rho}\right)^2 - \frac{\nabla^2\rho}{\rho}\right], \tag{18}$$

and

$$\vec{\Omega} = \frac{\partial\boldsymbol{u}}{\partial t} + (\boldsymbol{v}\cdot\nabla)\,\boldsymbol{u} - (\boldsymbol{u}\cdot\nabla)\,\boldsymbol{v} + \nu\,\Delta\,\boldsymbol{v}\,.$$

In deriving (18) we have observed that $\frac{1}{\nu}\nabla\boldsymbol{u} = -\frac{\Delta\rho}{\rho} + (\frac{\nabla\rho}{\rho})^2$. Furthermore, we have introduced $\frac{d\boldsymbol{v}}{dt}$ as the "convective (or hydrodynamic) acceleration" which in the present context merely represents an abbreviation

$$\frac{d\boldsymbol{v}(\boldsymbol{r},t)}{dt} = \frac{\partial\,\boldsymbol{v}}{\partial t} + \boldsymbol{v}\cdot\nabla\boldsymbol{v}\,. \tag{19}$$

The "stochastic potential" $V_{stoch}(\boldsymbol{r},t)$ depends on ν^2 whereas $\vec{\Omega}(\boldsymbol{r},t)$ is proportional to ν.

The latter constant is associated with the occurrence of the stochastic forces which - in the absence of an external force $\boldsymbol{F}(\boldsymbol{r})$ - would slow down the particle within a characteristic time τ.

Since the physical vacuum does not represent an embedding medium whose stochastic forces can cause a particle to slow down completely, we modify the character of the stochastic forces by assuming that they change periodically after a laps of $\approx \tau$ sec from down-slowing "Brownian" to motion enhancing "anti-Brownian" and vice versa. The "anti-Brownian" forces act as if the kinematic viscosity would have a negative sign. Hence, the corresponding equation of motion has the form

$$m_0\frac{d}{dt}\boldsymbol{v}(\boldsymbol{r},t) = \boldsymbol{F}(\boldsymbol{r}) - \nabla V_{stoch}(\boldsymbol{r},t) - \vec{\Omega}(\boldsymbol{r},t)\,. \tag{20}$$

In Section 15 we give an example of an embedding medium that acts on a test particle by alternating Brownian/anti-Brownian forces.

If we now additionally assume that the temporal changes that occur with all quantities in the two Eqs.(17) and (20) are slow on a scale of unit length τ - which is the standard requirement also in diffusion theory - the motion of the ensemble will be governed by the arithmetic mean of these equations , that is by

$$m_0 \frac{d}{dt}\boldsymbol{v}(\boldsymbol{r},t) = \boldsymbol{F}(\boldsymbol{r}) - \nabla V_{stoch}(\boldsymbol{r},t) . \tag{21}$$

A more detailed definition of the stochastic forces that ensure "conservative diffusion" will be given in Section 12. One might suspect that our subdivision into a Brownian "B"-ensemble and an anti-Brownian "A"-ensemble is unnecessarily clumsy and could be avoided at the outset by assuming vacuum forces that neither possess down-slowing components nor counterparts that effect motion enhancement, but rather consist of random (Gaussian) forces whose components form a normal distribution. However, from Einstein's theory of Brownian motion the kinematic viscosity (or "diffusion constant") emerges as

$$\nu = \frac{k_B T \tau}{m_0} \quad \text{(Einstein: } \overline{\Delta x_i \, \Delta x_j} = 2\,\delta_{ij}\,\nu\,\Delta t ; \quad i,j = 1,2,3 \text{)} \tag{22}$$

where m_0 is the mass of the particle under study, Δx_i , Δx_j are displacements of its position and T is the effective temperature of the embedding medium. This temperature enters into the derivation as the width of the distribution of the random (Gaussian) forces that act on the particle apart from the **directional** down-slowing force. Because of the latter there is a down-slowing motion that we have already alluded to. The associated time constant is denoted by τ. Equating the down-slowing forces to zero amounts to $\tau \to \infty$ which would yield infinite kinematic viscosity. Hence, there is no alternative to our approach.

Obviously, the physical dimension of the numerator of the above fraction in Eq.(22) is that of an action, i.e. energy$\times$time. As ν appears via $V_{stoch}(\boldsymbol{r},t)$ in Eq.(21) which is constructed to describe dissipationless motion in a "stochastic vacuum" whose effect on a particle can only be associated with a new constant of nature, one is justified in equating $k_B\, T\, \tau$ with $\frac{1}{2}\,\hbar$ where $h = 2\pi\,\hbar$ is Planck's constant. Of course, instead of $1/2$ there could be any other dimensionless prefactor in front of $\hbar$, but it turns out that the numerical results of all quantum mechanical calculations that follow from Eq.(21) are only consistent with the above choice. *Clearly, that choice can only be made once and for all.*

Having thus calibrated the "vacuum-ν"we rewrite Eq.(21) in the form

$$m_0 \frac{d}{dt}\boldsymbol{v}(\boldsymbol{r},t) = \boldsymbol{F}(\boldsymbol{r}) - \nabla V_{QP}(\boldsymbol{r},t) , \tag{23}$$

where we have substituted the subscript of $V_{stoch}(\boldsymbol{r},t)$ by "QP"

$$V_{QP} = \frac{\hbar^2}{4\,m_0} \left[\frac{1}{2} \left(\frac{\nabla \rho}{\rho} \right)^2 - \frac{\nabla^2 \rho}{\rho} \right] \text{ "quantum potential"} , \tag{24}$$

and we have set

$$\frac{\hbar}{2m_0} = \nu = \frac{k_B T \tau}{m_0} . \tag{25}$$

The "quantum potential" has first been introduced by de Broglie [30] and later been taken up again by David Bohm [28]. Obviously Eq.(23) may be viewed as a modification of Newton's second law. If not stated differently, we shall assume $\boldsymbol{F}(\boldsymbol{r})$ to be conservative:

$$\boldsymbol{F}(\boldsymbol{r}) = -\nabla V(\boldsymbol{r})\,. \tag{26}$$

The assumption made above, viz. that all changes of the ensemble properties have to be sufficiently slow on a time scale of unit length τ may raise questions about the validity of such a constraint. Eqs.(23) and (24) will prove equivalent to the time-dependent Schrödinger equation whose validity is unquestioned at the non-relativistic level. Hence, τ is obviously sufficiently small within the experimentally tested range of the Schrödinger equation. Conversely, as one may conclude then from Eq.(25) that the "effective temperature" of the vacuum must be very high compared to those temperatures commonly considered in applied thermodynamics and astrophysics.

Fundamentally different from our approach Bohm [28] derives Eqs.(23) and (24) by choosing the opposite direction starting from the time-dependent Schrödinger equation which he just considers given. Hence, he does not offer any new insight into what makes the motion of a microscopic particle different from what classical mechanics predicts. In the context of Bohm's mechanics Eq.(23) is frequently cast such that it resembles the Hamilton-Jacobi equation. To this end one sets

$$\boldsymbol{v}(\boldsymbol{r},t) = \frac{1}{m_0}\nabla S(\boldsymbol{r},t) \quad \text{and} \quad \rho(\boldsymbol{r},t) = R^2(\boldsymbol{r},t)\,.$$

This implies that $\boldsymbol{v}(\boldsymbol{r},t)$ is irrotational, which is at best plausible, but remains unproven.

Eq.(23) in conjunction with (24) then attains the form

$$\frac{1}{m_0}\nabla\left[\frac{\partial S}{\partial t} + \frac{(\nabla S)^2}{2\,m_0} + V(\boldsymbol{r}) - \frac{\hbar^2}{2\,m_0}\frac{\Delta R}{R}\right] = 0\,.$$

This is equivalent to

$$\frac{\partial S}{\partial t} + \frac{(\nabla S)^2}{2\,m_0} + V(\boldsymbol{r}) - \frac{\hbar^2}{2\,m_0}\frac{\Delta R}{R} = 0\,,$$

and becomes identical with the Hamilton-Jacobi equation in the limit $\hbar \to 0$. However, the connection to classical mechanics is far more evident from Eq.(23), which reduces to Newton's second law

$$\boldsymbol{F} = m_0\,\frac{d}{dt}\boldsymbol{v}$$

as $\hbar$ tends to zero. In addition, Eq.(23) lends itself to a thought-experiment that is particularly illustrative of the quantum character of particle motion.

One starts with setting $\hbar = 0$ and assumes that all particles of the ensemble commence their motion under identical initial conditions. Their positions and trajectories will coincide then at any later time. One now lets $\hbar$ take on a finite value. As a consequence of the now occurring stochastic forces whose action on some particle is statistically independent from that on any other particle, the particle positions start diverging and form a cloud around the formerly common position along the trajectory. The particles of the ensemble now reach positions that are not accessible under energy conservation. It is hence obvious that the vacuum provides an embedding medium of a "universal noise" consisting of energy fluctuations which cause shifts of the individual particle trajectories such that the classical momentum and the energy are conserved on the average. This is reflected in the expectation value of the "vacuum force" $\boldsymbol{F}_{QP} = -\nabla V_{QP}(\boldsymbol{r}, t)$ which equals zero:

$$\int \rho(\boldsymbol{r}, t)\, \boldsymbol{F}_{QP}(\boldsymbol{r}, t)\, d^3r = 0\,. \tag{27}$$

We shift the proof of this equation to Section 13. Eq.(27) may be interpreted in the sense that the particles undergo only reversible scatterings. Figuratively speaking, the vacuum keeps track of the energy balance and remembers at later positions of a particle departures from its classical momentum and energy that occurred at previous positions. The undulatory properties of the **probability density** reside in this memory effect which gives rise to an unfamiliar non-locality. Hence, from our point of view it is illegitimate to correlate these properties with a **wave-like character of the particle**. We definitely side with Nevill Mott (1964) who argues:

"Students should not be taught to doubt that electrons, protons and the like are particles....The waves cannot be observed in any way than by observing particles."

5. The Time-independent Schrödinger Equation

As a first application we discuss the stationary state of a particle that is bound to a potential without symmetry elements. Hence, the real-space dependence of the potential does not display any distinct direction. That means, when a particle of the ensemble appears with a velocity $\boldsymbol{v}_i(t)$ in the elementary volume $\Delta^3 r$ around $\boldsymbol{r}$ there will always be another particle in that volume with approximately the opposite velocity, so that

$$\boldsymbol{v}(\boldsymbol{r}, t) = \frac{1}{n(\boldsymbol{r}, t)} \sum_{i}^{n(\boldsymbol{r}, t)} \boldsymbol{v}_i(t) \equiv 0\,. \tag{28}$$

Hence, if one recalls (26) Eq.(23) reduces to

$$\nabla \left(\frac{\hbar^2}{4\,m_0} \left[-\frac{1}{\rho} \nabla^2 \rho + \frac{1}{2} \left(\frac{\nabla \rho}{\rho} \right)^2 \right] + V(\boldsymbol{r}) \right) = 0\,.$$

This is equivalent to:

$$\frac{\hbar^2}{4\,m_0} \left[-\frac{1}{\rho} \nabla^2 \rho + \frac{1}{2} \left(\frac{\nabla \rho}{\rho} \right)^2 \right] + V(\boldsymbol{r}) = E\,, \tag{29}$$

where E denotes a constant. Eq.(29) represents a **non-linear** partial differential equation in $\rho(\boldsymbol{r})$.

On substituting $\rho(\boldsymbol{r})$ by a function $\psi(\boldsymbol{r})$ defined through

$$\rho(\boldsymbol{r}) = \psi^2(\boldsymbol{r}) \tag{30}$$

one obtains because of

$$\nabla\rho = 2\,\psi\,\nabla\psi\,; \quad \frac{1}{2}\left(\frac{\nabla\rho}{\rho}\right)^2 = 2\left(\frac{\nabla\psi}{\psi}\right)^2$$

and

$$\nabla^2\rho = 2\,\psi\,\nabla^2\psi + 2\,(\nabla\psi)^2$$
$$-\frac{1}{\rho}\,\nabla^2\rho = -2\,\frac{\nabla^2\psi}{\psi} - 2\left(\frac{\nabla^2\psi}{\psi}\right)^2$$

a **linear** differential equation

$$\frac{\hbar^2}{2\,m_0}\left[-\frac{1}{\psi}\nabla^2\psi\right] + V(\boldsymbol{r}) = E \quad \text{that is}$$
$$-\frac{\hbar^2}{2\,m_0}\,\nabla^2\psi + V(\boldsymbol{r})\,\psi = E\,\psi \tag{31}$$

which constitutes the time-independent Schrödinger equation.

6. Including Currents

For the familiar problem of a particle in a box Eq.(31) reduces in the one-dimensional case to

$$\left[\frac{d^2}{dx^2} + k^2\right]\psi(x) = 0 \tag{32}$$

where we have set

$$k^2 = \tfrac{2m_0}{\hbar^2}\,E\,; \quad E = \tfrac{m_0}{2}\,v^2$$

and

$$V(x) = \begin{cases} 0 & \text{for } 0 \le x \le a \\ \infty & \text{else} \end{cases}$$

The solutions

$$\psi(x) = \frac{1}{\sqrt{a/2}}\,\sin k_n x \;\text{ where } k_n = \frac{\pi}{a}\,n\,; \quad n = 1, 2, 3.. \tag{33}$$

may be recast as

$$\psi(x) = \tfrac{1}{\sqrt{2}}\,[\psi_+(x) + \psi_-(x)]$$

where

$$\psi_\pm(x) = \tfrac{1}{\sqrt{a}}\,e^{\pm\, i\varphi(x)}\,; \quad \varphi(x) = k_n x + \tfrac{\pi}{2}\,.$$

In the spirit of our approach the two independent solutions to the differential equation (32), $\psi_\pm(x)$, refer to the particle moving at a velocity $v_n = \frac{\hbar\,k_n}{m_0}$ either to the right or (after reflection at $x = a$) to the left where it is reflected again at $x = 0$.

We are thus led to surmise that we have in the general case of a freely moving particle

$$\psi(\boldsymbol{r}) = |\psi(\boldsymbol{r})|\, e^{i\varphi(\boldsymbol{r})} \quad \text{and} \quad \boldsymbol{v}(\boldsymbol{r}) = \frac{\hbar}{m_0}\, \nabla\varphi(\boldsymbol{r})\,. \tag{34}$$

The validity of this conjecture will be shown in Section 7.

In a stationary state of the one-particle system in which $\frac{\partial}{\partial t}\boldsymbol{v} = 0$ but $\boldsymbol{v}(\boldsymbol{r}) \neq 0$ we have according to Eq.(19) $\frac{d}{dt}\boldsymbol{v} = \boldsymbol{v}\cdot\nabla\boldsymbol{v} = \frac{1}{2}\nabla\boldsymbol{v}^2$ where we have exploited in advance that, according to Eq.(34), $\boldsymbol{v}(\boldsymbol{r})$ is irrotational. Hence, in the presence of a stationary current Eq.(29) contains the kinetic energy $\frac{m_0}{2}\,\boldsymbol{v}^2$ as an additional term, that is

$$\frac{\hbar^2}{4\,m_0}\left[-\frac{1}{\rho}\nabla^2\rho + \frac{1}{2}\left(\frac{\nabla\rho}{\rho}\right)^2\right] + V(\boldsymbol{r}) + \frac{m_0}{2}\,\boldsymbol{v}^2 = E\,. \tag{35}$$

If one now makes use of Eq.(34) instead of Eq.(30)

$$\rho(\boldsymbol{r}) = |\psi(\boldsymbol{r})|^2 = \left(\psi(\boldsymbol{r})\, e^{-i\varphi(\boldsymbol{r})}\right)^2 \tag{36}$$

and substitutes $\boldsymbol{v}(\boldsymbol{r})$ by $\frac{\hbar}{m_0}\,\nabla\varphi(\boldsymbol{r})$ the bracketed term in Eq.(35) becomes

$$\begin{aligned}\frac{\hbar^2}{4\,m_0}\left[-\frac{1}{\rho}\nabla^2\rho + \frac{1}{2}\left(\frac{\nabla\rho}{\rho}\right)^2\right] &= -\frac{\hbar^2}{2\,m_0}\,\frac{1}{\psi}\nabla^2\psi + \underbrace{\frac{\hbar^2}{2m_0}\,(\nabla\varphi)^2}_{=\frac{m_0}{2}\,\boldsymbol{v}^2} \\ &\quad \underbrace{+\,i\left[\frac{\hbar^2}{2\,m_0}\,\nabla^2\varphi + \frac{\hbar^2}{2\,m_0}\left(2\,\nabla\varphi\cdot\frac{\nabla\psi}{\psi}\right)\right]}_{=\frac{i\hbar}{2}\left[\nabla\cdot\boldsymbol{v}+2\boldsymbol{v}\cdot\frac{\nabla\psi}{\psi}\right]}\,.\end{aligned}$$

Invoking the equation of continuity in the form

$$\nabla\cdot\boldsymbol{j} = \nabla\cdot\rho\,\boldsymbol{v} = \rho\,\nabla\cdot\boldsymbol{v} + \boldsymbol{v}\cdot\nabla\rho = 0$$

it can readily be shown that the term $i[...]$ on the right-hand side equals $-m_0\,\boldsymbol{v}^2$. Hence we have from Eq.(35)

$$-\frac{\hbar^2}{2\,m_0}\,\frac{1}{\psi}\nabla^2\psi + V(\boldsymbol{r}) = E\,,$$

that is

$$-\frac{\hbar^2}{2\,m_0}\,\nabla^2\psi + V(\boldsymbol{r})\,\psi = E\,\psi \tag{37}$$

as before without a current.

It should be noticed that φ may well be time-dependent even when $\nabla\varphi$ is not, that is, we have in general

$$\varphi(\boldsymbol{r},t) = \varphi_0(\boldsymbol{r}) + f(t)$$

where $f(t)$ is a real-valued function. In this case the wave function $\psi(\boldsymbol{r},t)$ attains the form

$$\psi(\boldsymbol{r},t) = \hat{\psi}(\boldsymbol{r})\, e^{i\, f(t)} \quad \text{where} \quad \hat{\psi}(\boldsymbol{r}) = |\hat{\psi}(\boldsymbol{r})|\, e^{i\,\varphi_0(\boldsymbol{r})} \tag{38}$$

and hence, its time-derivative may be cast as

$$i\hbar\,\frac{\partial}{\partial t}\psi(\boldsymbol{r},t) = -\hbar\,\dot{f}\,\psi(\boldsymbol{r},t)\,. \tag{39}$$

Since $f(t)$ is primarily unspecified and $-\hbar\dot{f}$ possesses the dimension of an energy the latter may justifiably be identified with the energy E which is the only energy-related constant characterizing the wave function of the system:

$$-\hbar\,\dot{f} = E\,; \quad \text{that is} \quad i f(t) = -\frac{i}{\hbar}E\,t\,. \tag{40}$$

As a result, we have from Eq.(38)

$$\psi(\boldsymbol{r},t) = \hat{\psi}(\boldsymbol{r})\, e^{-\frac{i}{\hbar}E\,t} \tag{41}$$

for a wave function in a stationary state. Furthermore, we have from Eqs.(37), (39) and (40)

$$-\frac{\hbar^2}{2\,m_0}\,\nabla^2\psi(\boldsymbol{r},t) + V(\boldsymbol{r})\,\psi(\boldsymbol{r},t) = i\hbar\,\frac{\partial}{\partial t}\psi(\boldsymbol{r},t) \tag{42}$$

which constitutes the time-dependent Schrödinger equation. Its validity is here still restricted to stationary systems, but it will be shown in Section 10 that it retains this form also for non-stationary systems. However, in order to achieve this consistency, one has to introduce the negative sign in Eq.(40) which seems to lack reason and can actually not be justified without reference to Section 10.

7. The Velocity Potential and Phase Uniqueness

We rewrite Eq.(23) in the form

$$m_0\frac{d}{dt}\boldsymbol{v}(\boldsymbol{r},t) = -\nabla P(\boldsymbol{r},t) \tag{43}$$

where

$$P(\boldsymbol{r},t) = \frac{1}{m_0}\left[V(\boldsymbol{r}) + V_{PQ}(\boldsymbol{r},t)\right],$$

and we have made use of Eq.(19) defining the “hydrodynamic” or convective acceleration

$$\frac{d}{dt}\boldsymbol{v}(\boldsymbol{r},t) = \frac{\partial}{\partial t}\boldsymbol{v} + (\boldsymbol{v}\cdot\nabla)\,\boldsymbol{v}\,.$$

In hydrodynamics Eq.(43) corresponds to the Euler equation of perfect (frictionless) fluids and constitutes the starting point of Helmholtz’s theory of vortices. Thomson’s more elaborate analysis on vortices [33] builds on Helmholtz’s considerations. We confine ourselves here to reporting only the general ideas as far as they directly concern the present theory.

If we set $\vec{\omega} = \nabla \times \boldsymbol{v}$ for the curl of the ensemble average of the particle velocity, we have from Eq.(16)

$$(\boldsymbol{v} \cdot \nabla)\, \boldsymbol{v} = \nabla \frac{\boldsymbol{v}^2}{2} - \boldsymbol{v} \times \vec{\omega}\,.$$

We now form the curl of Eq.(43) and use this expression together with Eq.(19). The result may be cast as

$$\frac{\partial}{\partial t}\vec{\omega}(\boldsymbol{r},t) - \nabla \times [\boldsymbol{v}(\boldsymbol{r},t) \times \vec{\omega}(\boldsymbol{r},t)] = 0\,, \tag{44}$$

where we have used $\nabla \times \nabla P = 0$ and $\nabla \times \nabla \boldsymbol{v}^2 = 0$. One recognizes from Eq.(44) that $\frac{\partial}{\partial t}\vec{\omega}(\boldsymbol{r},t)|_{t=0}$ becomes zero for some chosen time, which we here equate to zero for convenience, if $\vec{\omega}(\boldsymbol{r},t)|_{t=0} = 0$ at that time. Forming the time derivative of Eq.(44) and setting again $t = 0$ we see that the second time derivative of $\vec{\omega}(\boldsymbol{r},t)$ vanishes as well. This can be carried further to any higher order of the time derivative. Hence, the system stays irrotational if it is irrotational at $t = 0$. We now consider an ensemble of free particles ($\boldsymbol{F}(\boldsymbol{r}) \equiv 0$) when $\hbar = 0$. They may start their motion at $t = 0$ at the same point in real-space and with the same momentum $\boldsymbol{p}_0 = m_0\, \boldsymbol{v}_0$. If one allows $\hbar$ to attain its natural value, the particle positions diverge and form a point cloud. Outside this cloud there are no particles and therefore $\boldsymbol{v}(\boldsymbol{r},t) \equiv 0$. Since the ensemble does not exchange momentum with the vacuum on the average and consequently no angular momentum, we have everywhere within the space of normalization

$$\nabla \times \boldsymbol{v}(\boldsymbol{r},t) = \vec{\omega}(\boldsymbol{r},t) \equiv 0 \quad \forall \boldsymbol{r}, t\,. \tag{45}$$

If one now turns on some (physically realistic) potential $V(\boldsymbol{r})$, weighting it with a smooth switch function from zero to one, starting at $t = t_0$, the velocity distribution $\boldsymbol{v}(\boldsymbol{r},t)$ for $t > t_0$ will now change differently, of course, but because of Eqs.(44) and (45) for $t = t_0$, we have as before $\vec{\omega}(\boldsymbol{r},t_0) \equiv 0$ and $\frac{\partial}{\partial t}\vec{\omega}(\boldsymbol{r},t)|_{t=t_0} \equiv 0$ which again applies to any higher order time-derivative at $t = t_0$. We thus arrive at the conclusion that an ensemble whose equation of motion is given by Eq.(43) is irrotational. In other words, $\boldsymbol{v}(\boldsymbol{r},t)$ possesses a potential $\varphi(\boldsymbol{r},t)$ which we express in the form

$$\boldsymbol{v}(\boldsymbol{r},t) = \frac{\hbar}{m_0} \nabla\varphi(\boldsymbol{r},t)\,. \tag{46}$$

Because of the prefactor $\hbar/m_0$ the function $\varphi(\boldsymbol{r},t)$ becomes dimensionless. Eq.(46) may equivalently be cast as

$$\varphi(\boldsymbol{r}) = \frac{m_0}{\hbar} \int_{\boldsymbol{r}_0}^{\boldsymbol{r}} \boldsymbol{v}(\boldsymbol{r}') \cdot d\boldsymbol{r}' \tag{47}$$

where we have omitted the time-dependence in confining ourselves to a stationary state situation. As in the theory of perfect fluids there may be singular vortex lines which occur if $V(\boldsymbol{r})$ possesses axial or spherical symmetry. A vortex line then defines an axis of quantization. The latter may be regarded as the boundary line of a semi-plane. Even in the presence of a vortex line, can $\varphi(\boldsymbol{r})$ be defined such that it remains unique if one only stipulates that the starting point of the line integral in Eq.(47), $\boldsymbol{r}_0$, lies on one side of this semi-plane and that the path along which the integral is performed never crosses that semi-plane. The point $\boldsymbol{r}_0$ may be chosen at will. In general, $\varphi(\boldsymbol{r})$ will now be discontinuous at the semi-plane. The ensuing section deals with this particular problem.

8. Quantization of Angular Momentum

The primary objective of this section is to disprove Wallstrom's notable objection [11] against Madelung's conviction, also held by other theorists of this school of thought, that Newton's modified second law (23) is equivalent to the time-dependent Schrödinger equation which we shall derive below. In so doing we have to exploit the uniqueness of the velocity potential shown in the preceding section. By contrast, in standard quantum mechanics the time-dependent Schrödinger equation is regarded as given. It is customarily converted into the equation of continuity

$$\dot{\rho} + \nabla \cdot \left[\tfrac{\hbar}{2i\, m_0}\{\psi^* \nabla \psi - \psi \nabla \psi^*\}\right] = 0$$

to show that the bracketed expression has to be interpreted as the current density $\boldsymbol{j}(\boldsymbol{r}, t)$. This conclusion is only legitimate if $\boldsymbol{j}(\boldsymbol{r}, t)$ has been proven to be irrotational which, however, is only tacitly presupposed. Inserting

$$\psi(\boldsymbol{r}, t) = |\psi(\boldsymbol{r}, t)|\, e^{i\varphi(\boldsymbol{r}, t)} \tag{48}$$

into the bracketed expression yields

$$\boldsymbol{j}(\boldsymbol{r}, t) = |\psi(\boldsymbol{r}, t)|^2 \underbrace{\frac{\hbar}{m_0} \nabla\varphi(\boldsymbol{r}, t)}_{=\boldsymbol{v}(\boldsymbol{r},t)},$$

as a consequence of which one obtains Eq.(46). If one is dealing with a stationary state whose velocity field contains a vortex line, e. g. an excited state of a hydrogen electron possessing an orbital momentum, we have

$$\oint \boldsymbol{v}(\boldsymbol{r}) \cdot d\boldsymbol{r} \neq 0 \tag{49}$$

for any path encircling the vortex line (=quantization axis). On inserting here $\boldsymbol{v} = \frac{\hbar}{m_0} \nabla\varphi$ one obtains

$$\int_{\boldsymbol{r}_0}^{\boldsymbol{r}} \nabla\varphi(\boldsymbol{r}) \cdot d\boldsymbol{r} = \varphi(\boldsymbol{r}) - \varphi(\boldsymbol{r}_0) \neq 0 \tag{50}$$

where $\boldsymbol{r}$ and $\boldsymbol{r}_0$ are two points facing each other across the semi-plane, introduced in Section 7, at an infinitesimal distance. Thus, in general the phase of the wave function, and consequently the wave function itself, will be discontinuous at the semi-plane as opposed to $\rho(\boldsymbol{r})$ and $\boldsymbol{j}(\boldsymbol{r})$ which may be presupposed to be smooth functions everywhere.

Clearly, as follows from Eq.(48), $\psi(\boldsymbol{r})$ remains continuous at the semi-plane if

$$\varphi(\boldsymbol{r}) - \varphi(\boldsymbol{r}_0) = 2m\,\pi \quad \text{where} \quad m = \text{integer}. \tag{51}$$

But there is no immediately obvious reason why one should require $\psi(\boldsymbol{r})$ to be continuous because only $\rho(\boldsymbol{r})$ and $\boldsymbol{j}(\boldsymbol{r})$ can be regarded as reflecting physical properties of the system.

We are hence led to conclude that without an additional argument **neither** our derivation **nor** standard quantum mechanics yields a justification of the proven relation

$$m_0 \oint \boldsymbol{v}(\boldsymbol{r}) \cdot d\boldsymbol{r} = 2m\pi\,\hbar = m\,h \quad \text{where } m = \text{integer} \tag{52}$$

which comprises Eqs.(46), (49) to (51). This has already been pointed out more than 75 years ago by Pauli [26] and Born and Jordan [27]. As opposed to these considerations Wallstrom states in his paper [11]: *"To the best of my knowledge, this condition* (Eq.(52)) *has not yet found any convincing explanation* **outside the context of the Schrödinger equation***"*.

This is definitely incorrect: within that context the assumption of continuity (Eq.(51)) has to be justified by an additional argument as well.

What else necessitates then the continuity of $\psi(\boldsymbol{r})$ everywhere?

To keep the formalism as simple as possible we confine the considerations to a two-dimensional one-particle system in which the potential $V(\boldsymbol{r})$ is cylindrically symmetric. For this case the time-independent Schrödinger equation (31) attains the form

$$\left(-\frac{\hbar^2}{2\,m_0}\left[\frac{\partial^2}{\partial r^2}+\frac{1}{r}\frac{\partial}{\partial r}+\frac{1}{r^2}\frac{\partial^2}{\partial\varphi^2}\right]+V(r)\right)\phi(r,\varphi)=E\,\phi(r,\varphi)\,.$$

If one introduces

$$\phi(r,\varphi) = R(r)\,e^{i\tilde{\varphi}} \quad \text{where} \quad \tilde{\varphi} = k\,\varphi \quad \text{and} \quad k \in \Re\,,$$

the Schrödinger equation becomes

$$\left[\frac{d^2}{dr^2}+\frac{1}{r}\frac{d}{dr}-\frac{k^2}{r^2}+\varepsilon-v(r)\right]R(r)=0 \quad \text{where} \quad \varepsilon - v(r) = \tfrac{2m_0}{\hbar^2}[E-V(r)]\,. \tag{53}$$

For any choice of k and for an appropriate value of ε one can always find a normalizable solution to this differential equation which is regular at $r = 0$ and vanishes exponentially for $r \to \infty$, provided that $V(r)$ is not ill-behaved and allows bound states. In this case one can always think of performing a numerical integration of this differential equation to obtain a bound state $R(r)$. If $R(r)$ satisfies Eq.(53), then $\phi(r,\varphi)$ satisfies the Schrödinger-Gleichung everywhere even if k is **non-integer**. True, $\phi(r,\varphi)$ is discontinuous within the interval $0 < \varphi \le 2\pi$ for $\varphi = 0$ since

$$e^{i\,k\,2\pi} \neq 1\,,$$

however $|\phi(r,\varphi)|^2$ and

$$\boldsymbol{v}(r,\varphi) = \tfrac{\hbar}{m_0}\tfrac{1}{r}\tfrac{\partial}{\partial\varphi}\,\tilde{\varphi}(\varphi)\,\boldsymbol{e}_\varphi = \tfrac{\hbar}{m_0}\tfrac{k}{r}\,\boldsymbol{e}_\varphi$$

remain smooth functions everywhere, and the associated angular momentum is

$$\frac{1}{2\pi}\oint m_0\,\boldsymbol{v}(r,\varphi)\cdot d\boldsymbol{r} = \frac{\hbar}{2\pi}\int_0^{2\pi}\frac{k}{r}\,r\,d\varphi = k\,\hbar\,.$$

We now consider two functions R_{k_1} and R_{k_2} which solve Eq.(53) for two non-integer values k_1 and k_2, respectively. We choose an appropriate normalization

$$2\pi \int_0^\infty R_{k_1}^2(r)\, r\, dr = 1 \qquad 2\pi \int_0^\infty R_{k_2}^2(r)\, r\, dr = 1$$

and cast their energy eigenvalues $\varepsilon_{k_1}, \varepsilon_{k_2}$, obtained from numerical integration, for example, as

$$\varepsilon_{k_1} = \hbar\, \omega_{k_1} \quad \text{und} \quad \varepsilon_{k_2} = \hbar\, \omega_{k_2}\,.$$

As already alluded to in Section 6 the time-dependent Schrödinger equation which we are going to derive in Section 10 , has the form

$$\left(-\frac{\hbar^2}{2\, m_0}\left[\frac{\partial^2}{\partial r^2} + \frac{1}{r}\frac{\partial}{\partial r} + \frac{1}{r^2}\frac{\partial^2}{\partial \varphi^2}\right] + V(r)\right)\phi(r,\varphi,t) = i\,\hbar\,\frac{\partial}{\partial t}\,\phi(r,\varphi,t)\,.$$

Since it constitutes a linear partial differential equation it will be satisfied also by a linear combination of the two functions

$$\begin{aligned}\phi(\boldsymbol{r},t) = \tfrac{1}{\sqrt{2\pi}}[&c_{k_1}\, R_{k_1}(r)\, e^{i(\,k_1\,\varphi - \omega_{k_1}\,t)} \\ &+ c_{k_2}\, R_{k_2}(r)\, e^{i(\,k_2\,\varphi - \omega_{k_2}\,t)}]\,.\end{aligned}$$

Without loss of generality the two coefficients c_{k_1}, c_{k_2} may be chosen as real-valued. We now form the expression for the norm of $\phi(\boldsymbol{r},t)$

$$\begin{aligned}\int \underbrace{|\phi(\boldsymbol{r},t)|^2}_{=\rho(\boldsymbol{r},t)}\, d^2r \overset{!}{=} 1 = c_{k_1}^2 + c_{k_2}^2 + c_{k_1}\, c_{k_2}\, I_{k_1\,k_2} \\ \times \left[\int_0^{2\pi} e^{i(k_2-k_1)\varphi}\, d\varphi\ e^{i(\omega_{k_1}-\omega_{k_2})\,t} + \text{c.c.}\right]\,.\end{aligned}$$

Here $I_{k_1\,k_2}$ denotes

$$I_{k_1\,k_2} = \int_0^\infty R_{k_1}(r)\, R_{k_2}(r)\, r\, dr\,.$$

If $k_2 - k_1$ is non-integer, $I_{k_1\,k_2}$ does not vanish, and the norm of $\phi(\boldsymbol{r},t)$ becomes time-dependent which is inadmissible, of course. Hence $k_2 - k_1$ has to be integer. Since the groundstate of the system, associated with ε_{k_1}, for example, is definitely associated with zero current, i. e. $k_1 = 0$, it follows immediately that k_2 must be integer for any state with angular momentum ($k_2 \neq 0$).

9. An Instructive Objection, Quantum Beats and a Possible Which-Way Detection

An apparently serious objection against a stochastic foundation of quantum mechanics along the lines of the preceding sections goes back to Mielnik and Tengstrand [31]. The authors refer to an experimental setup as sketched in Figure 1 where the test particle enters from a distant source on the left-hand side and is kept within a tube that extends up to a screen on the right. The tube contains an impermeable partition that completely seals off the upper part (A) from the lower part (B). It possesses a limited, but macroscopic length of, say, 10 cm. The authors argue that according to conventional quantum mechanics the incoming wave would split up into an upper and totally independent lower portion. Yet both portions retain their capability of interfering with each other when they merge again within the area C and beyond. However, if the wave portions are replaced by the set of irregular trajectories which stochastic quantum mechanics claims to be an equivalent of, it seems to be very unlikely that stochastic-force controlled trajectories can preserve information over so long a distance as well as waves. This criticism amounts to perceiving the preceding derivation of the Schrödinger equation from Eq.(23) as ill-founded or even erroneous. It is just the solution to the Schrödinger equation for the particular setup around which the present authors' consideration revolve. On the other hand, it is easy to verify the validity of the derivation. There is simply no step where one may be in doubt. But one has to keep in mind that the solutions $\psi(\boldsymbol{r}) = |\psi(\boldsymbol{r})|\, e^{i\varphi(\boldsymbol{r})}$ to the Schrödinger equation provide only information on ensemble properties and not on a particular trajectory that is a member of the ensemble under study. For example, the velocity $\boldsymbol{v}(\boldsymbol{r}) = \frac{\hbar}{m_0}\nabla\varphi(\boldsymbol{r})$ at some point in the area marked C represents such an average over all trajectories of the ensemble running through that point. This ensemble defines the probability in which direction a particular

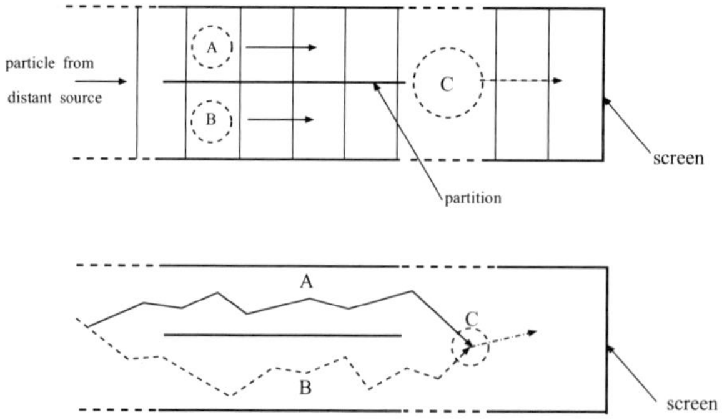

Figure 1. Interference of trajectories.

particle that has arrived at C, e. g. along the "A"-trajectory, will move further. (S. Figure 1, lower panel.) This is analogous to considerations we shall discuss in the context of the Smoluchowski equation (Section 33). The properties of the ensemble are just an image of the property of the vacuum fluctuations to ensure the absence of dissipation. This manifests itself in the fact that $\boldsymbol{v}(\boldsymbol{r})$ is irrotational as in ideal fluids. An individual particle that has moved along the "A"-trajectory and arrives at C "feels", so to speak, the possibility of a "B"-trajectory. As stated above, it continues its trajectory depending also on the family of "B"-trajectories running through C. If the partition in the tube would be elongated and the point C correspondingly shifted to the right, irrespective of how much, the "A"- and "B"subset of trajectories would now be different, but the scattering probability at C of a particle that has moved along an "A"(or "B")-trajectory would still be influenced by possible "B" (or "A")-trajectories. Furthermore, if one would place some spatially confined array into the upper part of the setup which would cause an accelerating electric field, the "A"-trajectories would change accordingly and give rise to a different interference pattern within the "C"-range.

To make the surprising content of this observation even more striking we consider a situation where one has particles enter the setup one by one from the left so that only one particle traverses the setup at a time. First, we switch the electrostatic array off so that there is no extra potential along the "A"-trajectory. If one has placed a detector, an electron multiplier, for example, at some position $\boldsymbol{r}_{screen}$ on the screen, it would monitor the incoming electrons at a certain rate. These electrons come either along an 'A"- or a "B"-trajectory. Once the extra field has been turned on, the count rate at $\boldsymbol{r}_{screen}$ changes. Although an electron may have moved along the unmodified "B"-portion of the setup, it feels the modification of the "A"-portion when it arrives at "C". As explained above, this is due to the change of the vacuum scattering probability at C. Electrons that have arrived at some elementary volume within C and have so far preferentially been scattered into $\boldsymbol{r}_{screen}$ are now also scattered to other positions on the screen, thereby changing the count rate at $\boldsymbol{r}_{screen}$.

If the array in the "A"-portion would simply consist of two planar parallel grids perpendicular to the average particle motion, and if one applies an accelerating voltage V between the grids, the particles' kinetic energy ϵ_0 increases by an amount $\Delta\epsilon = e\,V$ where e denotes the particle charge. The wave function $\psi_{screen}(\boldsymbol{r},t)$ at the screen is the sum of the "A"- and "B"-related contributions:

$$= \frac{1}{\sqrt{2}}\left[\hat{\psi}_A(\boldsymbol{r}_{screen})\,e^{-i\omega_A\,t} + \hat{\psi}_B(\boldsymbol{r}_{screen})\,e^{-i\omega_B\,t}\right] \tag{54}$$

where

$$\hat{\psi}_{A/B}(\boldsymbol{r}_{screen}) = \frac{1}{\sqrt{\mathcal{V}}}\,e^{ik_{A/B}\,r_{screen}}$$

and

$$\hbar\,\omega_A = \epsilon_0 + \Delta\epsilon\,;\;\hbar\,\omega_B = \epsilon_0\,,$$

and with $1/\sqrt{\mathcal{V}}$ denoting an appropriate normalization factor. The function $\psi(\boldsymbol{r},t)$ solves the time-dependent Schrödinger equation (42) for the particular array under study. If we introduce $\overline{\epsilon} = \epsilon_0 + \frac{1}{2}\,\Delta\epsilon$ we may cast $\hbar\,k_{A/B}$ as $\hbar\,k_{A/B} \approx \sqrt{2m_0\,\overline{\epsilon}}\,(1 \pm \frac{1}{2}\,\frac{\Delta\epsilon}{\overline{\epsilon}})$ if $\frac{\Delta\epsilon}{\overline{\epsilon}} << 1$ where m_0 denotes the rest mass of the particle. Eq.(54) can then be rewritten

$$\psi(\boldsymbol{r}_{screen},t) = \frac{1}{\sqrt{\mathcal{V}}}\,e^{i(\overline{k}\,r_{screen} - \overline{\omega}\,t)}$$
$$\times\frac{1}{\sqrt{2}}\left[e^{i(\Delta k\,r_{screen} - \Delta\omega\,t)} + e^{-i(\Delta k\,r_{screen} - \Delta\omega\,t)}\right]$$

where $\hbar\,\overline{k} = \sqrt{2m_0\,\overline{\epsilon}}$, $\Delta k = k_A - k_B$ and $\hbar\,\Delta\omega = \frac{1}{2}\,\Delta\epsilon$. Hence we have for the current density $\boldsymbol{j}(\boldsymbol{r},t) \propto$ count rate at $\boldsymbol{r}_{screen}$

$$\boldsymbol{j}(\boldsymbol{r}_{screen},t) = \frac{\hbar\overline{\boldsymbol{k}}}{m_0}\,|\psi(\boldsymbol{r}_{screen},t)|^2 = \frac{1}{4\mathcal{V}}\,\frac{\hbar\overline{\boldsymbol{k}}}{m_0}\,[1 + \cos(2\Delta k\,r_{screen} - 2\Delta\omega\,t)]\,. \tag{55}$$

That means: the count rate oscillates at a period of $T = \frac{2\pi\,\hbar}{\Delta\epsilon}$. This most surprising effect of "quantum beats" has, in fact, been observed by Rauch and collaborators (s. Badurek et al. [86]) who used spin polarized neutrons instead of electrons. The energy change $\Delta\epsilon$ in the "A"-section of the setup was in that case imparted to the respective neutron by flipping its spin within a spatially confined magnetic field along the "A"-trajectory. (In practice one used a spin flipper also in the "B"-portion of the setup where the corresponding magnetic field was slightly lower than in the "A"-portion so that $\Delta\epsilon$ referred to the difference of two spin flip energies in this case. Another spin-flip was necessary anyway to enable the two beams to interfere with each other.)

It is worth noticing that $\boldsymbol{j}(\boldsymbol{r}_{screen}, t)$ displays - apart from its oscillatory time dependence - an oscillatory behavior also in space, i. e. in the plane of the screen. This is due to the occurrence of $\Delta k\, r_{screen}$ in the argument of the cosine. Hence, in the plane of the screen the current density displays an interference pattern which moves perpendicular to the interference lines at a velocity $\Delta\omega/\Delta k$. If one would replace the screen by a one-slit diaphragm of adjustable width, a detector behind the slit would monitor the incoming electrons one by one. According to Eq.(55) the current density oscillates at the frequency $2\Delta\omega$ about the value $\frac{1}{4\mathcal{V}}\frac{\hbar \boldsymbol{k}}{m_0}$. If the capture width of the detector, i. e. the slit of the diaphragm comprises a bright and a dark interference line, there is no oscillation of the count rate any more. If the width is narrower than that, oscillations occur which indicate the presence of the interference pattern. This situation will (very likely) not be affected, that is, the oscillations will persist, if one places an energy analyzer between the diaphragm and the detector. The analyzer can be set such that only electrons that have the energy of the "A-trajectory" are allowed to pass. As the oscillations still occur, one is led to conclude then that tracing the electron's path does not destroy the interference.

10. The Time-Dependent Schrödinger Equation

In the most general case $\boldsymbol{v}$ and hence φ are time-dependent. As already pointed out in Section 6 the substitution of $\rho(\boldsymbol{r})$ has to be modified then in the form

$$\psi(\boldsymbol{r}, t) = \pm\sqrt{\rho(\boldsymbol{r}, t)}\, e^{i\varphi(\boldsymbol{r}, t)} \tag{56}$$

which was introduced by Madelung in 1926 [34] The $\pm$-sign requires a comment. As discussed in Section 3, $\rho(\boldsymbol{r}, t)$ will generally be presupposed as a smooth function. The zeros of $\rho(\boldsymbol{r}, t)$ pose a particular problem that occurred already in Section 6, but was not explicitly mentioned. The admissible type of zeros limits the set of functions $\rho(\boldsymbol{r})$ that can be mapped onto $\psi(\boldsymbol{r})$ according to Eq.(36). For simplicity we confine ourselves to the time-independent case and assume that the zeros of $\rho(\boldsymbol{r})$ lie on the faces of a rectangular parallelepiped defined by the equations $x_\nu = x_{\nu 0}$ with x_ν and $\nu = 1, 2, 3$ denoting Cartesian coordinates. Hence close to $x_\nu = x_{\nu 0}$ and perpendicular to the respective face the density varies as $(x_\nu - x_{\nu 0})^2$. Since we have everywhere $\rho(\boldsymbol{r}) \geq 0$ its square root varies as $|x_\nu - x_{\nu 0}|$ and thus would not be differentiable at $x_\nu = x_{\nu 0}$. In defining the map $\rho(\boldsymbol{r}) \to \psi(\boldsymbol{r})$ one is forced hence to choose the positive sign in front of $\sqrt{\rho(\boldsymbol{r})}$ outside the rectangular parallelepiped if one has chosen the minus sign inside (or vice versa) to ensure that $\psi(\boldsymbol{r})$ stays differentiable across the face of the rectangular parallelepiped. Hence, mapping functions $\rho(\boldsymbol{r})$ onto differentiable functions $\psi(\boldsymbol{r})$ is only possible if the zeros of $\rho(\boldsymbol{r})$ subdivide the space of volume $\mathcal{V}$ into cells without leaving empty space. At first sight it appears that this limitation in the set of admissible functions $\rho(\boldsymbol{r})$ constitutes a serious drawback of the entire concept. One has to bear in mind, however, that the functions $\psi(\boldsymbol{r})$ are not determined as a map of $\rho(\boldsymbol{r})$ but rather by solving the Schrödinger equation (31) which has been the objective of the derivation. Physical meaningful solutions to Eq.(31) have automatically the required spatial structure of their zeros.

We now move on to derive the time-dependent Schrödinger equation under the supposition that the above considerations apply to the time-dependent case as well.

If one uses instead of Eq.(23) the arithmetic mean of the original Eq.(7) and its "anti-Brownian" analogue where the sign of ν and $\boldsymbol{u}(\boldsymbol{r},t)$ is reversed, one obtains

$$\frac{\partial}{\partial t}\boldsymbol{v}+(\boldsymbol{v}\cdot\nabla)\,\boldsymbol{v}-(\boldsymbol{u}\cdot\nabla)\,\boldsymbol{u}+\frac{\hbar}{2\,m_0}\,\Delta\boldsymbol{u}=\frac{1}{m_0}\,\boldsymbol{F}(\boldsymbol{r})\,. \tag{57}$$

This can be simplified in the form:

$$\frac{\partial}{\partial t}\boldsymbol{v}=-\frac{1}{m_0}\,\nabla V-\frac{1}{2}\,\nabla\boldsymbol{v}^2+\frac{1}{2}\,\nabla\boldsymbol{u}^2-\frac{\hbar}{2\,m_0}\,\Delta\boldsymbol{u}\,, \tag{58}$$

where we have made use of the relations

$$\boldsymbol{v}\cdot\nabla\boldsymbol{v}=\frac{1}{2}\nabla\boldsymbol{v}^2\,;\quad \boldsymbol{u}\cdot\nabla\boldsymbol{u}=\frac{1}{2}\nabla\boldsymbol{u}^2\quad\text{and}\quad \nu=\frac{\hbar}{2m_0}\,.$$

With the first two equations it has been observed that $\boldsymbol{v}$ and $\boldsymbol{u}$ are irrotational. On differentiating $\boldsymbol{u}$ with respect to time and using Eq.(8) one obtains

$$\frac{\partial}{\partial t}\boldsymbol{u}=-\frac{\hbar}{2\,m_0}\,\nabla\,(\frac{\partial\,\rho}{\partial\,t}/\rho)\,, \tag{59}$$

Invoking the equation of continuity

$$\partial\,\rho/\partial\,t+\nabla\cdot(\rho\,\boldsymbol{v})=0 \tag{60}$$

that is

$$\partial\,\rho/\partial\,t+\rho\,\nabla\cdot\boldsymbol{v}+\boldsymbol{v}\cdot\nabla\rho=0$$

$\frac{\partial\,\rho}{\partial\,t}/\rho$ can be replaced with $-\nabla\cdot\boldsymbol{v}-\boldsymbol{v}\cdot\frac{1}{\rho}\,\nabla\rho$ which yields

$$-\frac{\hbar}{2\,m_0}\,\nabla\,(\frac{\partial\,\rho}{\partial\,t}/\rho)=\frac{\hbar}{2\,m_0}\,\nabla\,(\nabla\cdot\boldsymbol{v})-\nabla\left[\boldsymbol{v}\cdot\left(-\frac{\hbar}{2\,m_0}\,\frac{1}{\rho}\,\nabla\rho\right)\right]\,. \tag{61}$$

Using Eq.(8) we may substitute the expression in the [...]-brackets on the right-hand side by $\boldsymbol{v}\cdot\boldsymbol{u}$. Hence Eq.(61) takes the form

$$\frac{\partial}{\partial\,t}\boldsymbol{u}=\frac{\hbar}{2\,m_0}\,\nabla(\nabla\cdot\boldsymbol{v})-\nabla(\boldsymbol{u}\cdot\boldsymbol{v})\,. \tag{62}$$

On multiplying the equation of motion (58) by the imaginary unit i and subtracting Eq.(62) we obtain

$$\frac{\partial}{\partial\,t}(-\boldsymbol{u}+i\,\boldsymbol{v})=-\frac{i}{m_0}\,\nabla V-\frac{i}{2}\,\nabla\boldsymbol{v}^2+\frac{i}{2}\,\nabla\boldsymbol{u}^2-i\,\frac{\hbar}{2\,m_0}\,\Delta\boldsymbol{u}-\frac{\hbar}{2\,m_0}\,\nabla(\nabla\cdot\boldsymbol{v})$$
$$+\nabla(\boldsymbol{u}\cdot\boldsymbol{v})\,,$$

which after reordering the terms on the right-hand side becomes

$$\frac{\partial}{\partial t}(-\boldsymbol{u}+i\,\boldsymbol{v})=\frac{i}{2}\nabla(-\boldsymbol{u}+i\,\boldsymbol{v})^2+\frac{i\,\hbar}{2\,m_0}\nabla\left[\nabla\cdot(-\boldsymbol{u}+i\,\boldsymbol{v})\right]-\frac{i}{m_0}\nabla V\,. \tag{63}$$

Here we insert Eqs.(13), (46) and (56) in the form

$$-\boldsymbol{u}+i\,\boldsymbol{v}=\frac{\hbar}{m_0}\nabla\ln\left[\psi/\sqrt{\rho_0}\right]. \tag{64}$$

After interchanging the operators $\partial/\partial t$ and ∇ one obtains

$$\nabla\left(\frac{\hbar}{m_0}\frac{1}{\psi}\frac{\partial\psi}{\partial t}\right)=\nabla\left[\frac{i}{2}\frac{\hbar^2}{m_0^2}\left\{\left(\frac{1}{\psi}\nabla\psi\right)^2+\nabla\cdot\left(\frac{1}{\psi}\nabla\psi\right)\right\}-\frac{i}{m_0}V\right].$$

If the gradient of some function equals that of another function the two functions can only differ by a real-space independent function of time which we denote by $\beta(t)$. Hence, if one divides the above equation by the imaginary unit the result may be cast as

$$-i\,\frac{\hbar}{m_0}\frac{1}{\psi}\frac{\partial\psi}{\partial t}=\frac{1}{2}\frac{\hbar^2}{m_0^2}\left[\left(\frac{1}{\psi}\nabla\psi\right)^2+\nabla\cdot\left(\frac{1}{\psi}\nabla\psi\right)\right]-\frac{1}{m_0}V-i\,\beta(t)\,. \tag{65}$$

One can now make use of the identity

$$\nabla\cdot\left(\frac{1}{\psi}\nabla\psi\right)=-\left(\frac{1}{\psi}\nabla\psi\right)^2+\frac{1}{\psi}\nabla^2\psi$$

and multiply Eq.(65) by $-m_0\,\psi$. This yields

$$i\hbar\,\frac{\partial\psi}{\partial t}=-\frac{\hbar^2\nabla^2}{2\,m_0}\psi+V\,\psi+\gamma(t)\,\psi \tag{66}$$

where

$$\gamma(t)=i\,m_0\,\beta(t)\,.$$

If $\psi(\boldsymbol{r},t)$ is replaced with $\widehat{\psi}(\boldsymbol{r},t)$ defined through

$$\psi(\boldsymbol{r},t)=\widehat{\psi}(\boldsymbol{r},t)\,\exp\left[-\frac{i}{\hbar}\int_{t_0}^{t}\gamma(t')\,dt'\right],$$

Eq.(66) becomes an equation for $\widehat{\psi}(\boldsymbol{r},t)$:

$$i\hbar\,\frac{\partial\,\widehat{\psi}(\boldsymbol{r},t)}{\partial t}=\left[\frac{\widehat{\boldsymbol{p}}^2}{2\,m_0}+V(\boldsymbol{r})\right]\widehat{\psi}(\boldsymbol{r},t)\,, \tag{67}$$

where

$$\widehat{\boldsymbol{p}}\equiv-i\hbar\nabla\,. \tag{68}$$

The two functions $\psi(\boldsymbol{r},t)$ and $\widehat{\psi}(\boldsymbol{r},t)$ differ only by a time-dependent phase factor without physical relevance. Only the functions

$$\rho(\boldsymbol{r},t) = \psi^*(\boldsymbol{r},t)\,\psi(\boldsymbol{r},t) \qquad \text{(density)} \tag{69}$$

and the current density:

$$\boldsymbol{j}(\boldsymbol{r},t) = \rho(\boldsymbol{r},t)\,\frac{\hbar}{m_0}\,\nabla\,\varphi(\boldsymbol{r},t)\,, \tag{70}$$

refer to relevant quantities of the system which obviously do not depend on this phase factor. For this reason we may set $\gamma(t) \equiv 0$, that is replace $\widehat{\psi}(\boldsymbol{r},t)$ in Eq.(67) by $\psi(\boldsymbol{r},t)$ without loss of generality. To simplify the notation we introduce the so-called Hamiltonian defined by

$$\widehat{H} \equiv \frac{\widehat{\boldsymbol{p}}^2}{2\,m_0} + V(\boldsymbol{r})\,. \tag{71}$$

Eq.(67) then takes the familiar form of the Schrödinger equation

$$i\hbar\,\frac{\partial\,\psi(\boldsymbol{r},t)}{\partial\,t} = \widehat{H}(\boldsymbol{r})\,\psi(\boldsymbol{r},t)\,. \tag{72}$$

The first order time derivative on the left-hand side can be tracked down to the acceleration $(\partial/\partial t)\,\boldsymbol{v}$ in Newton's modified second law (57).

Using

$$\psi(\boldsymbol{r},t) = |\psi(\boldsymbol{r},t)|\,e^{i\,\varphi(\boldsymbol{r},t)}$$

and inserting this into Eqs.(69) and (70) one obtains the familiar expression

$$\boldsymbol{j}(\boldsymbol{r},t) = \rho(\boldsymbol{r},t)\,\boldsymbol{v}(\boldsymbol{r},t) = \frac{\hbar}{2i\,m_0}\,\left[\psi^*(\boldsymbol{r},t)\nabla\,\psi(\boldsymbol{r},t) - \psi(\boldsymbol{r},t)\nabla\,\psi^*(\boldsymbol{r},t)\right] \tag{73}$$

which on real-space integration and multiplication by m_0 yields

$$m_0\,\langle \boldsymbol{v}(t)\rangle = \int \psi^*(\boldsymbol{r},t)\,\widehat{\boldsymbol{p}}\,\psi(\boldsymbol{r},t)\,d^3r \equiv \langle\widehat{\boldsymbol{p}}\rangle \tag{74}$$

where $\psi(\boldsymbol{r},t)$ has been required to satisfy the usual boundary conditions at the surface of the normalization volume. Because of Eq.(74) one is justified in terming $\widehat{\boldsymbol{p}}$ "momentum operator".

In Bohm's version of quantum mechanics [28] Eq.(73) is recast to define the velocity field

$$\boldsymbol{v}(\boldsymbol{r},t) = \frac{\hbar}{m_0}\,\Im\left(\frac{\nabla\psi(\boldsymbol{r},t)}{\psi(\boldsymbol{r},t)}\right)\,.$$

The flowlines of this field are interpreted as true particle trajectories. From our point of view this appears to be rather absurd because the explicit $\boldsymbol{r}$-dependence of $\boldsymbol{v}$ comes about by forming the ensemble average over the (in principle infinite) family of true trajectories as defined in Eq.(3). Bohm's definition of $\boldsymbol{v}$ as describing the true velocity of the particle leads inescapably to strange results, notably with stationary real-valued wave functions $\psi(\boldsymbol{r})$ for which $\boldsymbol{v}(\boldsymbol{r}) \equiv 0$. Hence, the particle appears to be at rest although the kinetic energy of the particle

$$\langle \widehat{T} \rangle = \int \psi^*(\boldsymbol{r}) \, \frac{\widehat{\boldsymbol{p}}^2}{2\, m_0} \, \psi(\boldsymbol{r}) \, d^3 r \equiv \frac{\langle \widehat{\boldsymbol{p}}^2 \rangle}{2\, m_0} \tag{75}$$

is definitely different from zero.

The time-dependent Schrödinger equation represents the center of non-relativistic quantum mechanics. Fundamentally different from the present approach where it is derived from a new vacuum concept, in conventional quantum mechanics it falls out of the blue, and this applies to Bohm's theory as well. As the latter associates the pattern of smooth flowlines with the set of true particle trajectories, it is forced to explain the probabilistic character of the information contained in $\psi(\boldsymbol{r}, t)$ by an additional "quantum equilibrium"- hypothesis. It is therefore hard to see that anything can be gained by "going Bohmian". The "process of measurement" in which a particle moves from a source to the detector where it fires a counter, is in our view described by one of the irregular trajectories which is terminated at the detector. Due to the stochastic forces that cause this irregularity, the information on the ensemble properties is naturally probabilistic.

A frequently raised objection against Bohm's theory concerns the asymmetric way in which it deals with the particle's real-space position and its momentum. In fact, the real-space position $\boldsymbol{r}$ plays a pivotal role in Bohm's theory compared to the other observables which are "contextualized" by resorting to the wave function $\psi(\boldsymbol{r}, t)$ that solves the Schrödinger equation for the system under study. By contrast, in our approach the ensemble's i-th particle position $\boldsymbol{r}_i$ and its velocity $\boldsymbol{v}_i(t)$ enter into the theory as autonomous quantities. This is reflected in the occurrence of two independent functions $\rho(\boldsymbol{r}, t)$ and $\boldsymbol{v}(\boldsymbol{r}, t) = \frac{\hbar}{2\, m_0} \nabla \varphi(\boldsymbol{r}\, t)$. It is this pair of information $\rho(\boldsymbol{r}, t)\, ; \varphi(\boldsymbol{r}, t)$ that necessitates the description of the one-particle system by a **complex**-valued function

$$\psi(\boldsymbol{r}, t) = \pm \sqrt{\rho(\boldsymbol{r}, t)} \, e^{i\, \varphi(\boldsymbol{r}, t)} \, .$$

11. The Uncertainty Relation and the Issue of "Measurement"

By performing a Fourier transform on $\psi(\boldsymbol{r}, t)$

$$\psi(\boldsymbol{r}, t) = \frac{1}{(2\, \pi)^{3/2}} \int C(\boldsymbol{k}, t) \, e^{i\, \boldsymbol{k} \cdot \boldsymbol{r}} \, d^3 k \tag{76}$$

Eqs.(74) and (75) may alternatively be written

$$\begin{aligned} \langle \widehat{\boldsymbol{p}} \rangle &= \int \psi^*(\boldsymbol{r}, t) \, \widehat{\boldsymbol{p}} \, \psi(\boldsymbol{r}, t) \, d^3 r = \int C^*(\boldsymbol{k}, t) \, (\hbar\, \boldsymbol{k}) \, C(\boldsymbol{k}, t) \, d^3 k \\ \langle \widehat{\boldsymbol{p}}^2 \rangle &= \int \psi^*(\boldsymbol{r}, t) \, \widehat{\boldsymbol{p}}^2 \, \psi(\boldsymbol{r}, t) \, d^3 r = \int C^*(\boldsymbol{k}, t) \, (\hbar\, \boldsymbol{k})^2 \, C(\boldsymbol{k}, t) \, d^3 k \end{aligned} \tag{77}$$

where

$$C^*(\boldsymbol{k}, t) \, C(\boldsymbol{k}, t) \, \Delta^3 k = P(\boldsymbol{k}, t) \, \Delta^3 k \tag{78}$$

describes the probability of the particle possessing a momentum that lies within $\Delta^3 k$ around $\boldsymbol{k}$ in the $\boldsymbol{k}$-space. We temporarily label the coordinate-components of the particle in the two spaces by an index ν; $\nu = 1, 2, 3$. The mean square departures of the position coordinates x_ν and k_ν, respectively, from their arithmetic means $\bar{x}_\nu$ and $\bar{k}_\nu$ are given by

$$\langle (x_\nu - \bar{x}_\nu)^2 \rangle_t = \int \psi^*(\boldsymbol{r}, t)\, (x_\nu - \bar{x}_\nu)^2\, \psi(\boldsymbol{r}, t)\, d^3 r$$

and

$$\langle (k_\nu - \bar{k}_\nu)^2 \rangle_t = \int C^*(\boldsymbol{k}, t)\, (k_\nu - \bar{k}_\nu)^2\, C(\boldsymbol{k}, t)\, d^3 k\,.$$

Since $C(\boldsymbol{k}, t)$ is the Fourier transform of $\psi(\boldsymbol{r}, t)$ we have as a fundamental mathematical theorem

$$\langle (x_\nu - \bar{x}_\nu)^2 \rangle_t\, \langle (k_\nu - \bar{k}_\nu)^2 \rangle_t \geq \frac{1}{4}$$

that is

$$\langle (x_\nu - \bar{x}_\nu)^2 \rangle_t\, \langle (\hbar\, k_\nu - \hbar\, \bar{k}_\nu)^2 \rangle_t \geq \frac{\hbar^2}{4}\,. \tag{79}$$

Following the standard notation by setting $\Delta x_\nu = \sqrt{\langle (x_\nu - \bar{x}_\nu)^2 \rangle_t}$ and $\Delta p_\nu = \sqrt{\langle (\hbar\, k_\nu - \hbar\, \bar{k}_\nu)^2 \rangle_t} = \sqrt{\langle (\widehat{\boldsymbol{p}} - \langle \widehat{\boldsymbol{p}} \rangle)^2 \rangle_t}$ the latter relation may be cast as

$$\Delta x_\nu\, \Delta p_\nu \geq \frac{\hbar}{2} \tag{80}$$

which constitutes the celebrated uncertainty relation. It is commonplace to interpret this relation, loosely speaking, by saying: "momentum and position of a particle cannot be measured simultaneously with any desirable precision".

From our point of view it does in no ways refer to any measurement on the position or momentum of the particle in question. It is nothing more than the theorem Eq.(79) on the product of two quantities that are interconnected by a Fourier transform. Furthermore, since this relation is - besides the Schrödinger equation - just another consequence of our concept, it cannot possibly conflict with the existence of trajectories which constitute a fundamental element of that concept.

Eq.(80) is considered ground-laying for the Copenhagen interpretation of quantum mechanics. The latter is based on the conviction that it is the measurement that causes the indeterminacy in quantum mechanics and necessitates a probabilistic description of microscopic mechanical systems. In a highly respected article [35] Heisenberg gives a revealing example of such a measurement. To pinpoint an electron moving along the x-axis within an experimental setup he considers a γ-ray source, that illuminates the electron beam, and a hypothetical γ-ray microscope that possesses a sufficiently high resolution in detecting the position of that electron up to an error of Δx. He demonstrates that the γ-ray photon that "hits the electron" and is subsequently scattered into the microscope, transfers a momentum Δp_x to the electron so that

$$\Delta x\, \Delta p_x \approx \hbar\,. \tag{81}$$

The above result reflects only a property of the microscope

$$\Delta x \, \Delta k_x \approx 2\,\pi$$

which interrelates the resolved linear dimensions $\Delta x = \lambda / \sin\alpha$ of an object and the admissible maximum angle α required to ensure that the scattered wave (of wavelength λ) is still captured by the front lens of the microscope, and $\Delta k_x = k \, \sin\alpha$ which describes the k_x-change of the wave vector of the scattered wave. But this interrelation expresses only the content of Eq.(79) in a different form. The measurement, however, is completely fictional for two reasons. Firstly, imaging systems within that regime of wavelength are for fundamental reasons unfeasible. Secondly, different from the picture insinuated by Heisenberg's phrasing, the interaction does not take place as an instantaneous collision process where a point-like particle (the photon) hits another point-like particle, the electron. Instead the transition probability of the electron for attaining a different momentum is given by the mod squared of the transition matrix element M_{opt}, a real-space integral that extends over a range of many light wave lengths in diameter. Moreover, the transition is not instantaneous but rather takes some time of the order $\hbar/|M_{opt}|$. Within this transition time the electron moves a distance $\Delta x'$ which has nothing to do with Δx in Eq.(81). Other examples of "measurement", e. g. diffraction at slits of a certain width Δx show even more directly that the probabilistic information on the (non-relativistic) motion of a particle is exhaustively described by the Schrödinger equation and boundary conditions for $\psi(\boldsymbol{r})$, and hence this information merely reflects our vacuum concept, irrespective of whether or not results on the diffraction are verified by measurements.

The host of considerations invoking the uncertainty relation (80) refers+- to situations where a particle is located within an interval Δx and one interprets this confinement of the particle indiscriminately in terms of a "measurement" of its coordinate x with limited accuracy. One concludes then from the uncertainty relation that Δx correlates unavoidably with a variance $\overline{\Delta p_x^2}$ of its momentum such that $\Delta x \, \Delta p_x \overset{>}{\approx} \frac{\hbar}{2}$ where $\Delta p_x \overset{\text{def}}{=} \sqrt{\overline{\Delta p_x^2}}$. In reality neither a measurement on Δx nor on $\overline{\Delta p^2}$ is truly executable. The uncertainty relation merely states that a solution of the one-dimensional Schrödinger equation for a particle in a box of length Δx yields a ground state energy $\Delta E = \frac{\Delta p^2}{2\,m_0}$ where $\Delta p^2 = \left(\hbar \, \frac{\pi}{\Delta x}\right)^2$. Hence one obtains simply as a consequence of solving the Schrödinger equation for that case "without observer"(!) $\Delta x \, \Delta p = \pi\,\hbar$. One cannot help but quote John Bell's question phrased in his stirring article "Against Measurement"[36]

"What exactly qualifies some physical systems to play the role of 'measurer'?"

The above considerations are in line with a discussion of Heisenberg's paper by Wigner [37].

12. Averaging over the Total Ensemble

In forming the arithmetic mean of the two equations (17) and (20) we omitted to mention a problem that we wish to discuss here in more detail.

We temporarily decompose the entire ensemble considered so far into a "Brownian" and "anti-Brownian" sub-ensemble, each characterized by the associated stochastic forces and comprising an equally large number of members. Accordingly we distinguish the velocities $\boldsymbol{v}(\boldsymbol{r},t)$ and the densities $\rho(\boldsymbol{r},t)$ in the respective sub-ensembles by subscripts B (for "Brownian") and A (for "Anti-Brownian"). If the velocities in these two equations agree at a time t, they are definitely different at a later time $t+\Delta t$. Yet forming the arithmetic mean of the two equations can only lead to the same average - which we could recast as "Newton's modified second law", Eq.(23) - if the two velocities $\boldsymbol{v}_B, \boldsymbol{v}_A$ and the densities ρ_B, ρ_A agree also at $t+\Delta t$ and any later time. At first sight the latter appears to be irreconcilable with the former. One has to recall, however, that our subdivision of the entire ensemble into sub-ensembles B and A represents only a simplifying model for the actually occurring reversible scatterings. In the real system the stochastic forces of the B-type become automatically forces of the A-type and vice versa within the characteristic time τ so that the change of the velocity $\Delta \boldsymbol{v}$ in either sub-ensemble is $[\Delta \boldsymbol{v}_A + \Delta \boldsymbol{v}_B]/2$ within a time span $\Delta t \gg \tau$ which, however, must be small compared to time intervals within which the quantities of interest change sizeably. The situation is similar to that encountered in diffusion theory where we have

$$\frac{\partial \rho}{\partial t} = \nu \, \Delta \rho \, .$$

This equation is obtained from the equation of continuity for $\boldsymbol{v}_B = \boldsymbol{u}$ and $\boldsymbol{u} = -\nu \, \nabla \rho / \rho$ with the latter equation based on similar considerations as the derivation of Eq.(7) invoking Einstein's law (22) which implies $\Delta t \gg \tau$. The above equation of diffusion hence describes changes that are actually defined only on a coarse grain time scale and its validity is confined to changes that are sufficiently slow on that time scale. As we have already discussed in Section 4, this is also the assumption underlying our derivation of Newton's modified second law (23).

We temporarily rewrite the two equations (17) and (20) for an - in that sense - "appropriately long, but sufficiently short time interval" Δt in the form

$$\Delta \boldsymbol{v}_{B/A}(\boldsymbol{r}, t+\Delta t) = \boldsymbol{R}_{B/A}(\boldsymbol{r},t) \, \Delta t$$

where

$$\boldsymbol{R}_{B/A}(\boldsymbol{r},t) = \frac{1}{m_0} \left(-\nabla [V(\boldsymbol{r}) + V_{QP}(\boldsymbol{r},t)] \pm \vec{\Omega}(\boldsymbol{r},t) \right) . \tag{82}$$

and

$$\vec{\Omega} = \frac{\partial \boldsymbol{u}}{\partial t} + (\boldsymbol{v} \cdot \nabla) \, \boldsymbol{u} - (\boldsymbol{u} \cdot \nabla) \, \boldsymbol{v} + \nu \, \Delta \, \boldsymbol{v} \, .$$

Here we have already used V_{QP} instead of V_{stoch}, but still denoted the prefactor of $\Delta \boldsymbol{v}$ by ν to demonstrate that $\vec{\Omega}$ (and consequently $\boldsymbol{u}$) changes sign when ν changes sign. It should be noticed that according to Eq.(24) V_{QP} has the property
$V_{QP}(\rho_B(\boldsymbol{r},t)) = V_{QP}(\rho_A(\boldsymbol{r},t)) = V_{QP}(\rho(\boldsymbol{r},t))$ since at the time t under consideration we have $\rho_A(\boldsymbol{r},t) = \rho_B(\boldsymbol{r},t) = \frac{1}{2}\,\rho(\boldsymbol{r},t)$.

After the elapse of a time Δt within which each of the $N/2$ particles in the two subsystems has changed its affiliation (B from A or vice versa) we have

$$\boldsymbol{v}_{B/A}(\boldsymbol{r}, t+\Delta t) = \boldsymbol{v}_{B/A}(\boldsymbol{r},t) + \Delta \boldsymbol{v}(\boldsymbol{r}, t+\Delta t) \, .$$

where

$$\Delta \boldsymbol{v}(\boldsymbol{r}, t+\Delta t) = \overline{\Delta \boldsymbol{R}_{B/A}}(\boldsymbol{r}, t+\Delta t) = \frac{1}{m_0}\left(-\nabla[V(\boldsymbol{r}) + V_{PQ}(\boldsymbol{r}, t)]\right)\Delta t \tag{83}$$

with $\overline{\Delta \boldsymbol{R}_{B/A}}$ denoting a time average.

$\gg$ We consider Eqs.(82) and (83) as implicitly defining "motion under reversible scattering"$\ll$.

If we now form the average over the entire ensemble we get

$$\boldsymbol{v}(\boldsymbol{r}, t+\Delta t) = \frac{1}{2}\boldsymbol{v}_B(\boldsymbol{r}, t+\Delta t) + \frac{1}{2}\boldsymbol{v}_A(\boldsymbol{r}, t+\Delta t) = \boldsymbol{v}(\boldsymbol{r}, t) + \frac{1}{m_0}\left(-\nabla[V(\boldsymbol{r}) + V_{PQ}(\boldsymbol{r}, t)]\right)\Delta t\,.$$

Thus we have

$$\boldsymbol{v}_B(\boldsymbol{r}, t+\Delta t) = \boldsymbol{v}_A(\boldsymbol{r}, t+\Delta t) = \boldsymbol{v}(\boldsymbol{r}, t+\Delta t)\,.$$

We want to demonstrate that the densities behave analogously. For this reason we resort to the equation of continuity (60) which holds for each sub-ensemble

$$\frac{\partial \rho_{B/A}}{\partial t} + \nabla \cdot (\rho_{B/A}\, \boldsymbol{v}_{A/B}) = 0\,. \tag{84}$$

It describes the conservation of the number of particles in each of the two subsystems. We conclude from this equation that $\dot{\rho}_B(\boldsymbol{r}, t) = \dot{\rho}_A(\boldsymbol{r}, t)$, if $\rho_B(\boldsymbol{r}, t) = \rho_A(\boldsymbol{r}, t)$ and $\boldsymbol{v}_B(\boldsymbol{r}, t) = \boldsymbol{v}_A(\boldsymbol{r}, t)$. If one differentiates Eq.(84) with respect to time and uses $\dot{\boldsymbol{v}}_B(\boldsymbol{r}, t) = \dot{\boldsymbol{v}}_A(\boldsymbol{r}, t) = \dot{\boldsymbol{v}}(\boldsymbol{r}, t)$ as a result of the preceding considerations, we may conclude $\ddot{\rho}_B(\boldsymbol{r}, t) = \ddot{\rho}_A(\boldsymbol{r}, t)$. One can carry this conclusion further to any order of time derivative. Thus, the Taylor-expansions of $\rho_B(\boldsymbol{r}, t+\Delta t)$ and $\rho_A(\boldsymbol{r}, t+\Delta t)$ agree for any length of the time interval Δt if ρ_B, ρ_A and $\boldsymbol{v}_B, \boldsymbol{v}_A$ agree at time t.

13. Conservative Diffusion. Ehrenfest's Theorem

We want to prove the validity of Eq.(27) which constitutes a necessary condition for the preservation of classical motional behavior on the average. To see this more clearly, we first consider one particle (the i-th) in the cube $\Delta^3 r$ around $\boldsymbol{r}$ acted upon by the external force $\boldsymbol{F}(\boldsymbol{r})$ and the stochastic force $\boldsymbol{F}_{s\,i}(t)$. According to Newton's second law we have

$$\frac{d}{dt}\, m_0\, \boldsymbol{v}_i(t) = \boldsymbol{F}(\boldsymbol{r}_i) + \boldsymbol{F}_{s\,i}(t)\,.$$

If we sum this equation over the $n(\boldsymbol{r}, t)$ particles contained in $\Delta^3 r$, divide by N and form ensemble averages similar to Eqs.(1) and (3) we obtain

$$\frac{\partial}{\partial t}\, m_0 \frac{1}{N} \underbrace{\sum_{i=1}^{n(\boldsymbol{r},t)} \boldsymbol{v}_i(t)}_{=n(\boldsymbol{r},t)\,\boldsymbol{v}(\boldsymbol{r},t)} = \frac{1}{N} \underbrace{\sum_{i=1}^{n(\boldsymbol{r},t)} \boldsymbol{F}(\boldsymbol{r}_i)}_{=n(\boldsymbol{r},t)\,\boldsymbol{F}(\boldsymbol{r})} + \frac{1}{N} \underbrace{\sum_{i=1}^{n(\boldsymbol{r},t)} \boldsymbol{F}_{s\,i}(t)}_{=n(\boldsymbol{r},t)\,\boldsymbol{F}_s(\boldsymbol{r},t)}\,. \tag{85}$$

Here the summation runs over all particles in the cell irrespective of whether they belong to the first or second sub-ensemble.

The idea of "conservative diffusion" implies that the $N = \sum_{\boldsymbol{r}}^{N_r} n(\boldsymbol{r},t)$ particles of the entire ensemble do not feel a stochastic force on the average although $\boldsymbol{F}_s(\boldsymbol{r},t)$ does locally not vanish in general. Thus, $\boldsymbol{F}_s(\boldsymbol{r},t)$ is required to have the property

$$\sum_{\boldsymbol{r}}^{N_r} \frac{n(\boldsymbol{r},t)}{N}\, \boldsymbol{F}_s(\boldsymbol{r},t) = \int_{\mathcal{V}} \rho(\boldsymbol{r},t)\, \underbrace{\boldsymbol{F}_{QP}(\boldsymbol{r},t)}_{\equiv \boldsymbol{F}_s(\boldsymbol{r},t)}\, d^3r = 0\; \forall t\,, \tag{86}$$

as a result of which Eq.(85) yields after summation over all elementary cells

$$\frac{\partial}{\partial t} \sum_{\boldsymbol{r}}^{N_r} m_0\, \frac{n(\boldsymbol{r},t)}{N}\, \boldsymbol{v}(\boldsymbol{r},t) = \underbrace{\frac{d}{dt} \int_{\mathcal{V}} \rho(\boldsymbol{r},t)\, m_0\, \boldsymbol{v}(\boldsymbol{r},t)\, d^3r}_{\equiv \langle \boldsymbol{p}(t)\rangle} = \underbrace{\sum_{\boldsymbol{r}}^{N_r} \frac{n(\boldsymbol{r},t)}{N}\, \boldsymbol{F}(\boldsymbol{r})}_{=\int \rho(\boldsymbol{r},t)\, \boldsymbol{F}(\boldsymbol{r})\, d^3r = \langle \boldsymbol{F}\rangle}\quad .$$

We thus obtain as a consequence of the required property of $\boldsymbol{F}_s(\boldsymbol{r},t)$

$$\frac{d}{dt}\, \langle \boldsymbol{p}(t)\rangle = \langle \boldsymbol{F}\rangle \tag{87}$$

which is Ehrenfest's first theorem[38].

In case of a force-free particle for which $\langle \boldsymbol{F}\rangle = 0$, Eq.(87) yields

$$\langle \boldsymbol{p}(t)\rangle = const.$$

which demonstrates that a free particle exposed to Brownian/anti-Brownian stochastic forces does not change its momentum on the average, as opposed to a particle that moves in a classical "Brownian" environment.

We now want to show that the expectation value of "Newton's modified second law" that we have derived in the form of Eq.(23), attains, in fact, exactly the form of Eq.(87). To this end it is convenient to recast Eq.(24) as

$$V_{QP} = \frac{\hbar^2}{4\, m_0} \left[\frac{1}{2} \left(\frac{\nabla \rho}{\rho} \right)^2 - \frac{\nabla^2 \rho}{\rho} \right] = m_0 \left[-\frac{\boldsymbol{u}^2(\boldsymbol{r},t)}{2} + \frac{\hbar}{2m_0}\, \nabla \cdot \boldsymbol{u}(\boldsymbol{r},t) \right]$$

where we have used Eq.(8) defining $\boldsymbol{u}(\boldsymbol{r},t)$. Hence

$$\begin{aligned} \int \rho(\boldsymbol{r},t)\, \boldsymbol{F}_{QM}(\boldsymbol{r},t)\, d^3r == m_0 \int \Bigg[& \frac{1}{2} \rho(\boldsymbol{r},t) \nabla \boldsymbol{u}^2(\boldsymbol{r},t) \\ & - \frac{\hbar}{2m_0}\, \rho(\boldsymbol{r},t) \Delta \boldsymbol{u}(\boldsymbol{r},t) \Bigg]\, d^3r\,. \end{aligned} \tag{88}$$

We rewrite the integral over the second term on the right-hand side using Gauss' theorem

$$\int_{\mathcal{V}} \rho\, \underbrace{\nabla \cdot (\nabla \boldsymbol{u})}_{=\Delta \boldsymbol{u}}\, d^3r = \underbrace{\int_{\mathcal{V}} \nabla \cdot (\rho\, \nabla \boldsymbol{u})\, d^3r}_{=\int_{\mathcal{F}} \rho\, \nabla \boldsymbol{u} \cdot d^2\boldsymbol{r}} - \int_{\mathcal{V}} \nabla \rho \cdot \nabla \boldsymbol{u}\, d^3r\,.$$

We assume that $\rho(\boldsymbol{r},t)$ differs sizeably from zero only within a volume that lies completely within the finite space and drops sufficiently fast to zero toward infinity so that the surface integral vanishes. Using again Eq.(8) we hence arrive at

$$-\int_{\mathcal{V}} \nabla\rho\cdot\nabla\boldsymbol{u}\,d^3r = \frac{2m_0}{\hbar}\int_{\mathcal{V}}\rho\underbrace{(\boldsymbol{u}\cdot\nabla)\boldsymbol{u}}_{=\frac{1}{2}\nabla\boldsymbol{u}^2}d^3r$$

which shows that, in fact, the right-hand side of Eq.(88) equals zero. Thus, the expectation value of the right-hand side of "Newton's modified second law", Eq.(23), becomes equal to $\langle\boldsymbol{F}\rangle$. However, we have on the left-hand side $\langle\frac{d}{dt}\,m_0\,\boldsymbol{v}\rangle$ instead of $\frac{d}{dt}\,\langle m_0\,\boldsymbol{v}\rangle$. Nevertheless, the two expressions are equal as follows from multiplying $\frac{d}{dt}\,m_0\,\boldsymbol{v}$ by $\rho(\boldsymbol{r},t)$ and observing that $\boldsymbol{v}(\boldsymbol{r},t)$ is irrotational. Because of the latter we have

$$\frac{d}{dt}\boldsymbol{v} = \frac{\partial}{\partial t}\boldsymbol{v} + \frac{1}{2}\nabla\boldsymbol{v}^2$$

which can be recast as

$$m_0\,\rho\,\frac{d}{dt}\,\boldsymbol{v} = m_0\,\rho\,\frac{\partial}{\partial t}\boldsymbol{v} + \left[m_0\,\boldsymbol{v}\,\frac{\partial\rho}{\partial t} - m_0\,\boldsymbol{v}\,\frac{\partial\rho}{\partial t}\right] + \frac{m_0}{2}\,\rho\,\nabla\boldsymbol{v}^2$$

where we have added zero in the form of the bracketed expression. The real-space integral over this equation may be written after reordering

$$\underbrace{\frac{\partial}{\partial t}\int_{\mathcal{V}}\rho(\boldsymbol{r},t)\,m_0\,\boldsymbol{v}(\boldsymbol{r},t)\,d^3r}_{=\frac{d}{dt}\int_{\mathcal{V}}\rho(\boldsymbol{r},t)\,m_0\,\boldsymbol{v}(\boldsymbol{r},t)\,d^3r} = \langle\frac{d}{dt}\,m_0\,\boldsymbol{v}\rangle + \frac{m_0}{2}\int_{\mathcal{V}}\left[2\boldsymbol{v}\,\frac{\partial\rho}{\partial t} - \rho\,\nabla\boldsymbol{v}^2\right]d^3r\,. \quad (89)$$

The integral on the right-hand side vanishes because of the equation of continuity

$$\frac{\partial\rho}{\partial t} + \nabla\cdot(\rho\,\boldsymbol{v}) = 0\,. \quad (90)$$

This follows from multiplying this equation by $\boldsymbol{v}$ and performing a real-space integration. We then have

$$\begin{aligned}\int_{\mathcal{V}}\boldsymbol{v}\,\frac{\partial\rho}{\partial t}\,d^3\,r &= -\sum_{\nu=1}^{3}\boldsymbol{e}_\nu\int_{\mathcal{V}}v_\nu\,\nabla\cdot(\rho\,\boldsymbol{v})\,d^3\,r\\ &= -\sum_{\nu=1}^{3}\boldsymbol{e}_\nu\underbrace{\int_{\mathcal{V}}\nabla\cdot(v_\nu\,\rho\,\boldsymbol{v})\,d^3\,r}_{=\int_A v_\nu\,\rho\,\boldsymbol{v}\cdot d^2\,\boldsymbol{r}} + \underbrace{\sum_{\nu=1}^{3}\boldsymbol{e}_\nu\int_{\mathcal{V}}\rho\,\boldsymbol{v}\cdot\nabla v_\nu\,d^3\,r}_{=\int_{\mathcal{V}}\rho\,(\boldsymbol{v}\cdot\nabla)\,\boldsymbol{v}\,d^3\,r}\end{aligned} \quad (91)$$

with $\boldsymbol{e}_\nu$ denoting unit vectors. The surface integral as been obtained by invoking Gauss' theorem. It vanishes since we may assume $\rho\,|\boldsymbol{v}|$ to vanish sufficiently toward infinity. Again exploiting the property of $\boldsymbol{v}$ being irrotational the second integral on the right-hand side can be written

$$\int_{\mathcal{V}}\rho\,(\boldsymbol{v}\cdot\nabla)\,\boldsymbol{v}\,d^3\,r = \frac{1}{2}\int_{\mathcal{V}}\rho\,\nabla\boldsymbol{v}^2\,d^3\,r\,.$$

It follows then from Eq.(91) that the integral on the right-hand side of Eq.(89) is, in fact, equal to zero. Thus we have shown that the expectation value of the "vacuum force" $\boldsymbol{F}_{QM}(\boldsymbol{r},t)$ vanishes

$$\int \rho(\boldsymbol{r},t)\, \boldsymbol{F}_{QM}(\boldsymbol{r},t)\, d^3r = 0$$

which plays also a central role in information theory (s. e. g. Garbaczewski [39]).

14. The Time-Dependent Schrödinger Equation in the Presence of an Electromagnetic Field

In going through the various steps that led from Eq.(23) ("Newton's modified second law") to the time-dependent Schrödinger equation (72) one recognizes that we implied nowhere that $\boldsymbol{F}$ has to be time-independent. Hence one is justified in allowing $\boldsymbol{F}$ in Eq.(23) to be time-dependent and attain the particular form

$$\boldsymbol{F}(\boldsymbol{r},t) = -\nabla V_{cons}(\boldsymbol{r}) + e\,\widehat{\boldsymbol{E}}(\boldsymbol{r},t) + e\,\boldsymbol{v}(\boldsymbol{r},t) \times \boldsymbol{B}(\boldsymbol{r},t) \tag{92}$$

if the particle under study possesses the charge e and is acted upon by an electric field $\widehat{\boldsymbol{E}}(\boldsymbol{r},t)$ and a magnetic field $\boldsymbol{B}(\boldsymbol{r},t)$. The quantity $V_{cons}(\boldsymbol{r})$ denotes the potential of an additional conservative field (e. g. the gravitational field) which we include to ensure full generality, and $\boldsymbol{v}(\boldsymbol{r},t)$ is the ensemble average defined by Eq.(3). From $\boldsymbol{B} = \nabla \times \boldsymbol{A}$ and Faraday's law of induction we have $\nabla \times (\widehat{\boldsymbol{E}} + \dot{\boldsymbol{A}}) = 0$, and hence $\widehat{\boldsymbol{E}} + \dot{\boldsymbol{A}}$ may be expressed as a gradient of a scalar function which we denote by $-\frac{1}{e}\, V_{el}(\boldsymbol{r},t)$. Thus

$$e\,\widehat{\boldsymbol{E}}(\boldsymbol{r},t) = -e\,\dot{\boldsymbol{A}}(\boldsymbol{r},t) - \nabla V_{el}(\boldsymbol{r},t)\,. \tag{93}$$

If the magnetic field is switched on, it induces a voltage V_R along any circular path C

$$V_R = \oint_C \widehat{\boldsymbol{E}}_{ind.}(\boldsymbol{r}',t)\cdot d\boldsymbol{r}' = -\frac{\partial}{\partial t}\int_{\mathcal{A}} \boldsymbol{B}(\boldsymbol{r}',t)\cdot d^2\boldsymbol{r}'$$

where C is the rim of the surface $\mathcal{A}$. On multiplying this equation by e and observing that $e\,\widehat{\boldsymbol{E}}_{ind.}$ represents an additional force that changes the momentum of the particle, we obtain

$$\oint_C \dot{\boldsymbol{p}}(\boldsymbol{r}',t')\cdot d\boldsymbol{r}' = \oint_C e\,\widehat{\boldsymbol{E}}_{ind.}(\boldsymbol{r}',t')\cdot d\boldsymbol{r}' = -\frac{\partial}{\partial t'}\oint_C e\,\boldsymbol{A}(\boldsymbol{r}',t')\cdot d\boldsymbol{r}'\,.$$

Integrating this equation from t_0 to t and assuming $\boldsymbol{A}(\boldsymbol{r}',t_0) \equiv 0$ we obtain

$$-\oint_C \boldsymbol{p}(\boldsymbol{r}',t_0)\cdot d\boldsymbol{r}' + \oint_C \boldsymbol{p}(\boldsymbol{r}',t)\cdot d\boldsymbol{r}' = -\oint_C e\,\boldsymbol{A}(\boldsymbol{r}',t)\cdot d\boldsymbol{r}'\,.$$

where

$$\oint_C \boldsymbol{p}(\boldsymbol{r}',t_0)\cdot d\boldsymbol{r}' = m_0 \oint_C \boldsymbol{v}(\boldsymbol{r}',t_0)\cdot d\boldsymbol{r}' = 0\,,$$

which follows from Eq.(45). Thus

$$\oint_C \left[\boldsymbol{v}(\boldsymbol{r}',t) + \frac{e}{m_0}\,\boldsymbol{A}(\boldsymbol{r}',t)\right]\cdot d\boldsymbol{r}' = 0 \quad \forall\, t$$

which means that the curl of the integrand vanishes:

$$\nabla \times \left[\boldsymbol{v}(\boldsymbol{r},t) + \frac{e}{m_0}\,\boldsymbol{A}(\boldsymbol{r},t) \right] \equiv 0\,. \tag{94}$$

Consequently, it can be expressed as a gradient of a scalar function which we denote by $(\hbar/m_0)\,\varphi(\boldsymbol{r},t)$. Hence we arrive at

$$\boldsymbol{v}(\boldsymbol{r},t) + \frac{e}{m_0}\,\boldsymbol{A}(\boldsymbol{r},t) = \frac{\hbar}{m_0}\,\nabla\varphi(\boldsymbol{r},t)\,. \tag{95}$$

which now stands in place of Eq.(46).

We note here only in passing that we have because of $\psi(\boldsymbol{r}) = |\psi(\boldsymbol{r})|e^{i\varphi(\boldsymbol{r})}$

$$\frac{1}{2i}\left[\psi^*\nabla\psi - \psi\nabla\psi^*\right] = |\psi(\boldsymbol{r})|^2\,\nabla\varphi\,.$$

Using Eq.(95) one can recast this as

$$\frac{\hbar}{2m_0}\left[\psi^*\nabla\psi - \psi\nabla\psi^*\right] = \rho\,\boldsymbol{v} + \frac{e}{m_0}\,|\psi|^2\boldsymbol{A}$$

or equivalently

$$\rho\,\boldsymbol{v} = \frac{1}{2m_0}\,\left[\psi^*\widehat{\boldsymbol{P}}\psi + c.c.\right]$$

where $\widehat{\boldsymbol{P}}$ is short-hand for $\widehat{\boldsymbol{p}} - e\,\boldsymbol{A}$. After real-space integration and an integration by parts one arrives at

$$\langle \boldsymbol{v} \rangle = \frac{1}{m_0}\int \psi^*(\boldsymbol{r},t)\widehat{\boldsymbol{P}}\,\psi(\boldsymbol{r},t)\,. \tag{96}$$

Because of Eq.(94) the expression $(\boldsymbol{v}\cdot\nabla)\,\boldsymbol{v}$ which appears in

$$m_0\,\frac{d}{dt}\,\boldsymbol{v}(\boldsymbol{r},t) = m_0\left[\frac{\partial}{\partial t}\,\boldsymbol{v} + (\boldsymbol{v}\cdot\nabla)\,\boldsymbol{v}\right] = \boldsymbol{F}(\boldsymbol{r},t) + \boldsymbol{F}_{QP}(\boldsymbol{r},t) \tag{97}$$

cannot be replaced by $\frac{1}{2}\nabla\boldsymbol{v}^2$ any more. Because of the generally valid relation

$$(\boldsymbol{a}\cdot\nabla)\,\boldsymbol{a} = \nabla\,\frac{\boldsymbol{a}^2}{2} - \boldsymbol{a}\times(\nabla\times\boldsymbol{a})$$

and because of Eq.(94) we now have

$$(\boldsymbol{v}\cdot\nabla)\,\boldsymbol{v} = \frac{1}{2}\nabla\boldsymbol{v}^2 - \boldsymbol{v}\times(\nabla\times\boldsymbol{v}) = \frac{1}{2}\nabla\boldsymbol{v}^2 + \frac{e}{m_0}\boldsymbol{v}\times(\nabla\times\boldsymbol{A})\,.$$

Using $\nabla\times\boldsymbol{A} = \boldsymbol{B}$ we may recast this as

$$(\boldsymbol{v}\cdot\nabla)\,\boldsymbol{v} = \frac{1}{2}\nabla\boldsymbol{v}^2 + \frac{e}{m_0}\boldsymbol{v}\times\boldsymbol{B}\,.$$

Inserting this result together with Eq.(92) and $\boldsymbol{F}_{QP} = -\nabla V_{QP}$ into Eq.(97) we notice that the Lorentz-force $e\,\boldsymbol{v}(\boldsymbol{r},t) \times \boldsymbol{B}(\boldsymbol{r},t)$ drops out in favor of $\boldsymbol{A}(\boldsymbol{r},t)$, and we get

$$\frac{\partial}{\partial t}(\boldsymbol{v} + \frac{e}{m_0}\boldsymbol{A}) = -\frac{1}{m_0}\nabla V - \frac{1}{2}\nabla \boldsymbol{v}^2 + \frac{1}{2}\nabla \boldsymbol{u}^2 - \frac{\hbar}{2\,m_0}\Delta \boldsymbol{u} \tag{98}$$

where we have introduced

$$V(\boldsymbol{r},t) = V_{cons}(\boldsymbol{r}) + V_{el}(\boldsymbol{r},t)\,. \tag{99}$$

We now multiply Eq.(98) by the imaginary unit i and subtract Eq.(62) which gives in complete analogy to Eq.(63)

$$\frac{\partial}{\partial t}\left[-\boldsymbol{u} + i(\boldsymbol{v} + \frac{e}{m_0}\boldsymbol{A})\right] = \frac{i}{2}\nabla(-\boldsymbol{u} + i\,\boldsymbol{v})^2 + \frac{i\,\hbar}{2\,m_0}\nabla\left[\nabla\cdot(-\boldsymbol{u} + i\,\boldsymbol{v})\right] - \frac{i}{m_0}\nabla V \tag{100}$$

We mention here only in passing that Eq.(62) is equivalent to Fick's law and is hence not affected by the presence of an electromagnetic field as long as Einstein's law (22) remains unchanged which is obvious from his derivation. (S. also Fritsche and Haugk [14].)

As in the case without electromagnetic field we absorb the two independent scalar informations $\rho(\boldsymbol{r},t)$ and $\varphi(\boldsymbol{r},t)$ into one complex-valued function

$$\psi(\boldsymbol{r},t) = \pm\sqrt{\rho(\boldsymbol{r},t)}\,e^{i\varphi(\boldsymbol{r},t)}\,. \tag{101}$$

As $\boldsymbol{v}$ is no longer equal to $\frac{\hbar}{m_0}\nabla\varphi$ we have now in place of Eq.(64)

$$\frac{\hbar}{m_0}\nabla(\ln\psi/\sqrt{\rho_0}) = -\boldsymbol{u} + i(\boldsymbol{v} + \frac{e}{m_0}\boldsymbol{A})\,. \tag{102}$$

The left-hand side of Eq.(100) is obviously the time-derivative hereof. It will be useful to notice that

$$\frac{\partial}{\partial t}\frac{\hbar}{m_0}\nabla(\ln\psi/\sqrt{\rho_0}) = \nabla(\frac{1}{m_0\,\psi}\hbar\frac{\partial}{\partial t}\psi)\,. \tag{103}$$

We may also use Eq.(102) to recast the first expression on the right-hand side of Eq.(100)

$$\frac{i}{2}\nabla(-\boldsymbol{u} + i\,\boldsymbol{v})^2 = \frac{i}{2}\frac{\hbar^2}{m_0^2}\nabla[\nabla\ln(\psi/\sqrt{\rho_0})]^2$$
$$+\frac{\hbar}{m_0}\frac{e}{m_0}\nabla[\boldsymbol{A}\cdot\nabla\ln(\psi/\sqrt{\rho_0})] - \frac{i}{2}\nabla(\frac{e}{m_0}\boldsymbol{A})^2\,.$$

If one observes that

$$\nabla[\nabla\ln(\psi/\sqrt{\rho_0})] = \frac{1}{\psi}\Delta\psi - [\nabla\ln(\psi/\sqrt{\rho_0})]^2\,, \tag{104}$$

the second expression on the right-hand side of Eq.(100) can be written

$$i\frac{\hbar}{2\,m_0}\nabla\left[\nabla\cdot(-\boldsymbol{u} + i\,\boldsymbol{v})\right] = -\frac{i}{2}\frac{\hbar^2}{m_0^2}\nabla[\nabla\ln(\psi/\sqrt{\rho_0})]^2$$
$$+i\nabla\left[\frac{1}{\psi}(\frac{1}{2}\frac{\hbar^2}{m_0^2}\Delta\psi - \frac{\hbar}{2\,m_0}\frac{e}{m_0}\underbrace{\psi\,\nabla\cdot\boldsymbol{A}}_{=(\nabla\cdot\boldsymbol{A})\psi-(\boldsymbol{A}\cdot\nabla)\psi}\quad)\right]\,.$$

Hence we obtain

$$\frac{i}{2}\nabla(-\boldsymbol{u}+i\,\boldsymbol{v})^2+\frac{i\,\hbar}{2\,m_0}\nabla\left[\nabla(-\boldsymbol{u}+i\,\boldsymbol{v})\right]=-i\,\nabla\left[\frac{1}{m_0\,\psi}\mathcal{G}\,\psi\right]$$

where

$$\mathcal{G}=-\frac{\hbar^{\in}}{\in\Updownarrow_{,}}\cdot\psi+\rangle\,\frac{\hbar}{\in}\frac{]}{\Updownarrow_{,}}\nabla\cdot\mathcal{A}+\rangle\in\frac{\hbar}{\in}\frac{]}{\Updownarrow_{,}}\mathcal{A}\cdot\nabla+\frac{\infty}{\in\Updownarrow_{,}}(\rceil\,\mathcal{A})^{\in}\,.$$

The right-hand side of t his equation may be compactified by using the momentum operator (68) as a convenient short-hand notation

$$\frac{i}{2}\nabla(-\boldsymbol{u}+i\,\boldsymbol{v})^2+\frac{i\,\hbar}{2\,m_0}\nabla\left[\nabla(-\boldsymbol{u}+i\,\boldsymbol{v})\right]=-i\,\nabla\left[\frac{1}{m_0\,\psi}\frac{(\widehat{\boldsymbol{p}}-e\,\boldsymbol{A})^2}{2\,m_0}\,\psi\right].$$

Inserting this result into Eq.(100) which derives from Eq.(97) ("Newton's modified second law") and Eq.(9) (≡ Fick's law) and exploiting the Eqs.(103) and (104) we arrive at

$$i\hbar\,\frac{\partial\,\psi(\boldsymbol{r},t)}{\partial\,t}=\widehat{H}(\boldsymbol{r},t)\,\psi(\boldsymbol{r},t) \tag{105}$$

where

$$\widehat{H}(\boldsymbol{r},t)=\frac{\widehat{\boldsymbol{P}}^{\,2}}{2\,m_0}+V(\boldsymbol{r},t)\quad\text{and}\quad\widehat{\boldsymbol{P}}=\widehat{\boldsymbol{p}}-e\,\boldsymbol{A}(\boldsymbol{r},t)\,.$$

15. A Model for Non-Markovian Diffusion Illustrating the Origin of Non-Locality

It is instructive to consider a model illustrating "conservative diffusion". The latter is a consequence of forming the arithmetic mean of Eqs.(17) and (20) which leads to Eq.(21). If one were to follow the motion of an individual particle, just one member out of the total ensemble, one would directly see the effect of stochastic forces changing back and forth from "Brownian" to "anti-Brownian" with the latter causing a motion enhancement after the former have effected a slow down of the particle motion. Figure 2 shows three situation of the (free) particle which moves within a two-dimensional frame where a two-slit diaphragm has been inserted on the left-hand side. The "walls" of the frame are assumed elastically reflecting. The stochastic forces acting on the particle are simulated by a two-dimensional gas of N identical point masses ($N \gg 1$) that interact via Lennard-Jones pair-potentials with each other and with the particle under study as well. The latter will henceforth be referred to as "test particle". It is this situation which the original derivation of Einstein's law (22) refers to where the motion of the test particle is described by a Langevin equation into which the embedding of the particle enters through a stochastic force. (The practical calculations have been performed with slightly modified Lennard-Jones potentials that were truncated at twice the average particle distance.) In our model the particle motion of the embedding gas results from a molecular dynamics simulation which one starts by first keeping the test particle fixed at the point $\boldsymbol{r}_A$ and letting the N gas particles start from some corner of the frame with equal absolute values of their momenta. Thereby one defines

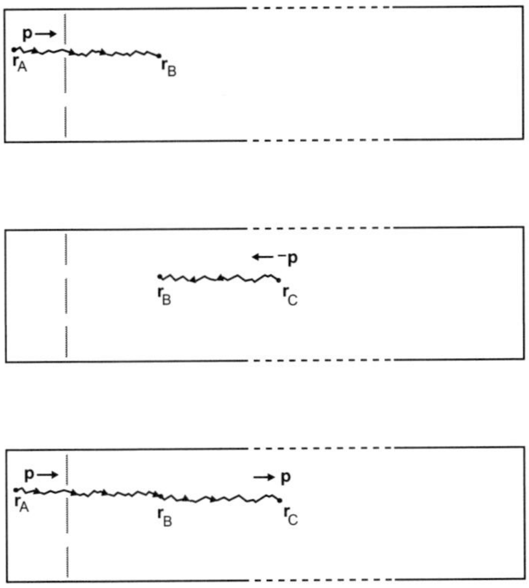

Figure 2. Trajectory of the test particle undergoing reversible scatterings.

a certain value of their total kinetic energy E_{kin}^{gas}. After a short simulation time the gas particles are uniformly distributed within the frame and their distribution in the momentum space has become Maxwellian. The latter is associated with a certain temperature such that the thermodynamical expectation value of the kinetic energy equals E_{kin}^{gas}. It is this temperature which finally shows up in Einstein's law (22). After the embedding gas has "thermalized" one imparts a certain momentum $\boldsymbol{p}$ on the test particle and continues the molecular dynamics simulation with the test particle now included. As indicated in the upper panel of the figure, it performs an irregular (Brownian) motion and loses momentum to the embedding gas whose particles are not shown in the figure. We have chosen the starting point $\boldsymbol{r}_A$ such that the particle moves through the upper slit of the diaphragm and reaches the point $\boldsymbol{r}_B$ after a simulation time Δt of the order of τ which is the time constant of a freely moving particle in a gaseous medium with friction. We now look for a point $\boldsymbol{r}_C$ further to the right in the forward direction of the test particle (s. panel in the middle of the figure). At this point we impart a momentum $-\boldsymbol{p}$ on the particle (after thermalization of the embedding gas), i.e. just the reverse of the momentum at $\boldsymbol{r}_A$.

The point $\boldsymbol{r}_C$ is chosen such that the trajectory ends - again after an identical simulation time of Δt seconds - at point $\boldsymbol{r}_B$. At this point the test particle has lost its original momentum $-\boldsymbol{p}$ almost completely. If one now turns the velocities of **all** particles around by 180^0 and starts the simulation again with the time running forward as before, the test particle continues its motion from point $\boldsymbol{r}_B$ and moves exactly along the trajectory it had formerly followed in the opposite direction coming from $\boldsymbol{r}_C$. When it has reached $\boldsymbol{r}_C$ again, it has regained the previously lost momentum, but this time with the sign reversed. Hence, in moving from $\boldsymbol{r}_A$ to $\boldsymbol{r}_C$ the particle undergoes scattering processes that are in alternating succession Brownian and anti-Brownian within a time interval of the order τ. Thereby the average momentum of the particle is conserved. This is illustrated in the third panel (bottom). A striking feature of the momentum reconstruction by the above scattering processes is the occurrence of non-locality. This can be demonstrated by repeating the procedure that led to the trajectory portion from r_C to r_B with a crucial modification: If one closes the lower slit of the diaphragm and starts then with **the same position/velocity configuration** of all particles as before, the trajectory of the test particle evolves now differently and does no longer join the previously generated trajectory portion at r_B. This is what the molecular dynamics simulation clearly yields. On the other hand, this is to be expected anyway because every momentum transfer from the test particle to the gas spreads with sound velocity throughout the entire structure and probes the change that has been introduced. The stochastic forces acting on the test particle are modified by such a change when these sound waves are reflected back on the particle. If one wants the modified trajectory to join the first trajectory portion at r_B again, one has to choose a different starting point r'_C. Once the test particle has arrived at r_B, one inverts all the velocities as before, and the particle will now recover the momentum p on its modified trajectory toward r'_C. Note: this change in the course of the particle motion results just from closing the **lower** slit although the particle definitely traverses the **upper** slit. One is tempted to surmise that this mechanism of probing the environment "in real life" as the particle exchanges temporarily momentum with the **vacuum**, occurs at light velocity. The latter would impose a limit on the distance beyond which a previously passed potential structure can no longer affect the evolution of the particle's trajectory at its current position.

If one were dealing with Brownian scattering only, the succession of scattering events could be classed as "Markovian". (Shorthand definition: given the presence, future and past are independent.) However, the overall character of the combined Brownian/anti-Brownian scattering processes is obviously non-Markovian. It is true that the particle has almost completely lost its memory of its original momentum when it arrives at r_B, but its future time evolution while moving toward r_C reconstructs, so to speak, past scattering events. The particle's momentum $p(0)$ when it is at r_A, and its momentum $p(t)$ at r_C are **strongly correlated**.

This does not apply to the positions $r_A(0)$ and $r_C(t)$: if one repeats the experiment and lets the particle start at r_A with the same momentum $p(0)$ as before, but with the thermalization process of the embedding gas started some time interval earlier, the particle's trajectory will now be different and lead to a point different from r_C though it regains its original momentum after (approximately) the same traveling time.

Obviously, the non-Markovian (reversible) character of particle motion which results from such a combination of scattering processes can only show up on a coarse grain time scale which is the crucial assumption underlying our derivation of Newton's modified second laws Eqs.(23) and (97) together with (92).

16. Operators and Commutators

An important advantage of our approach may be seen in the derivability of Hermitian operators which in standard quantum mechanics can merely be obtained from educated guessing employing Jordan's replacement rules. In Section 10 we have already derived the momentum operator

$$\widehat{p} = -i\hbar\nabla$$

exploiting our expression (46) for $\boldsymbol{v}(\boldsymbol{r},t)$ and $\boldsymbol{j}(\boldsymbol{r},t) = \rho(\boldsymbol{r},t)\,\boldsymbol{v}(\boldsymbol{r},t)$. The same arguments used in deriving $\widehat{\boldsymbol{p}}$ apply to the angular momentum operator $\widehat{\boldsymbol{L}}$ which occurs on forming the expectation value of the angular momentum of a particle with respect to a center located at $\boldsymbol{r} = 0$. This expectation value $\langle \boldsymbol{L} \rangle$ is primarily defined as a real-space integral over the angular momentum density $\boldsymbol{r} \times m_0\, \boldsymbol{j}$:

$$\langle \boldsymbol{L}(t) \rangle = \int \boldsymbol{r} \times m_0\, \boldsymbol{j}(\boldsymbol{r},t)\, d^3r \tag{106}$$

If one here inserts $\boldsymbol{j}$ from Eq.(73), integrates by parts and requires $\psi(\boldsymbol{r},t)$ to vanish sufficiently fast toward infinity, the result may be written

$$\langle \boldsymbol{L}(t) \rangle = \int \psi^*(\boldsymbol{r},t)\,(\boldsymbol{r} \times \widehat{\boldsymbol{p}})\,\psi(\boldsymbol{r},t)\, d^3r \tag{107}$$

which justifies terming $\widehat{\boldsymbol{L}} \equiv \boldsymbol{r} \times \widehat{\boldsymbol{p}}$ "angular momentum operator".

The kinetic energy of an individual particle, labeled by the index j, is defined as the work performed on that particle by the external force $\boldsymbol{F}$ in accelerating it from zero velocity at time $t = 0$ to its velocity $\boldsymbol{v}_{cj}(t)$ at time t, which yields

$$E^j_{kin} = \frac{m_0}{2}\, \boldsymbol{v}^2_{cj}(t)\,.$$

Forming the ensemble average according to Eq.(3) one obtains $E_{kin} = \frac{m_0}{2}\, \boldsymbol{v}^2_c(\boldsymbol{r},t)$. Thus, the density of the kinetic energy is given by

$$\epsilon_{kin}(\boldsymbol{r},t) = \frac{m_0}{2}\, \rho(\boldsymbol{r},t)\, [\boldsymbol{v}_c(\boldsymbol{r},t)]^2\,.$$

In the two subsystems "B" and "A" we are considering the velocity $\boldsymbol{v}$ is always the same, whereas the convective velocity $\boldsymbol{v}_c$ is different, and therefore we distinguish $\boldsymbol{v}^B_c$ from $\boldsymbol{v}^A_c$ and form the ensemble average over the two subensembles:

$$\epsilon_{kin}(\boldsymbol{r},t) = m_0\, \tfrac{\rho(\boldsymbol{r},t)}{2}\, \tfrac{1}{2} \left(\left[\boldsymbol{v}^B_c(\boldsymbol{r},t)\right]^2 + \left[\boldsymbol{v}^A_c(\boldsymbol{r},t)\right]^2 \right) \tag{108}$$

According to Eq.(6) which still refers to the "B"-system, we have

$$\boldsymbol{v}^B_c = \boldsymbol{v} - \boldsymbol{u} \quad \text{and therefore} \quad \boldsymbol{v}^A_c = \boldsymbol{v} + \boldsymbol{u}\,.$$

Consequently Eq.(108) may be cast

$$\epsilon_{kin}(\boldsymbol{r},t) = m_0\, \frac{\rho(\boldsymbol{r},t)}{2} \left([\boldsymbol{v}(\boldsymbol{r},t)]^2 + [\boldsymbol{u}(\boldsymbol{r},t)]^2 \right)\,. \tag{109}$$

From Eq.(64) we have

$$-\boldsymbol{u} + i\,\boldsymbol{v} = \frac{\hbar}{m_0}\, \frac{1}{\psi}\, \nabla\psi\,.$$

The modulus square of this equation times $m_0\, \rho/2$ is equal to the right-hand side of Eq.(109), that is

$$\epsilon_{kin.}(\boldsymbol{r},t) = \frac{\hbar^2}{2\, m_0}\, |\nabla\psi(\boldsymbol{r},t)|^2\,.$$

Taking the real-space integral of this expression one obtains the kinetic energy

$$E_{kin} \equiv \langle T(t) \rangle = \int_{\mathcal{V}} \frac{\hbar^2}{2\, m_0} \nabla \psi^*(\boldsymbol{r}, t) \cdot \nabla \psi(\boldsymbol{r}, t)\, d^3 r \tag{110}$$

which by employing Green's theorem may be given the familiar form

$$\int_{\mathcal{V}} \frac{\hbar^2}{2\, m_0} \nabla \psi^*(\boldsymbol{r}, t) \cdot \nabla \psi(\boldsymbol{r}, t)\, d^3 r = \int_{\mathcal{V}} \psi^*(\boldsymbol{r}, t) \left[-\frac{\hbar^2\, \nabla^2}{2\, m_0} \right] \psi(\boldsymbol{r}, t)\, d^3 r \,,$$

and hence

$$\langle T(t) \rangle = \int_{\mathcal{V}} \psi^*(\boldsymbol{r}, t)\, \frac{\widehat{\boldsymbol{p}}^2}{2\, m_0}\, \psi(\boldsymbol{r}, t)\, d^3 r \,,$$

which justifies terming $\widehat{\boldsymbol{p}}^2 / 2\, m_0$ "kinetic energy operator".

In practical calculations one often benefits from the fact that E_{kin} may alternatively be cast as in Eq.(110) where the integrand is real-valued and may immediately be interpreted as "kinetic energy density".

The statistical operator is a particular example of derivability from a simple concept. We confine ourselves here to the case of a quantum mechanical system of a bound particle in contact with a heat bath of temperature T. In a stationary state the latter constantly exchanges energy with the system, in the simplest case photons. Hence, the wave function of that system cannot be one of its eigenstates any more, but rather represents a solution to the time-dependent Schrödinger equation and can be expanded in terms of eigenfunctions $\psi_n(\boldsymbol{r})$

$$\psi(\boldsymbol{r}, t) = \sum_n c_n(t)\, \psi_n(\boldsymbol{r})\, e^{-\frac{i}{\hbar} E_n\, t} \tag{111}$$

where E_n denotes eigenvalues of the unperturbed one-particle Hamiltonian $\widehat{H}$.

To make the external system classifiable as a heat bath, the time-averaged coupling energy of the two systems must be negligibly small compared to the difference $E_{n'} - E_n$ of any two eigenvalues. The particle's thermodynamical expectation value of its energy (indicated by double brackets) is given by

$$\langle\langle \widehat{H} \rangle\rangle \equiv U = \frac{1}{\tau} \int_t^{t+\tau} \left[\int \psi^*(\boldsymbol{r}, t')\, \widehat{H}\, \psi(\boldsymbol{r}, t')\, d^3 r \right] dt' \tag{112}$$

where τ (not to be confused with the slow-down time in Section 4) has to be chosen sufficiently large such that U does not depend on t any more. Quantities that derive from U like the specific heat, are only defined as time-averages of this kind.

Inserting Eq.(111) into (112) we obtain

$$U = \sum_n E_n \left\{ \frac{1}{\tau} \int_t^{t+\tau} |c_n(t')|^2\, dt' \right\} . \tag{113}$$

The expression in curly brackets may be interpreted as the relative frequency of the system of being in the n-th eigenstate.

Straight-forward thermodynamics yields for a system that possesses energy levels E_n

$$U = \sum_n E_n \frac{1}{\sigma} e^{-\beta E_n}, \quad \beta = \frac{1}{k_B T} \tag{114}$$

where

$$\sigma = \sum_n e^{-\beta E_n}.$$

Thus, we have from Eq.(113)

$$\frac{1}{\tau} \int_t^{t+\tau} |c_n(t')|^2 \, dt' = \frac{1}{\sigma} e^{-\beta E_n}.$$

If one defines a statistical operator

$$\widehat{\rho} = \frac{1}{\sigma} e^{-\beta \widehat{H}}$$

Eq.(114) can alternatively be cast as

$$U = \sum_n \langle \psi_n | \widehat{\rho} \widehat{H} | \psi_n \rangle \equiv \mathrm{Tr}(\widehat{\rho} \widehat{H}).$$

Commutation rules for the operators apply when the potential $V(\boldsymbol{r})$ in the time-independent Schrödinger equation (31) possesses a certain symmetry. If $V(\boldsymbol{r})$ is spherically symmetric, for example, one verifies simply by performing partial differentiations that

$$\left(\widehat{H}\widehat{\boldsymbol{L}}^2 - \widehat{\boldsymbol{L}}^2\widehat{H}\right) \psi_n(\boldsymbol{r}) \equiv [\widehat{H}, \widehat{\boldsymbol{L}}^2] \, \psi_n(\boldsymbol{r}) = 0$$

and similarly

$$[\widehat{H}, \widehat{\boldsymbol{L}}_z] \, \psi_n(\boldsymbol{r}) = 0$$

if $\psi_n(\boldsymbol{r})$ is an eigenfunction of $\widehat{H}$.

If one is dealing with some operator $\widehat{A}$ which represents just some analytical expression in $\boldsymbol{r}$ and $\widehat{\boldsymbol{p}}$, the time dependence of its expectation value $\langle \widehat{A} \rangle$ can be determined by employing the time-dependent Schrödinger equation which gives

$$\frac{d}{dt} \int \psi^*(\boldsymbol{r}, t) \, \widehat{A} \, \psi(\boldsymbol{r}, t) \, d^3 r = \int \psi^*(\boldsymbol{r}, t) \frac{i}{\hbar} [\widehat{H}, \widehat{A}] \, \psi(\boldsymbol{r}, t) \, d^3 r,$$

in short-hand notation

$$\frac{d}{dt} \widehat{A} = \frac{i}{\hbar} [\widehat{H}, \widehat{A}].$$

Commutation rules of the above kind, again in short-hand notation

$$[\widehat{H}, \widehat{\boldsymbol{L}}^2] = 0; \quad [\widehat{H}, \widehat{\boldsymbol{L}}_z] = 0,$$

similarly

$$[\widehat{H}, \widehat{\boldsymbol{p}}] = 0 \ \text{ if } \ V(\boldsymbol{r}) = \text{const.},$$

but also

$$[\widehat{p}_j, x_k] = \frac{\hbar}{i}\,\delta_{j\,k} \quad \text{where} \quad j = 1,2,3\,;\; k = 1,2,3$$

constitute fundamental elements of standard quantum mechanics and are discussed as pivotal in the context of measurement. From our point of view they are just byproducts of the Schrödinger equation and do not contain any more physics than has already gone into the derivation of the Schrödinger equation. In practice it is impossible to find quantum systems where eigenvalues of $\widehat{H}$ and $\widehat{L}_z$, for example, can be measured simultaneously although there is a widespread belief to the contrary. It is not even possible, for example, to measure the eigenvalues of $\widehat{H}$ for a hydrogen atom which - in clamped proton approximation - represents the archetypal one-particle system and the starting point of quantum mechanics. The lines one observes in its discrete optical spectrum refer to eigenvalue differences and possess - different from true eigenvalues - a natural line width which goes to zero only in the hypothetical case of zero radiation coupling, that is when the lines cannot be observed any more.

There is a remark by Wigner [41] which reveals exactly that lack of stringency and consistency in the foundation of orthodox quantum mechanics:

" All these are concrete and clearly demonstrated limitations on the measurability of operators. They should not obscure the other, perhaps even more fundamental weakness of the standard theory, that it postulates the measurability of operators but does not give directions as to how the measurement should be carried out."

17. Collaps of the Wave Function and the Node Problem

A vital point of the Copenhagen interpretation consists in the notion that the wave function of a stationary one-particle state collapses on performing a measurement on the position of the particle, for example. Within our approach a phenomenon of this kind cannot occur. First of all, in our view "measurement" is not a process of something foreign intruding the realm of quantum mechanics but is rather a part of it. If one calculates, for example, the time-independent wave function $\psi(\boldsymbol{r})$ for a stationary situation where electrons in a diffraction chamber leave a tunneling cathode, sufficiently far behind each other, run through a two-slit diaphragm and finally hit a fluorescent screen, $|\psi(\boldsymbol{r})|^2$ will display the familiar diffraction pattern behind the diaphragm and in particular on the screen. But clearly, the structure of this pattern reflects the distribution of the entire ensemble of electrons that leave the cathode, and a particular electron, that hits the screen some place, is only one member out of this ensemble. Hence its capture on the screen does not destroy the properties of the ensemble. The electron capture by an atom of the screen constitutes a process that has only marginally to do with the diffraction state in that the latter determines the probability of the electron being at that particular atom. Otherwise the capture process is governed by the time dependent Schrödinger equation and the perturbation caused by the electromagnetic field of the outgoing photon. All this is completely independent of the possible presence of an "observer" who might see that photon.

Despite deceptive similarities the situation becomes conceptually different when one replaces the tunneling tip in the otherwise unchanged diffraction chamber by a light source that emits, again in sufficiently large time intervals, photons of the same wave length as the previously considered electrons. Since the space-time structure of the wave (in principle $\propto \cos[\boldsymbol{k} \cdot \boldsymbol{r} - \omega t]$) with which each photon is associated, is not defined as the property of an ensemble of mechanical objects but rather by classical electrodynamics (Maxwell's equations), it will, in fact, disappear on the disappearance of the photon in question. There are a couple of properties by which photons differ crucially from massive particles: They move always at light velocity along straight lines in vacuo and the associated waves are vector-valued functions. By contrast, the de Broglie waves of massive particles are in general complex-valued functions, and the average velocity of the particles is given by the gradient of the functions' phase. Their interaction with other particles and parts of an experimental setup is described by the Schrödinger equation and the potentials therein. On the other hand, the interaction of photons with polarizers, mirrors, quaterwave plates, filters etc. is governed by classical electrodynamics. Malus' law, for example, constitutes a law of classical optics. Because of these rather fundamental differences any analysis of photon correlation experiments, for example, should critically be scrutinized whether a transfer to analogous experiments with massive particles is truly justified. In Section 30 we shall draw on the familiar example of the Stern-Gerlach experiment to demonstrate that the selection mechanism for up-spin and down-spin particles in the Stern-Gerlach magnet has nothing to do with the mechanism separating horizontally and vertically polarized photons in a polarizing beam spitter.

According to Mielnik and Tengstrand [31] excited stationary states appear to pose a serious problem in that $\psi(\boldsymbol{r})$ possesses nodal surfaces at which the normal derivative $\frac{\partial}{\partial n}\rho(\boldsymbol{r})$ vanishes but the normal component of the osmotic velocity

$$\boldsymbol{u}_n(\boldsymbol{r}) = -\frac{\hbar}{2m_0} \frac{\frac{\partial}{\partial n}\rho(\boldsymbol{r})}{\rho(\boldsymbol{r})}\, \boldsymbol{e}_n$$

becomes formally infinite. Moreover, at surfaces across which $\rho(\boldsymbol{r})$ attains a maximum, $\frac{\partial}{\partial n}\rho(\boldsymbol{r})$ vanishes as well, but $\boldsymbol{u}_n(\boldsymbol{r})$ becomes now zero. If $\psi(\boldsymbol{r})$ is real-valued then $\boldsymbol{v}(\boldsymbol{r})$ vanishes everywhere, and therefore we have on such surfaces with maximum probability density

$$\boldsymbol{v} = \boldsymbol{u}_n = 0\,.$$

In the 2s-state of a hydrogen electron, for example, one has a spherical surface of this kind. Hence, it seems that this sphere separates two regions of space that are mutually inaccessible for the electron. But the above velocities are only ensemble averages or - in the spirit of the definition (5) - averages of non-vanishing velocities $\boldsymbol{v}(t_i)$, $\boldsymbol{u}_n(t_i)$ of different directions over a sufficiently long time T.

As for $|\boldsymbol{u}_n(\boldsymbol{r})|$ going to infinity as one crosses a nodal surface of $\psi(\boldsymbol{r})$, one has to keep in mind that stationary excited states (excited eigenstates) are highly fictional and do actually not exist in nature. Because of $\Delta E\,\Delta t \approx \hbar$ and $\Delta E = 0$ for an eigenstate it would take an infinite time to prepare them. Hence, truly existing excited states do not possess nodal surfaces where $\psi(\boldsymbol{r})$ vanishes exactly. But even if one would allow them to exist, the kinetic energy density $\frac{m_0}{2}\,\rho(\boldsymbol{r})\,\boldsymbol{u}^2(\boldsymbol{r}) = \frac{\hbar^2}{2m_0}|\nabla\psi(\boldsymbol{r})|^2$ remains finite and hence ensures a physically meaningful behavior even for this idealized situation.

18. The Feynman Path Integral

As our concept builds on the existence of particle trajectories one might surmise that there should be some affinity to Feynman's path integral method [42] which also relates to possible paths a particle might take. We shall outline that there is neither any formal kinship nor does Feynman name any cause for the possible occurrence of non-classical trajectories. In so doing we limit ourselves, as Feynman in his article, to the one-dimensional case of a particle that moves non-relativistically in a potential $V(x)$. Feynman's considerations are based on two hypotheses that may be summarized by stating that the wave function $\psi(x, t + \Delta t)$ of the particle at some point x and time $t + \Delta t$ is connected with the wave function $\psi(x - \sigma, t)$ at a previous point $x - \sigma$ and earlier time t by an integral equation similar to the Smoluchowski equation (278) of the ensuing section, viz.

$$\psi(x, t + \Delta t) = \int \psi(x - \sigma, t)\, F(x, x - \sigma, t, \Delta t)\, d\sigma \tag{115}$$

where $F(x, x - \sigma, t, \Delta t)$ is the function that brings in classical mechanics. It is defined as

$$F(x, x - \sigma, t, \Delta t) = \frac{1}{A}\, e^{\frac{i}{\hbar} S(x, x-\sigma, t, \Delta t)}$$

where

$$A = \left(\frac{2\pi\hbar\, i\, \Delta t}{m_0}\right)^{\frac{1}{2}} .$$

Here $S(x, x - \sigma, t, \Delta t)$ denotes Hamilton's first principle function for a particle moving classically in a potential $V(x)$ along a trajectory from a point $x - \sigma$ to x within an infinitesimally small time span Δt. Hence

$$S(x, x - \sigma, t, \Delta t) = \\ \text{Min.} \int_t^{t+\Delta t} \left[\frac{m_0}{2}\, \dot{\sigma}^2 - V(x - \sigma(t'))\right] dt'$$

where

$$L(\dot{\sigma}(t), \sigma(t)) = \frac{m_0}{2}\, \dot{\sigma}^2 - V(x - \sigma(t))$$

denotes the Lagrangean.

As Δt is infinitesimally small $S(x, x - \sigma, t, \Delta t)$ may be approximated

$$S = \Delta t \left[\frac{m_0}{2} \left(\frac{\sigma}{\Delta t}\right)^2 - V(x)\right] .$$

Hence one has

$$F = \frac{1}{A} \left[e^{\frac{i\, m_0}{2\hbar\, \Delta t}\, \sigma^2} \cdot e^{-\frac{i\, V(x)\, \Delta t}{\hbar}}\right] .$$

The first exponential oscillates rapidly as a function of σ because of the prefactor $1/\Delta t$ in the exponent, whereas, by comparison, $\psi(x - \sigma, t)$ may be assumed slowly varying as a function of σ. The value of the integral in Eq.(115) depends therefore only on a small interval of σ around the point x. Within this interval $\psi(x-\sigma,t)$ may be expanded as

$$\psi(x-\sigma,t) = \psi(x,t) - \frac{d\psi}{dx}\,\sigma + \frac{1}{2}\frac{d^2\psi}{dx^2}\,\sigma^2 \,.$$

If one inserts this into Eq.(115), observes

$$\frac{1}{A}\int_{-\infty}^{\infty} e^{\frac{i\,m_0}{2\hbar\,\Delta t}\,\sigma^2}\, d\sigma = 1\,;$$

(note that this equation defines A!), further

$$\frac{1}{A}\int_{-\infty}^{\infty} e^{\frac{i\,m_0}{2\hbar\,\Delta t}\,\sigma^2}\,\sigma\, d\sigma = 0\,,$$

$$\frac{1}{A}\int_{-\infty}^{\infty} e^{\frac{i\,m_0}{2\hbar\,\Delta t}\,\sigma^2}\,\sigma^2\, d\sigma = \frac{i\hbar}{m_0}\,\Delta t$$

and uses

$$e^{-\frac{i\,V(x)\,\Delta t}{\hbar}} \approx 1 - \frac{i\,V(x)\,\Delta t}{\hbar}\,,$$

one obtains

$$\psi(x,t+\Delta t) = \psi(x,t)\left(1 - \frac{i}{\hbar}\,V(x)\,\Delta t\right) + \frac{1}{2}\frac{d^2\psi}{dx^2}\cdot\frac{i\hbar}{m_0}\,\Delta t\left(1 - \frac{i\,V(x)\,\Delta t}{\hbar}\right).$$

Multiplying this equation by $\frac{i\hbar}{\Delta t}$ and letting Δt tend to zero one arrives at the time dependent Schrödinger equation

$$i\hbar\frac{\partial}{\partial t}\,\psi(x,t) = \left[-\frac{\hbar^2}{2\,m_0}\frac{\partial^2}{\partial x^2} + V(x)\right]\psi(x,t)\,.$$

Though the Schrödinger equation is obviously recovered following this line of argument, it remains unclear why Hamilton's classical first principle function should appear in the exponent of $F(x, x-\sigma, t, \Delta t)$. Feynman's considerations lean closely on arguments of measurement typical of the Copenhagen school of thought, cast into an axiomatic framework notably by v. Neumann [43]. But $\psi(x,t)$ may, for example, describe the motion of a harmonic oscillator in the absence of any measurement. In fact, if one were to perform a measurement on the harmonic oscillator the Schrödinger equation would contain a perturbative extra term that would give rise to a different wave function. Clearly, as already stated in Section1 the probabilistic character of $\psi(x,t)$ does not originate from indeterminacies caused by the process of measurement. The complex-valuedness of the wave function in the form $\psi(x,t) = |\psi(x,t)|\,e^{i\varphi(x,t)}$ comes about by incorporating two autonomous real-valued informations: the probability density $|\psi(x,t)|^2$ of the particle being at x and time t and the ensemble average $v(x,t) = \frac{\hbar}{m_0}\frac{d}{dx}\,\varphi(x,t)$ of its velocity. For that reason our derivation of the Schrödinger equation requires two Smoluchowski equations for the real-valued functions $\rho(\boldsymbol{r},t)$ and $\boldsymbol{v}(\boldsymbol{r},t)$ instead of Feynman's single Eq.(115). We believe, therefore, that in our derivation the connection to classical mechanics becomes definitely more transparent and convincing.

19. Spontaneous Light Emission

Understanding the discrete spectrum of light emitting atoms had been the primary motivation for developing a theory beyond classical mechanics and electrodynamics. It was far from being likely that Planck's constant h which he introduced to explain the continuous spectrum of light emitting incandescent "black bodies" could have anything to do with those discrete spectra. By hindsight it must be seen as a surprise when Niels Bohr [44] could explain the well studied spectrum of the hydrogen atom by requiring the associated electron to orbit around the nucleus on concentric circles where its angular momentum L equals integer multiples of Planck's reduced constant $\hbar = \frac{h}{2\pi}$: $L = n\,\hbar$ and $n = 1, 2, \ldots$ with h denoting Planck's originally introduced constant. In each of the orbitals the electron was considered to be in a stable state, but it was allowed to jump spontaneously to another orbital of lower energy and convert the energy difference into light. Those "quantum jumps" still belong to the vocabulary of present-day quantum mechanics (see e. g.[45]) although their existence lacks any foundation as we shall demonstrate.

In Bohr's theory the electron always possesses a non-vanishing angular momentum so that the centrifugal force keeps it well separated from the nucleus and thereby ensures a well defined size of the hydrogen atom for $n = 1$, its state of lowest energy. In reality, i. e. according to our approach, the particle under study, the electron, is driven by the combined action of the static classical Coulomb force exercised by the nucleus and by the stochastic forces of the vacuum. As a consequence, its probability amplitude obeys the Schrödinger equation, the time-independent solutions of which, $\hat{\psi}_{nlm}(\boldsymbol{r})$, are characterized by integer quantum numbers n, l, m where $\boldsymbol{r}$ is referenced to the position of the nucleus. The latter is considered to be a clamped point charge for simplicity. The state $\hat{\psi}_{100}(\boldsymbol{r})$ refers to the groundstate where $\langle \boldsymbol{L} \rangle = 0$, distinctly different from Bohr's theory. Only excited states ($n > 1$) for which l equals $|m|$ display a toroidal probability density and resemble diffuse circular Bohr orbitals.

The energies that are associated with the eigensolutions $\hat{\psi}_{nlm}(\boldsymbol{r})$ are given by

$$E_{nlm} = -\frac{R}{n^2} \quad \text{where} \quad n > l + |m| \quad l = 0, 1, 2, \ldots\,, \; m = -l, \ldots 0 \ldots + l \qquad (116)$$

with n being an integer itself, and R stands for

$$R = \tfrac{1}{2}\,\alpha^2 m_0\, c^2 = \text{Rydberg constant}\,; \quad R = 13.6059\, eV\,.$$

Here m_0 denotes the electron's rest mass, c the velocity of light in vacuo and α the fine structure constant:

$$\alpha = \frac{e^2}{4\pi\varepsilon_0\,\hbar c} = \frac{1}{137.036}\,; \qquad \hbar = \frac{h}{2\pi}\,.$$

The quantity ε_0 represents, as before, the permittivity of the vacuum and e is the elementary charge.

As has already been discussed in Section 11, it is one of the fundamental credos of conventional quantum mechanics that eigenvalues of the energy as in (116) constitute results of appropriate measurements, more precisely, as Mermin [46] states in a widely recognized article:

"...quantum mechanics requires that the result of measuring an observable be an eigenvalue of the corresponding Hermitian operator....."

Although this statement belongs to the seemingly ineradicable rituals in conversing about quantum mechanics, it is void of meaning. Of course, there has always been the discrete hydrogen spectrum in the back of the minds of the founding fathers, and that spectrum seemed to be clearly some map of the eigenvalues (116). But in actual fact one commonly measures the wave length of the emitted light by a spectrometer about 10^{10} atomic diameters away from the emitter. When the packet of the light wave enters the "measurement process", i e. the spectrometer, the atom has long left its original state. The phraseology "measuring eigenvalues" invites the impression as if it would compare to measuring somebody's collar size. Moreover, as we shall show, in the emission process the energy of the atom attains all values between the eigenvalues that are involved in the transition, but the frequency and the measured associated wavelength of the emitted light remain constant.

Before we go into the details of our approach we want to emphasize that its basic idea is almost identical with what has been popularized by E. T. Jaynes already in 1963 [47] and in the following years [48], [49] under the name "neoclasical theory (NCT)". The terminology is rather misleading. In the present theory (and in Jaynes' theory as well) the electromagnetic field is generated by an oscillating electronic current density which sets up a vector potential and this, in turn, appears in the kinetic energy operator of the electronic Hamiltonian. What should be considered "neoclassical" in using this interrelation? Quantum mechanical current densities are foreign objects to classical electrodynamics.

Although we arrive at a time evolution of the light emission that differs in crucial details from purely exponential decay as obtained by Weisskopf and Wigner [50], the decay times are in agreement with each other. This applies to Jaynes' theory as well. Most of the criticism voiced against Jaynes' work revolves around his non-exponential time evolution. He tries to defend this result by considering ineffective excitations in which the atom acquires only a small portion of the full excitation energy. He argues that this will normally happen in reality, and therefore only the tail of his transition curve can show up in the experiment. This tail is essentially an exponential function.

We fundamentally disagree with this notion. If the atom re-emits light with a frequency ω the associated wave packet contains a photon of energy $\hbar\omega$ and not a fraction of it. Hence the atom must have definitely been in its excited state prior to the emission process.

How does spontaneous light emission fit into the framework that we have developed so far? Our point of departure from the standard approach consists in questioning the assumption that eigenstates "can be prepared". How should such a "preparation" be achieved? By definition, an eigenstate is associated with zero variance of its energy. Hence, because of

$$\Delta E\,\Delta t \approx \hbar \quad \text{where} \quad \Delta E = \sqrt{\overline{(E - E_{nlm})^2}} \tag{117}$$

the preparation time Δt is infinite for an eigenstate.[1] That is, in reality, an excited state can

[1]A similar situation occurs if one wants to excite a superconducting cavity in one of its modes. If the cavity can lose energy to the outside at a small rate it behaves as if it were slightly attenuated. The time it takes to arrive at a stationary state grows longer and longer the weaker the energy loss becomes. The excitation time tends to infinity as the loss rate tends to zero.

only be a solution to the **time-dependent** Schrödinger equation and hence may be cast as

$$\psi(\boldsymbol{r},t) = \sum_{n,l,m} c_{nlm}\,\hat{\psi}_{nlm}(\boldsymbol{r})\,e^{-i\frac{E_{nlm}}{\hbar}t} \tag{118}$$

where

$$\int \hat{\psi}^*_{n'l'm'}(\boldsymbol{r})\,\hat{\psi}_{nlm}(\boldsymbol{r})\,d^3r = \delta_{n'n}\,\delta_{l'l}\,\delta_{m'm} \tag{119}$$

and because of

$$\int |\psi(\boldsymbol{r},t)|^2\,d^3r = 1 \quad \text{one has} \quad \sum_{n,l,m} |c_{nlm}|^2 = 1\,. \tag{120}$$

Each term under the sum in Eq.(118) satiesfies individually the time-dependent Schrödinger equation of the hydrogen electron since by definition

$$\hat{H}\hat{\psi}_{nlm}(\boldsymbol{r}) = E_{nlm}\,\hat{\psi}_{nlm}(\boldsymbol{r}) \quad \text{where} \quad \hat{H} = \frac{\widehat{\boldsymbol{p}}^{\,2}}{2\,m_0} - \frac{e^2}{4\pi\varepsilon_0\,r} \quad \text{and} \quad \widehat{\boldsymbol{p}} = -i\hbar\,\nabla\,. \tag{121}$$

A realistic "eigenstate" is characterized by the property that the square modulus of one the coefficients c_{nlm} in Eq.(118) is close to unity, that of the others correspondingly small.

In the following we consider the situation in which the hydrogen atom has been excited from the ground state 1s to the state 2p where $m = 0$. The excitation may have been caused by absorbing linearly polarized light. As stated above, it is, as a matter of fact, impossible that the atom in the excitation process really ends up in the eigenstate 2p. Its state will rather have the form

$$\psi(\boldsymbol{r},t) = c_0\,\hat{\psi}_{1s}(\boldsymbol{r})\,e^{-i\frac{E_{1s}}{\hbar}t} + c_1\,\hat{\psi}_{2p}(\boldsymbol{r})\,e^{-i\frac{E_{2p}}{\hbar}t} \quad \text{where} \quad 0 < |c_0| \ll |c_1| < 1\,. \tag{122}$$

Using Eq.(122) we obtain

$$\int \psi^*(\boldsymbol{r},t)\,\hat{H}\,\psi(\boldsymbol{r},t)\,d^3r = \underbrace{|c_1|^2}_{=1-|c_0|^2}\;E_{2p} + |c_0|^2\,E_{1s} \quad \text{that is} \quad E = E_{2p} - \widetilde{E}\,|c_0|^2 \tag{123}$$

where

$$\widetilde{E} = E_{2p} - E_{1s}\,. \tag{124}$$

The expression $\widetilde{E}\,|c_0|^2$ represents obviously the uncertainty ΔE with which the 2p-state has been "prepared".

From Eq.(122) we may form the electronic charge density $\rho(\boldsymbol{r},t) = e\,|\psi(\boldsymbol{r},t)|^2$ which we cast as

$$\rho(\boldsymbol{r},t) = \rho_0(\boldsymbol{r}) + \widetilde{\rho}(\boldsymbol{r},t) \tag{125}$$

where

$$\rho_0(\boldsymbol{r}) = e\left[|c_0|^2\,\hat{\psi}^2_{1s}(\boldsymbol{r}) + |c_1|^2\,\hat{\psi}^2_{2p}(\boldsymbol{r})\right] \tag{126}$$

and

$$\widetilde{\rho}(\boldsymbol{r},t) = |c_0^*\, c_1|\, e\, \hat{\psi}_{1s}(\boldsymbol{r})\, \hat{\psi}_{2p}(\boldsymbol{r}) \left[e^{i\left[\frac{E_{2p}-E_{1s}}{\hbar}\, t+\varphi\right]} + e^{-i\left[\frac{E_{2p}-E_{1s}}{\hbar}+\varphi\right]} \right] . \tag{127}$$

where φ is defined through

$$c_0^* c_1 = |c_0^* c_1|\, e^{i\varphi} . \tag{128}$$

Here we have exploited the fact that $\hat{\psi}_{1s}(\boldsymbol{r})$ and $\hat{\psi}_{2p}(\boldsymbol{r})$ are real-valued functions. Eq.(125) may hence be rewritten

$$\rho(\boldsymbol{r},t) = \rho_0(\boldsymbol{r}) + 2\,|c_0^*\, c_1|\, \widetilde{\rho}_0(\boldsymbol{r}) \cos(\omega t + \varphi) \tag{129}$$

where

$$\widetilde{\rho}_0(\boldsymbol{r}) = e\, \hat{\psi}_{1s}(\boldsymbol{r})\, \hat{\psi}_{2p}(\boldsymbol{r}) \quad \text{and} \quad \omega = \frac{E_{2p} - E_{1s}}{\hbar} \quad \text{i.e.} \quad \hbar\, \omega = E_{2p} - E_{1s} \tag{130}$$

The following figure shows four snapshots of the time evolution of $\rho(\boldsymbol{r},t)$.

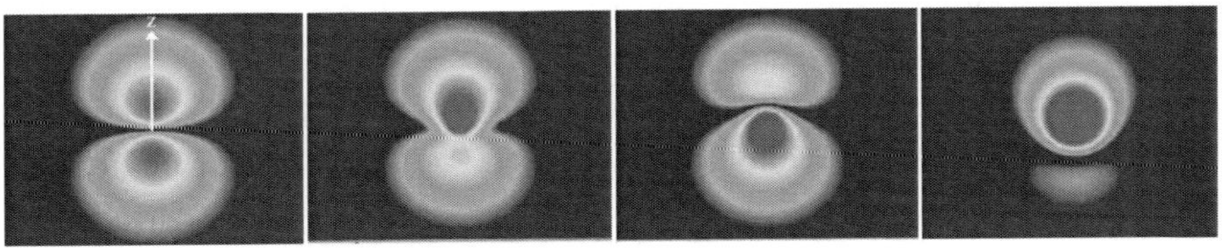

Figure 3. Four snapshots of the color coded density of a H-atom in the transition 2p→1s.

As one can see from Eq.(123):

$$E = |c_0|^2\, E_{1s} + |c_1|^2\, E_{2p} \quad \text{and} \quad 0 \le |c_0|\, ; \, 0 \le |c_1| \quad |c_0|^2 + |c_1|^2 = 1$$

the energy of the electron can have any value between E_{2p} and E_{1s}. **Regardless of the value of E its charge density oscillates sharply at ω!** In the following calculation we shall derive a transition time which the electron takes to change its state from (122) with $0 < |c_0| \ll |c_1| < 1$ to a form of $\psi(\boldsymbol{r},t)$ where $|c_0| \approx 1$ and $|c_1| \approx 0$. This transition time turns out to be of the order of $10^{-9}\, s$, that is, **there are no quantum jumps**! At this point it should be remembered that the article by Dehmelt and collaborators [45] "..on the observation of quantum jumps" shows a plot of these jumps on a time scale of 20 s unit length!

The function $\psi(\boldsymbol{r},t)$ defined by Eq.(122) is a solution to the time-dependent Schrödinger equation

$$\hat{H}\psi(\boldsymbol{r},t) = i\,\hbar \frac{\partial}{\partial t}\, \psi(\boldsymbol{r},t) \tag{131}$$

only as long as the coefficients c_0, c_1 are constant. However, even when c_0 is very small compared to unity, $\rho(\boldsymbol{r},t)$ oscillates at the frequency ω and thus gives rise to the emission of an electromagnetic wave. The latter is polarized in the direction of the quantization axis of $\hat{\psi}_{2p}(\boldsymbol{r})$ which also defines the symmetry axis of $\hat{\psi}_{1s}(\boldsymbol{r})\,\hat{\psi}_{2p}(\boldsymbol{r})$ in Eq.(130). As the atom loses energy in building up the electromagnetic wave, E in Eq.(123) decreases, and hence c_0 must now increase as a function of time. This is a consequence of the fact that the radiation field acts back on the atom, and therefore the Hamiltonion in Eq.(131) is now modified:

$$\hat{H}'\psi(\boldsymbol{r},t) = i\hbar\frac{\partial}{\partial t}\,\psi(\boldsymbol{r},t) \quad \text{where} \quad \hat{H}' = \frac{(\hat{\boldsymbol{p}} - e\boldsymbol{A}(\boldsymbol{r},t))^2}{2\,m_0} - \frac{e^2}{4\pi\varepsilon_0\,r} \tag{132}$$

where $\boldsymbol{A}(\boldsymbol{r},t)$ denotes the vector potential of the radiation field. It is set up by the quantum mechanical current density $\boldsymbol{j}(\boldsymbol{r},t)$. As we expect and what the calculation actually comes up with is an outgoing wave packet with a thickness of $c\,\hat{\tau}$ where $\hat{\tau}$ is the transition time. Moreover, the current density of the energy flow, $\boldsymbol{S}(\boldsymbol{r},t)$, will display the characteristic feature of a Hertzian dipole.

In the article by Weisskopf and Wigner [50] the radiation field is quantized which suggests that their derivation qualifies to be more fundamental. The authors consider a transition between two states, an initial state defined by the electrons of the atom being in an excited eigenstate with no photon present and a final state that describes the electrons being in the lower eigenstate and a linear combination of quantized electromagnetic modes in a cube of volume $\mathcal{V}$ with ideally reflecting walls. Hence, these modes represent standing waves. The sum of the mod squared of the coefficients in the linear combination equals unity. All modes that pertain to the final state are assumed to have equal weight and are associated with essentially the same frequency $\omega = \frac{E_i - E_f}{\hbar}$. Here E_i and E_f refer to the energy of the electronic initial and final state, respectively.

From a principal point of view this field-theoretical description appears to be rather absurd. The final state does not contain an outgoing electromagnetic wave of a certain thickness which forms a hollow sphere around the emitter, as with our theory, but rather a set of standing waves which penetrate the emitter undisturbed. They penetrate the atom completely unmodified also during the entire transition process. It seems to border on magic that a calculation of this kind still arrives at a result that is in agreement with the experiment, although the time evolution is hardly accessible and therefore still open to discussion.

We now turn back to our derivation. To obtain $\boldsymbol{A}(\boldsymbol{r},t)$ we first determine the current density

$$\boldsymbol{j}(\boldsymbol{r},t) = \frac{e\,\hbar}{2i\,m_0}\left[\psi^*(\boldsymbol{r},t)\nabla\,\psi(\boldsymbol{r},t) - \psi(\boldsymbol{r},t)\nabla\,\psi^*(\boldsymbol{r},t)\right]. \tag{133}$$

On inserting $\psi(\boldsymbol{r},t)$ from Eq.(122) the current density takes the form

$$\boldsymbol{j}(\boldsymbol{r},t) = \frac{e\,\hbar}{m_0}\,|c_0^*\,c_1|\,[\hat{\psi}_{1s}(\boldsymbol{r})\nabla\,\hat{\psi}_{2p}(\boldsymbol{r}) - \hat{\psi}_{2p}(\boldsymbol{r})\nabla\,\hat{\psi}_{1s}(\boldsymbol{r})]\,\sin(\omega\,t + \varphi)\,. \tag{134}$$

The vector potential $\boldsymbol{A}(\boldsymbol{r},t)$ and $\boldsymbol{j}(\boldsymbol{r},t)$ are interconnected by

$$\boldsymbol{A}(\boldsymbol{r},t) = \frac{\mu_0}{4\pi}\int\frac{\boldsymbol{j}(\boldsymbol{r}',t - \frac{|\boldsymbol{r}'-\boldsymbol{r}|}{c})}{|\boldsymbol{r}' - \boldsymbol{r}|}\,d^3r' \quad \text{where} \quad \mu_0 = \frac{1}{\varepsilon_0\,c^2}\,. \tag{135}$$

Because of Eq.(133) $\boldsymbol{A}(\boldsymbol{r},t)$ is a functional of $\psi(\boldsymbol{r},t)$. It follows then from inspection of Eq.(132) that this modified Schrödinger equation constitutes now a **non-linear** partial differential equation since the Hamiltonian $\hat{H}'$ depends on $\psi(\boldsymbol{r},t)$. Below we shall derive a detailed solution to this equation. If one is not interested in the details of the time dependence one can take a short-cut:

First we give the expressions

$$\hat{\psi}_{1s}(\boldsymbol{r})\nabla\,\hat{\psi}_{2p}(\boldsymbol{r}) \quad \text{and} \quad \hat{\psi}_{2p}(\boldsymbol{r})\nabla\,\hat{\psi}_{1s}(\boldsymbol{r})$$

a different form by using the identity (which is just an application of the chain rule):

$$[\hat{H}\,\boldsymbol{r} - \boldsymbol{r}\,\hat{H}]\,\hat{\psi}(\boldsymbol{r}) = -i\,\frac{\hbar}{m_0}\,\hat{\boldsymbol{p}}\,\hat{\psi}(\boldsymbol{r}) = -\frac{\hbar^2}{m_0}\,\nabla\,\hat{\psi}(\boldsymbol{r})\,. \tag{136}$$

This yields

$$\hat{\psi}_{1s}(\boldsymbol{r})\nabla\,\hat{\psi}_{2p}(\boldsymbol{r}) = \frac{m_0}{\hbar^2}\,[\hat{\psi}_{1s}(\boldsymbol{r})\,\boldsymbol{r}\,\hat{H}\,\hat{\psi}_{2p}(\boldsymbol{r}) - \hat{\psi}_{1s}(\boldsymbol{r})\,\hat{H}\,\boldsymbol{r}\,\hat{\psi}_{2p}(\boldsymbol{r})]$$

and

$$\hat{\psi}_{2p}(\boldsymbol{r})\nabla\,\hat{\psi}_{1s}(\boldsymbol{r}) = \frac{m_0}{\hbar^2}\,[\hat{\psi}_{2p}(\boldsymbol{r})\,\boldsymbol{r}\,\hat{H}\,\hat{\psi}_{1s}(\boldsymbol{r}) - \hat{\psi}_{2p}(\boldsymbol{r})\,\hat{H}\,\boldsymbol{r}\,\hat{\psi}_{1s}(\boldsymbol{r})]\,.$$

Forming the integral of Eq.(134), exploiting the hermitiaty of $\hat{H}$, using Eq.(181) and $\hbar\,\omega = E_{2p} - E_{1s}$ we thus obtain

$$I(t)\,\boldsymbol{e}_z = 2\,|c_0^*\,c_1|\,e\,\omega \int \hat{\psi}_{1s}(\boldsymbol{r})\,\boldsymbol{r}\,\hat{\psi}_{2p}(\boldsymbol{r})\,d^3r\,\sin(\omega\,t+\varphi) \quad \text{where} \quad \boldsymbol{e}_z || z-\text{axis}\,. \tag{137}$$

The quantity $I(t)$ denotes the alternating current that is set up in the atom as a result of c_0 not being zero. The quantization axis of $\hat{\psi}_{2p}(\boldsymbol{r})$ is taken along the z-axis. We may rewrite the above integral

$$\int \hat{\psi}_{1s}(\boldsymbol{r})\,\boldsymbol{r}\,\hat{\psi}_{2p}(\boldsymbol{r})\,d^3r = \int \rho_{dipole}(\boldsymbol{r})\,\boldsymbol{r}\,d^3r = \overline{\boldsymbol{r}}$$

where we have expressed the fact that $\hat{\psi}_{1s}(\boldsymbol{r})\,\hat{\psi}_{2p}(\boldsymbol{r})$ represents a dipole-type probability density. Hence, Eq.(137) can be cast as

$$I(t)\,\boldsymbol{e}_z = -\frac{d}{dt}\,\boldsymbol{p}(t) \quad \text{where} \quad \boldsymbol{p}(t) = g(t)\,|e|\,\overline{\boldsymbol{r}}\,\cos\omega\,t \quad \text{and} \quad g(t) = 2\,|c_0^*(t)\,c_1(t)|\,.$$

As the emission of the electromagnetic wave proceeds, the coefficient c_0 becomes larger and will finally attain its largest value 1 at the end of the transition. According to Eq.(118) it will be equal to $\frac{1}{\sqrt{2}}$ in the middle of the transition. The coefficient c_1 changes in reverse since the sum of the square of the coefficients must be unity at any time. Hence, in the middle of the transition the function $g(t)$ defined above attains its maximum value 1 and drops asymptotically to zero on either side. To serve the purpose of the present short-cut, we approximate the actually bell-shape time-dependence of $g(t)$ by a rectangle of width $\hat{\tau}$ and height unity.

We now invoke Hertz's result on the density $\boldsymbol{S}(\boldsymbol{r},t)$ of the energy flow from an oscillating dipole:

$$\boldsymbol{S}(r,\theta,t) = \frac{1}{16\pi^2\,\varepsilon_0\,c^3}\,\frac{\sin^2\theta}{r^2}\,\left[\frac{d^2}{dt^2}\,\boldsymbol{p}(t)\right]^2$$

where θ is the angle that $\boldsymbol{r}$ encloses with the dipole axis.

Forming a surface integral with $\boldsymbol{S}(r,\theta,t)$ over a concentric sphere of radius r and averaging over one oscillation period one arrives at

$$S(t) = [g(t)]^2 \, \frac{e^2\,\omega^4}{6\pi\,\varepsilon_0\,c^3}\,|\boldsymbol{p}_{n'n}|^2 \quad \text{where} \quad S(t) = \pi\,r^2 \int |\boldsymbol{S}(r,\theta,t)|\sin 2\,\theta\,d\,\theta$$

$$\text{and} \quad n' = 1s\,;n = 2p$$

and

$$\boldsymbol{p}_{n'n} = \int \hat{\psi}_{n'}(\boldsymbol{r})\,\boldsymbol{r}\,\hat{\psi}_{n}(\boldsymbol{r})\,d^3r\,; \qquad [g(t)]^2 = \begin{cases} 1 & \text{for} \quad |t| \le \hat{\tau}/2 \\ 0 & \text{for} \quad |t| > \hat{\tau}/2 \end{cases}. \tag{138}$$

Integration of $S(t)$ over the transition time $\hat{\tau}$ must yield $E_{2p} - E_{1s} = \hbar\,\omega$:

$$\int_0^{\hat{\tau}} S(t)\,dt = \hbar\,\omega = \frac{e^2\,\omega^4}{6\pi\,\varepsilon_0\,c^3}\,|\boldsymbol{p}_{n'n}|^2\hat{\tau}\,.$$

From this we obtain an expression for the inverse of the transition time

$$\tfrac{1}{\hat{\tau}} = \frac{e^2\,\omega^3}{6\pi\,\varepsilon_0\,c^3\,\hbar}\,|\boldsymbol{p}_{n'n}|^2 \quad \text{or} \quad \tfrac{1}{\hat{\tau}} = \alpha\,\frac{2\,\omega^3}{3\,c^2}\,|\boldsymbol{p}_{n'n}|^2 \quad \text{where} \quad \alpha = \frac{e^2}{4\pi\varepsilon_0\,\hbar} \tag{139}$$

in agreement with the result of the standard calculation (s. e. g. [50]) which is based on a remarkably different concept, as already mentioned above. It should be observed, however, that this calculation yields an expression for the transition rate $\frac{1}{\tau'}$ which is identified with $\frac{d}{dt}\,|c_0|^2(t)|_{t=0} = \frac{2}{\hat{\tau}}$. Hence

$$\tfrac{1}{\tau'} = \frac{e^2\,\omega^3}{3\pi\,\varepsilon_0\,c^3\,\hbar}\,|\boldsymbol{p}_{n'n}|^2\,.$$

It is worth noting that the problem of spontaneous light emission has for the first time been treated by Fermi [51] in 1927. He chose an approach very similar to ours, but used the classical expression for the radiation back action $\propto \frac{d^3}{dt^3}\,p$ which led to a frequency shift of the emitted light depending on the transition time. However, this is at variance with the observation.

We now turn back to the problem of calculating the detailed time-dependence of $|c_0(t)|^2, |c_1(t)|^2$. To this end we first observe that

$$\hat{H}' = \hat{H} + \hat{H}_{int} = \frac{(\hat{\boldsymbol{p}} - e\boldsymbol{A}(\boldsymbol{r},t))^2}{2\,m_0} - \frac{e^2}{4\pi\varepsilon_0\,r} = \hat{H} - \frac{e}{m_0}\,\boldsymbol{A}(\boldsymbol{r},t)\cdot\hat{\boldsymbol{p}} + \ldots \tag{140}$$

where the dots stand for $\frac{(e\boldsymbol{A}(\boldsymbol{r},t))^2}{2\,m_0}$ which will be neglected for the term linear in $\boldsymbol{A}(\boldsymbol{r},t)$. Inserting (140) and $\psi(\boldsymbol{r},t)$ from (122) into the time-dependent Schrödinger equation Eq.(132), multiplying this equation by $\hat{\psi}_{2p}(\boldsymbol{r})\,e^{i\frac{E_{2p}}{\hbar}\,t}$ or alternatively by $\hat{\psi}_{1s}(\boldsymbol{r})\,e^{i\frac{E_{1s}}{\hbar}\,t}$ and performing a real-space integration one arrives at

$$i\hbar\,\dot{c}_1 = c_0\,M_{10}\,e^{i\omega\,t} + c_1\,M_{11} \tag{141}$$

where $M_{10} = \int \hat{\psi}_{2p}(\boldsymbol{r})\,\hat{H}_{int}\,\hat{\psi}_{1s}(\boldsymbol{r})\,d^3r$ and $M_{11} = \int \hat{\psi}_{2p}(\boldsymbol{r})\,\hat{H}_{int}\,\hat{\psi}_{2p}(\boldsymbol{r})\,d^3r$

and

$$\hat{H}_{int} = i\,\frac{e\hbar}{m_0}\boldsymbol{A}(\boldsymbol{r},t)\cdot\nabla\,. \tag{142}$$

Analogously we have, in obvious notation

$$i\hbar\,\dot{c}_0 = c_1\,M_{01}\,e^{-i\omega\,t} + c_0\,M_{00}\,. \tag{143}$$

According to Eqs.(134) and (135) one has

$$\boldsymbol{A}(\boldsymbol{r},t) = \tag{144}$$

$$|c_0^*c_1|\,\frac{e\hbar}{4\pi\varepsilon_0\,m_0c^2}\int\frac{[\hat{\psi}_{1s}(\boldsymbol{r}')\nabla'\hat{\psi}_{2p}(\boldsymbol{r}') - \hat{\psi}_{2p}(\boldsymbol{r}')\nabla'\,\hat{\psi}_{1s}(\boldsymbol{r}')]\sin[\omega\,(t-\frac{|\boldsymbol{r}-\boldsymbol{r}'|}{c}) + \varphi]}{|\boldsymbol{r}-\boldsymbol{r}'|}\,d^3r'\,.$$

The retardation in the time dependence of the sine-function is crucial for the occurrence of an outgoing wave. If one were to neglect retardation the atom would not undergo a change of its energy, averaged over an oscillation period.

For light frequencies $\nu = \frac{\omega}{2\pi} \approx 10^{15}s^{-1}$ and $|\boldsymbol{r}-\boldsymbol{r}'| \lessapprox 4\cdot 10^{-8}cm$ for points within the atomic volume we have $\frac{\omega}{c}|\boldsymbol{r}-\boldsymbol{r}'| \lessapprox 10^{-2}$, and hence we may approximate:

$$\sin[\omega\,(t-\frac{|\boldsymbol{r}-\boldsymbol{r}'|}{c}) + \varphi] \approx \sin(\omega\,t+\varphi) - \omega\,\frac{|\boldsymbol{r}-\boldsymbol{r}'|}{c}\cos(\omega\,t+\varphi)$$

Inserting this into Eq.(144) one obtains

$$\boldsymbol{A}(\boldsymbol{r},t) = \boldsymbol{A}_1(\boldsymbol{r},t) + \boldsymbol{A}_2(\boldsymbol{r},t) = \hat{\boldsymbol{A}}_1(\boldsymbol{r})\sin(\omega\,t+\varphi) + \hat{\boldsymbol{A}}_2(\boldsymbol{r})\cos(\omega\,t+\varphi) \tag{145}$$

where

$$\hat{\boldsymbol{A}}_1(\boldsymbol{r}) = |c_0^*c_1|\,\frac{e\hbar}{4\pi\varepsilon_0\,m_0c^2}\int\frac{[\hat{\psi}_{1s}(\boldsymbol{r}')\nabla'\hat{\psi}_{2p}(\boldsymbol{r}') - \hat{\psi}_{2p}(\boldsymbol{r}')\nabla'\,\hat{\psi}_{1s}(\boldsymbol{r}')]}{|\boldsymbol{r}-\boldsymbol{r}'|}\,d^3r' \tag{146}$$

and

$$\hat{\boldsymbol{A}}_2(\boldsymbol{r}) = -|c_0^*c_1|\,\frac{e\,\hbar\,\omega}{4\pi\varepsilon_0\,m_0c^3}\int[\hat{\psi}_{1s}(\boldsymbol{r}')\nabla'\hat{\psi}_{2p}(\boldsymbol{r}') - \hat{\psi}_{2p}(\boldsymbol{r}')\nabla'\,\hat{\psi}_{1s}(\boldsymbol{r}')]\,d^3r' \tag{147}$$

The integral in Eq.(147) can be rewritten by using the identity (136):

$$\begin{aligned}\boldsymbol{A}_2(\boldsymbol{r},t) &= \hat{\boldsymbol{A}}_2\,\cos(\omega\,t+\varphi) = \\ &-2|c_0^*c_1|\,\frac{e\,\omega^2}{4\pi\varepsilon_0\,c^3}\underbrace{\int\hat{\psi}_{1s}(\boldsymbol{r}')\boldsymbol{r}'\,\hat{\psi}_{2p}(\boldsymbol{r}')\,d^3r'}_{=\boldsymbol{p}_{n'n}}\cos(\omega t+\varphi)\,.\end{aligned} \tag{148}$$

We now form the matrix element of $\hat{H}_{int} = i\,\frac{e\hbar}{m_0}\boldsymbol{A}(\boldsymbol{r},t)\cdot\nabla$ according to Eq.(141) using again the identity (136).

$$M_{10} = \frac{i\hbar}{m_0}\,2|c_0^*\,c_1|\,\frac{e^2\,\omega^2}{4\pi\,\varepsilon_0\,c^2}\,\boldsymbol{p}_{n'n}\cdot\underbrace{\int\hat{\psi}_{2p}\nabla\,\hat{\psi}_{1s}\,d^3r}_{=-\omega\,\frac{m_0}{\hbar}\boldsymbol{p}_{n'n}}\cos(\omega t+\varphi) \tag{149}$$

$$+\frac{i\hbar}{m_0}\int \hat{\psi}_{2p}(\boldsymbol{r})\,\frac{e}{m_0}\hat{\boldsymbol{A}}_1(\boldsymbol{r})\cdot\nabla\,\hat{\psi}_{1s}(\boldsymbol{r})\,d^3r\,\sin(\omega\,t+\varphi)$$

We multiply Eq.(141) by c_1^* and form the sum with its complex conjugate. The result may be written:

$$\begin{aligned}\frac{\partial}{\partial t}|c_1(t)|^2 &= -4\gamma\,|c_0^*(t)c_1(t)|^2\cos^2(\omega\,t+\varphi) \qquad (150)\\ &+2|c_0^*(t)c_1(t)|\int \hat{\psi}_{2p}(\boldsymbol{r})\,\frac{e}{m_0}\hat{\boldsymbol{A}}_1(\boldsymbol{r})\cdot\nabla\hat{\psi}_{1s}(\boldsymbol{r})\,d^3r\,\cos(\omega\,t+\varphi)\sin(\omega\,t+\varphi)\\ &+2|c_1(t)|^2\int \hat{\psi}_{2p}(\boldsymbol{r})\,\frac{e}{m_0}\hat{\boldsymbol{A}}_1(\boldsymbol{r})\cdot\nabla\hat{\psi}_{2p}(\boldsymbol{r})\,d^3r\,\sin(\omega\,t+\varphi)\\ &+2|c_1(t)|^2\int \hat{\psi}_{2p}(\boldsymbol{r})\,\frac{e}{m_0}\hat{\boldsymbol{A}}_2(\boldsymbol{r})\cdot\nabla\hat{\psi}_{2p}(\boldsymbol{r})\,d^3r\,\cos(\omega\,t+\varphi)\,.\end{aligned}$$

The quantity γ in the first line of this equation stands for

$$\gamma = \frac{e^2\,\omega^3}{4\pi\,\varepsilon_0\,c^3\,\hbar}\,|\boldsymbol{p}_{n'n}|^2\,. \qquad (151)$$

We now perform a time average on the right-hand side of Eq.(150) over successive oscillation periods $T = \frac{2\pi}{\omega}$ of the emitted light. Since the emission time $\hat{\tau}$ is many orders of magnitude larger than T, one may approximate $|c_{0/1}(t)|$ by $|c_{0/1}(\bar{t}_\nu)|$ where $\nu = 1, 2, \ldots$ counts successive oscillation intervals and $\bar{t}_\nu$ denotes an appropriately chosen time in the respective interval. On performing the time average all terms on the right-hand side of Eq.(150) now drop out except for the first one. Hence we arrive at

$$\frac{\partial}{\partial t}|c_1(t)|^2 = -2\gamma\,|c_0(t)|^2|c_1(t)|^2 \qquad (152)$$

where we have used $\overline{\cos^2(\omega\,t+\varphi)} = \frac{1}{2}$ with the bar denoting time averaging. We have, furthermore, replaced the histogram-type functions of time $|c_{0/1}(\bar{t}_\nu)|^2$ on the right-hand side by their smooth least mean-square fits.

In complete analogy we obtain

$$\frac{\partial}{\partial t}|c_0(t)|^2 = 2\gamma\,|c_0(t)|^2|c_1(t)|^2\,. \qquad (153)$$

Since $|c_0(t)|^2 + |c_1(t)|^2 = 1$, the time derivative of this sum must vanish. This is obviously ensured by the above two coupled equations (152) and (153). It can readily be verified that their two solutions are

$$|c_0(t)|^2 = \tfrac{1}{2}\,(1+\tanh\tfrac{2t}{\tau}) \quad\text{and}\quad |c_1(t)|^2 = \tfrac{1}{2}\,(1-\tanh\tfrac{2t}{\tau})\,. \qquad (154)$$

On multiplying these two functions one gets

$$|c_0(t)|^2|c_1(t)|^2 = \tfrac{1}{4}\,\frac{1}{\cosh^2\frac{2t}{\tau}}\,.$$

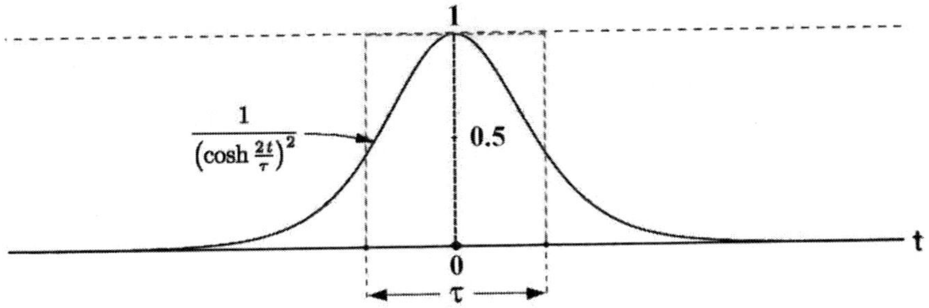

Figure 4. Optical transition: time-dependence of the "driving force" $\propto \boldsymbol{A}_2^2$.

From Eq.(152) we have

$$\frac{\partial}{\partial t}|c_1(t)|^2 = -2\gamma\,|c_0(t)|^2|c_1(t)|^2 = -\tfrac{1}{2}\,\gamma\,\frac{1}{\cosh^2\frac{2t}{\tau}}\,.$$

On the other hand it follows from Eq.(154) on differentiating $|c_1(t)|^2$

$$\frac{\partial}{\partial t}|c_1(t)|^2 = -\frac{1}{\tau}\,\frac{1}{\cosh^2\frac{2t}{\tau}}\,.$$

That means, the functions $|c_1(t)|^2$ and $|c_0(t)|^2$ fulfill Eqs.(152) and (153) if

$$\tfrac{1}{\tau} = \tfrac{1}{2}\,\gamma = \frac{e^2\,\omega^3}{8\pi\,\varepsilon_0\,c^3\,\hbar}\,|\boldsymbol{p}_{n'n}|^2\,.$$

Comparing this result with our "short-cut calculation" (139) we see that it is 25% smaller than the latter:

$$\frac{1}{\tau} = 0.75\,\frac{1}{\hat{\tau}}\,.$$

This difference originates in the simplification of the time dependence of $|c_1(t)|^2$ and $|c_0(t)|^2$ in taking the short-cut.

The following figure illustrates the time dependences according to Eq.(154).

We have marked two points A and B on the left-hand side of the curve for $|c_1(t)|^2$. As explained in connection with Eq.(123), the quantity $[E_{2p} - E_{1s}]\,|c_0|^2$ represents the energy uncertainty with which the state $\hat{\psi}_{2p}(\boldsymbol{r})$ has been "prepared" as a result of the finite preparation time Δt. With the aid of Eq.(117) and $|c_1(t)|^2 = 1 - |c_0(t)|^2$ this can be recast

$$1 - |c_1(t)|^2 = \frac{\hbar}{[E_{2p} - E_{1s}]\,\Delta t}\,.$$

The shorter the excitation time, the more $|c_1(t)|^2$ departs from unity. Hence, point A refers to a longer excitation time than point B. Corresondingly, if the system has landed at A after the excitation process, it takes a longer time to reach the transition interval (marked by two vertical dashed lines) than it would take if it would start at B. One may refer to these residence times prior to emission as "dead times". It should be noticed, however, that the

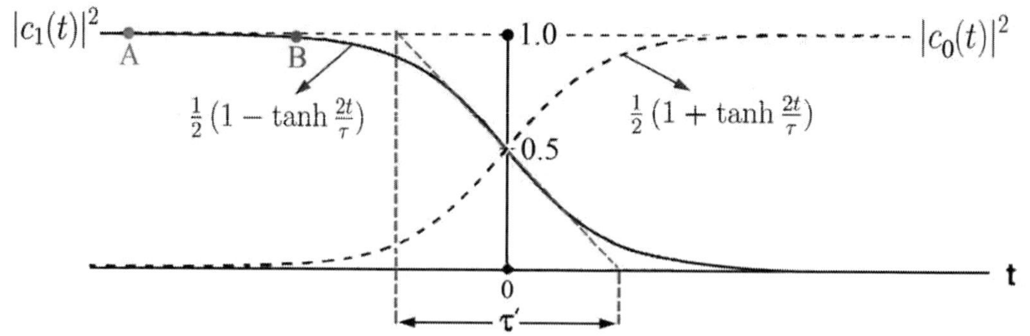

Figure 5. Optical transition between two states: Time dependence of the modulus square of the two-state related coefficients.

"emission time", limited by the two vertical dashed lines, remains largely unaffected by the different lengths of the dead times. That is to say, largely independent of the form of the excitation one observes a spectral line of a natural width that is only determined by the two states of the atom under study.

As already alluded to, the time evolution of the transition obtained above contrasts remarkably with that of Weisskopf and Wigner [50]. Their article is still considered groundlaying for the theory of spontaneous light emission. However, these authors arrive at $|c_1(t)|^2 = e^{-\frac{t}{\tau'}}$ which appears to be plausible at first sight, but is inconsistent with a solution to the time-dependent Schrödinger equation as follows from our derivation.

Another point of misunderstanding concerns "the measurement" of that exponential decay law $e^{-\frac{t}{\tau'}}$. Clearly, neither $|c_1(t)|^2$ nor any related quantity, for example $E = |c_0(t)|^2 E_{1s} + |c_1(t)|^2 E_{2p}$, is experimentally accessible in any ways since one can only detect the light when it has been fully emitted. What is actually done in the experiments is measuring the time-dependence of the photon-capture rate at which a photon detector fires after a large number of identical atoms has been excited by a flash. In the ensuing process the atoms re-emit the light spontaneously. The time point of the flash serves as a reference point with respect to which the detector records the flow of the incoming photons, and this flow is exponentially decaying in time. But each atom contributes only one single photon. Obviously, the photons are emitted at different times from different atoms. What has the time-dependence of their flow to do with the Weisskopf-Wigner decay law which refers to the time evolution of a single atom? According to their theory all atoms start decaying immediately after the excitation by the flash. Since the photons can only be detected after the asssociated wave packets have been fully emitted, they are monitored by the detector also at approximately the same time. Our derivation relates the re-emission of photons at different times after excitation of the atoms by a flash to different deadtimes which occur because of the spectral width of the light-flash.

To show that we assume for simplicity that the spectrum of the flash is rectangular around the frequency ω_0 for which the atom under study would be completely transferred from its ground state 1s to the excited state 2p if the atom would be exposed to the associated electromagnetic wave for an infinitely long time. Hence

$$\hbar\,\omega_0 = E_{2p} - E_{1s} = \Delta E \quad \text{and} \quad |c_1(t)|^2 = 1\,.$$

If the atom absorbs a photon of lower frequency from the flash ($\omega < \omega_0$) it ends up in a state where $|c_1(t)|^2$ is now smaller than unity and we have because of Eq.(123)

$$\hbar\,\Delta\omega = \Delta E\,[1 - |c_1(t)|^2] \quad \text{which can be recast} \quad \hbar\,\Delta\omega = \Delta E\,|c_0(t)|^2 \tag{155}$$

with $\Delta\omega$ denoting $\omega_0 - \omega$. We temporarily abbreviate $\frac{2t}{\tau}$ into x and observe

$$\tanh x = \frac{e^{x} - e^{-x}}{e^{x} + e^{-x}} \approx -1 + e^{2x} \quad \text{if} \quad x \ll -1 \quad \text{and hence} \quad \tfrac{1}{2}\,(1 + \tanh x) \approx \tfrac{1}{2}\,e^{2x}\,.$$

Eq.(155) may therefore be rewritten

$$\hbar\,\Delta\omega = \tfrac{\Delta E}{2}\,e^{\frac{4t}{\tau}}\,. \tag{156}$$

Here $t \ll -\tau$ is referenced to the middle of the transition interval. If one wants to find the position of some point like A or B in the above figure for some atom that has absorbed a photon of energy $\hbar(\omega_0 - \Delta\omega)$, one has to insert the particular $\Delta\omega$ in Eq.(156) and one obtains the associated t that gives the distance of that point from the middle of the transition interval. If the spectrum of the flash is rectangular the probability that an atom absorbs such a photon is equal for all frequencies of the spectrum. That means that the length of the dead times t associated with the various values of $\Delta\omega$ are ordered in an exponential fashion. As follows from Eq.(156) the decay constant in that exponential function is given by the transition time τ. Deviations from that exponential form are connected with a departure from the rectangular form of the flash spectrum.

As for the validity of the above assumption $x \ll -1$, that is $-t \gg \tau$, one has to keep in mind that there are experimental limitations which allow monitoring the incoming photons only many transition times after the flash.

In concluding this section we want to hint to a particular feature of our treatment that might encourage an interesting experiment: One could use the experimental setup by Dehmelt and associates [45] in which a single Ba^{+}-ion is kept in a Paul-trap. The ion contains one outer electron which behaves very similar to a hyrogen electron considered so far. One can transfer this electron from its 6s-groundstate to a 6p-state by the absorption of linearly polarized light of the appropiate energy. According to our theory the ion will spontaneously emit light then which is polarized in the same plane as the light that was previously absorbed. A detector monitors the emitted photon at some fixed distance d. It can be moved on a sphere of radius d. If the excitation has been repeated sufficiently often, the lateral distribution of detection events should display the characteristic feature of the Hertzian dipole radiation.

20. The Time-Dependent N-particle Schrödinger Equation

So far we have merely been concerned with a single particle whose stochastic behavior was described by regarding it as a member of N identically prepared, but statistically independent one-particle systems under the supposition that N be sufficiently large. To avoid confusion we shall henceforth rename that number by $\mathcal{N}$. Instead of a single particle we now consider N particles that interact via pair-forces. Each of these particles is individually a member of $\mathcal{N}$ statistically independent one-particle systems where the $N-1$ remaining particles appear at fixed positions $\boldsymbol{r}_2, \boldsymbol{r}_2, \ldots \boldsymbol{r}_N$ if the particle under consideration, picked at will, just happens to be "number 1". The considerations of Sections 10 and 14 carry over to this N-particle system. To see that one simply has to replace the 3-dimensional real-space of the single particle discussed as yet by a $3N$-dimensional space where the N particles appear as one point again. Instead of the probability density $\rho(\boldsymbol{r}, t)$ one is now dealing with

$$\rho(\boldsymbol{r}_1, \boldsymbol{r}_2, \ldots \boldsymbol{r}_N, t) = \rho(\boldsymbol{r}^N, t)\,;$$
$$\text{where} \quad \int \rho(\boldsymbol{r}^N, t)\, d^3r_1\, d^3r_2 \ldots d^3r_N = 1 \tag{157}$$

and

$$\boldsymbol{r}^N = (\boldsymbol{r}_1, \boldsymbol{r}_2, \ldots \boldsymbol{r}_N) = \sum_{j=1}^{N} \sum_{k=1}^{3} x_{j\,k}\, \boldsymbol{e}_{j\,k}$$

with $j = 1, 2, \ldots N$ numbering the particles and $x_{j\,k}$ denoting Cartesian coordinates which are associated with orthogonal unit vectors $\boldsymbol{e}_{j\,k}$. The quantities ∇^N, $\boldsymbol{u}^N$ and $\boldsymbol{v}^N$ are defined analogously.

Instead of $\varphi(\boldsymbol{r}, t)$ we now have $\phi(\boldsymbol{r}^N, t)$. Thus

$$\boldsymbol{v}^N(\boldsymbol{r}^N, t) = \frac{\hbar}{m_0}\, \nabla^N \phi(\boldsymbol{r}^N, t)\,. \tag{158}$$

Correspondingly, the $3N$-dimensional osmotic velocity has the form

$$\boldsymbol{u}^N(\boldsymbol{r}^N, t) = -\frac{\hbar}{2\,m_0}\, \nabla^N \ln[\rho(\boldsymbol{r}^N, t)/\rho_0]\,, \tag{159}$$

and hence we have similar to the single-particle case

$$\frac{\partial \boldsymbol{u}^N}{\partial t} = -\frac{\hbar}{2\,m_0}\, \nabla^N \frac{\partial}{\partial t}\, [\ln \rho/\rho_0] = -\frac{\hbar}{2\,m_0}\, \nabla^N \left[\frac{1}{\rho}\frac{\partial \rho}{\partial t}\right]. \tag{160}$$

Invoking the equation of continuity

$$\frac{\partial \rho}{\partial t} + \underbrace{\nabla^N \cdot (\rho\, \boldsymbol{v}^N)}_{=\rho\, \nabla^N \cdot \boldsymbol{v}^N + \boldsymbol{v}^N \cdot \nabla^N \rho} = 0$$

and using the definition (159), Eq.(160) can be cast as

$$\frac{\partial \boldsymbol{u}^N}{\partial t} = -\frac{\hbar}{2\,m_0}\, \nabla^N\, [(\nabla^N \cdot \boldsymbol{v}^N) - (\boldsymbol{u}^N \cdot \boldsymbol{v}^N)]\,. \tag{161}$$

In the following we first confine ourselves to time-independent conservative forces which - in the spirit of our notation - may be written

$$\boldsymbol{F}^N_{ext.}(\boldsymbol{r}^N) = \sum_{j=1}^{N} \sum_{k=1}^{3} F_k^{ext.}(\boldsymbol{r}_j)\, \boldsymbol{e}_{j\,k}$$

where

$$F_k^{ext.}(\boldsymbol{r}_j) = -\frac{\partial}{\partial x_{j\,k}} V_{ext.}(\boldsymbol{r}_j)$$

with $V_{ext.}(\boldsymbol{r})$ denoting an external potential. Hence $\boldsymbol{F}^N_{ext.}$ may alternatively be written

$$\boldsymbol{F}^N_{ext.}(\boldsymbol{r}^N) = -\nabla^N \widehat{V}_{ext.}(\boldsymbol{r}_1, \boldsymbol{r}_2, \ldots \boldsymbol{r}_N)$$

where

$$\widehat{V}_{ext.}(\boldsymbol{r}_1, \boldsymbol{r}_2, \ldots \boldsymbol{r}_N) = \sum_{j=1}^{N} V_{ext.}(\boldsymbol{r}_j)\,.$$

The force exerted on the j-th particle due to pair-interaction with the $N-1$ remaining particles is given by

$$F_{j\,k}^{inter}(\boldsymbol{r}_j) = -\frac{\partial}{\partial\, x_{j\,k}} \sum_{\substack{i=1\\ i\neq j}}^{N} V(|\boldsymbol{r}_j - \boldsymbol{r}_i|)\,,$$

where $V(|\boldsymbol{r}_j - \boldsymbol{r}_i|)$ denotes the interaction potential. The generalized total force in the $3N$-dimensional space may therefore be cast as

$$\boldsymbol{F}^N(\boldsymbol{r}^N) = -\nabla^N \widehat{V}(\boldsymbol{r}_1, \boldsymbol{r}_2, \ldots \boldsymbol{r}_N)\,,$$

where $\widehat{V}(\boldsymbol{r}_1, \boldsymbol{r}_2, \ldots \boldsymbol{r}_N)$ is defined by

$$\widehat{V}(\boldsymbol{r}_1, \boldsymbol{r}_2, \ldots \boldsymbol{r}_N) = \sum_{j=1}^{N} V_{ext.}(\boldsymbol{r}_j) + \frac{1}{2}\sum_{j=1}^{N}\sum_{\substack{i=1\\ i\neq j}}^{N} V(|\boldsymbol{r}_j - \boldsymbol{r}_i|)\,. \tag{162}$$

Newton's modified second law (58) hence attains the form

$$\frac{\partial\, \boldsymbol{v}^N}{\partial\, t} = -\nabla^N\left[\frac{1}{m_0}\,\widehat{V} + \frac{1}{2}\,(\boldsymbol{v}^N)^2 - \frac{1}{2}\,(\boldsymbol{u}^N)^2 + \frac{\hbar}{2\,m_0}\,\nabla^N\cdot\boldsymbol{u}^N\right]. \tag{163}$$

As in the one-particle case the two scalar functions $\rho^N(\boldsymbol{r}^N, t)$ and $\phi(\boldsymbol{r}^N, t)$ can be absorbed into a complex-valued function $\Psi(\boldsymbol{r}_1, \boldsymbol{r}_2, \ldots \boldsymbol{r}_N, t)$ defined by

$$\Psi(\boldsymbol{r}_1, \boldsymbol{r}_2, \ldots \boldsymbol{r}_N, t) = \pm\sqrt{\rho(\boldsymbol{r}_1, \boldsymbol{r}_2, \ldots \boldsymbol{r}_N, t)} \times \exp\left[i\,\phi(\boldsymbol{r}_1, \boldsymbol{r}_2, \ldots \boldsymbol{r}_N, t)\right].$$

This is equivalent to

$$-\boldsymbol{u}^N(\boldsymbol{r}^N, t) + i\,\boldsymbol{v}^N(\boldsymbol{r}^N, t) = \frac{\hbar}{m_0}\,\nabla^N(\ln\Psi(\boldsymbol{r}^N, t)/\sqrt{\rho_0}) = \frac{\hbar}{m_0}\,\frac{\nabla^N\Psi}{\Psi}$$

which is the analogue to Eq.(64), and we obtain accordingly

$$\frac{\partial}{\partial t}(-\boldsymbol{u}^N + i\,\boldsymbol{v}^N) = \nabla^N\left(\frac{\hbar}{m_0}\,\frac{1}{\Psi}\,\frac{\partial\Psi}{\partial t}\right).$$

If we here insert Eqs.(161) and(163) for $\frac{\partial \boldsymbol{v}^N}{\partial t}$ and $\frac{\partial \boldsymbol{u}^N}{\partial t}$ and proceed exactly as in the single-particle case we arrive at the N-particle Schrödinger equation

$$\underbrace{\left[\widehat{H}_0 + \frac{1}{2}\sum_{\substack{i,j\\ i\neq j}} V(|\boldsymbol{r}_j - \boldsymbol{r}_i|)\right]}_{=\widehat{H}} \Psi(\boldsymbol{r}_1, \boldsymbol{r}_2, \ldots \boldsymbol{r}_N, t) = i\hbar \frac{\partial}{\partial t}\Psi(\boldsymbol{r}_1, \boldsymbol{r}_2, \ldots \boldsymbol{r}_N, t) . \quad (164)$$

Here $\widehat{H}_0$ denotes the "free Hamiltonian"

$$\widehat{H}_0 = \sum_{j=1}^{N} \widehat{H}_j \quad \text{where} \quad \widehat{H}_j = \left[\frac{\widehat{\boldsymbol{p}}_j^2}{2\,m_0} + V_{ext.}(\boldsymbol{r}_j)\right] . \quad (165)$$

Because of Eq.(158) the phase of the wave function may still depend on time when ρ and $\boldsymbol{v}^N$ are time-independent:

$$\phi(\boldsymbol{r}^N, t) = \phi_0(\boldsymbol{r}^N) + f(t) .$$

Thus we have in this case

$$\begin{aligned} \Psi(\boldsymbol{r}^N, t) &= \Psi_0(\boldsymbol{r}^N)\, e^{-if(t)} ; \\ \Psi_0(\boldsymbol{r}^N) &= \pm\sqrt{\rho(\boldsymbol{r}^N)}\, \exp\left[i\,\phi_0(\boldsymbol{r}^N)\right] \end{aligned}$$

which on insertion into Eq.(164) yields

$$\widehat{H}\,\Psi_0(\boldsymbol{r}^N) = \hbar\,\dot{f}\,\Psi_0(\boldsymbol{r}^N) \quad \hookrightarrow \quad \hbar\,\dot{f} = const. = E \quad \hookrightarrow f(t) = \frac{E}{\hbar}\,t ,$$

whereby Eq.(164) becomes the time-independent Schrödinger equation

$$\widehat{H}\,\Psi_0(\boldsymbol{r}^N) = E\,\Psi_0(\boldsymbol{r}^N) . \quad (166)$$

21. States of Identical Particles and Entanglement

If the particles are non-interacting, one would naïvely expect their motions to be completely uncorrelated which means

$$\rho(\boldsymbol{r}_1, \boldsymbol{r}_2, \ldots \boldsymbol{r}_N, t) = \prod_{j=1}^{N} \rho_j(\boldsymbol{r}_j, t) , \quad (167)$$

and

$$\phi(\boldsymbol{r}_1, \boldsymbol{r}_2, \ldots \boldsymbol{r}_N, t) = \sum_{j=1}^{N} \varphi_j(\boldsymbol{r}_j, t) . \quad (168)$$

In that case Eq.(158) attains the form

$$(\boldsymbol{v}_1(\boldsymbol{r}_1, t), \boldsymbol{v}_2(\boldsymbol{r}_2, t), \ldots \boldsymbol{v}_N(\boldsymbol{r}_N, t)) = \frac{\hbar}{m_0}\,(\nabla_1\varphi_1(\boldsymbol{r}_1, t), \nabla_2\varphi_2(\boldsymbol{r}_2, t), \ldots \nabla_N\varphi_N(\boldsymbol{r}_N, t)) .$$

Likewise, Eq.(159) becomes

$$(\boldsymbol{u}_1(\boldsymbol{r}_1,t),\boldsymbol{u}_2(\boldsymbol{r}_2,t),\ldots\boldsymbol{u}_N(\boldsymbol{r}_N,t)) = -\frac{\hbar}{2\,m_0}\,(\nabla_1 \ln[\rho_1(\boldsymbol{r}_1,t)/\rho_{01}],$$
$$\nabla_2 \ln[\rho_2(\boldsymbol{r}_2,t)/\rho_{02}],\ldots\nabla_N[\ln\rho_N(\boldsymbol{r}_N,t)/\rho_{0N}])\,.$$

Newton's modified second law (163) decomposes accordingly into N analogous equations for single particles, as has to be expected. Each of these equations can be subjected to a Madelung transform which yields time-dependent one-particle Schrödinger equations solved by one-particle wave functions $\psi_j(\boldsymbol{r}_j,t)$. If one multiplies

$$\widehat{H}_j(\boldsymbol{r}_j)\,\psi_j(\boldsymbol{r}_j,t) = i\hbar\,\frac{\partial}{\partial t}\,\psi_j(\boldsymbol{r}_j,t) \tag{169}$$

by $\prod_{\substack{i=1\\ i\neq j}}^{N}\psi_i(\boldsymbol{r}_i,t)$ one obtains

$$\widehat{H}_j(\boldsymbol{r}_j)\prod_{i=1}^{N}\psi_i(\boldsymbol{r}_i,t) = \prod_{\substack{i=1\\ i\neq j}}^{N}\psi_i(\boldsymbol{r}_i,t)\,i\hbar\,\frac{\partial}{\partial t}\,\psi_j(\boldsymbol{r}_i,t)$$

which on forming the sum $\sum_{j=1}^{N}$ yields, in fact,

$$\widehat{H}_0\,\Psi(\boldsymbol{r}_1,\boldsymbol{r}_2,\ldots\boldsymbol{r}_N,t) = i\hbar\,\frac{\partial}{\partial t}\,\Psi(\boldsymbol{r}_1,\boldsymbol{r}_2,\ldots\boldsymbol{r}_N,t)$$
$$\text{where}\quad \Psi(\boldsymbol{r}_1,\boldsymbol{r}_2,\ldots\boldsymbol{r}_N,t) = \prod_{j=1}^{N}\psi_j(\boldsymbol{r}_j,t)\,. \tag{170}$$

Hence, the above time-dependent N-particle Schrödinger equation is solved by the product of individually time-dependent wave functions $\psi_j(\boldsymbol{r}_j,t)$.

Obviously, the density (167) that results from this wave function is **not** invariant against interchange of any two particles if they are in different states, say $\psi_{k_n}(\boldsymbol{r}_k,t)$ and $\psi_{l_m}(\boldsymbol{r}_l,t)$ where $k_n \neq l_m$.

It is not exactly physical wisdom but rather firm belief that even non-interacting massive particles, though non-existing in nature, do not perform an uncorrelated motion and can, therefore, not be described by the wave function (170). This belief is based on the idea that the particles cannot be tracked individually as they move (contrary to classical particles) because the uncertainty relation "forbids" the existence of trajectories. Our approach to the many-particle problem is characterized by the plausible assumption that each particle can be identified any time by an affix if it has been assigned to a certain number at some chosen instant since each particle follows an individual trajectory. The quantity $\boldsymbol{v}_1(\boldsymbol{r}_1,\boldsymbol{r}_2,\ldots\boldsymbol{r}_N,t)$, for example, represents the average over all particle velocities at $\boldsymbol{r}_1$ and time t of the ensemble associated with particle "number 1". In forming this average the positions $\boldsymbol{r}_2,\boldsymbol{r}_3,\ldots\boldsymbol{r}_N$ of the $N-1$ remaining particles are kept fixed, that is, the average results from the entire set of "number 1"-trajectories that occur in the "number 1"-ensemble while the particles "number 2, 3 ...N" are at fixed positions. Clearly, if one or more of those particles are kept at different positions and if all particles interact, $\boldsymbol{v}_1$ will in general be different at $\boldsymbol{r}_1$ and time t. Hence in our view there is no extra quantum phenomenon of indiscernibility. As in classical mechanics it is entirely sufficient to characterize identical particles merely by their property of having the same mass and charge.

If one insists, however, on "quantum indiscernibility" also for non-interacting particles, that is, on the invariance of $\rho(\boldsymbol{r}_1, \boldsymbol{r}_2, \ldots \boldsymbol{r}_N)$ against interchange of any two particles, one has to replace (170) with a renormalized linear combination of all $N!$ products that differ in the interchange of two particles

$$\Psi(\boldsymbol{r}_1, \boldsymbol{r}_2, \ldots \boldsymbol{r}_N) = \frac{1}{\sqrt{N!}} \sum_{P=1}^{N!} (\pm 1)^P \hat{P}(k,l) \prod_{j=1}^{N} \psi_{n_j}(\boldsymbol{r}_j) \,. \tag{171}$$

where $\hat{P}(k,l)$ is the permutation operator exchanging the particle referring to $j = k$ with that for $j = l$, and P numbers the permutations.

If the particles interact and are bound in an external potential or move in a parallelepiped where $\Psi(\boldsymbol{r}_1, \boldsymbol{r}_2, \ldots \boldsymbol{r}_N)$ is subjected to periodic boundary conditions, each particle is constantly scattered, and hence the probability density of some particle, say "number k", for being within an elementary volume $\Delta^3 r$ around $\boldsymbol{r}$ is given by:

$$P(\boldsymbol{r}) = \int |\Psi(\boldsymbol{r}_1, \ldots \boldsymbol{r}_{k-1}, \boldsymbol{r}, \boldsymbol{r}_{k+1} \ldots \boldsymbol{r}_N)|^2 \, d^3r_1 \ldots d^3r_{k-1} d^3r_{k+1} \ldots d^3r_N \,.$$

Indiscernibility means that $P(\boldsymbol{r})$ is the same for any particle one picks, that is, each particle appears at $\boldsymbol{r}$ with the same probability. Hence we have

$$\rho(\boldsymbol{r}) = N \, P(\boldsymbol{r})$$

with $\rho(\boldsymbol{r})\, \Delta^3 r$ denoting the probability of **any** of the N electrons being in $\Delta^3 r$.

The function $\rho(\boldsymbol{r}_1, \boldsymbol{r}_2, \ldots \boldsymbol{r}_N)$ is now naturally invariant against interchange of any two particles.

An important property of particles is their spin which will be discussed farther below in this article. In the present context it may be sufficient to introduce

$$\boldsymbol{x} = (\boldsymbol{r}, \sigma)$$

as a generalized particle coordinate where $\sigma = \pm 1$ denotes its discrete spin coordinate and refers to parallel or anti-parallel orientation with respect to a global axis. The wave function (171) for non-interacting particles then takes the form

$$\Psi(\boldsymbol{x}_1, \boldsymbol{x}_2, \ldots \boldsymbol{x}_N) = \frac{1}{\sqrt{N!}} \sum_{P=1}^{N!} (\pm 1)^P \hat{P}(k,l) \prod_{j=1}^{N} \psi_{n_j}(\boldsymbol{x}_j) \,. \tag{172}$$

The alternative in the sign under the sum is related to the two fundamentally different species of particles: The plus sign in $(\pm 1)^P$ characterizes bosons, the minus sign fermions. Hence the latter are associated with a wave function that changes sign on interchanging any two particles. This property persists when $\Psi(\boldsymbol{x}_1, \boldsymbol{x}_2, \ldots \boldsymbol{x}_N)$ describes N interacting fermions. Antisymmetry of the wave function gives rise to a peculiar behavior of the so-called pair-density

$$\rho_2(\boldsymbol{x}, \boldsymbol{x}') \overset{\text{Def}}{=} N(N-1) \int |\Psi(\boldsymbol{x}, \boldsymbol{x}', \boldsymbol{x}_3, \ldots \boldsymbol{x}_N)|^2 \, d^4x_3 \ldots d^4x_N$$

where

$$\int \ldots d^4x = \sum_{\sigma} \ldots d^3r \, .$$

Obviously

$$\Psi(\boldsymbol{x}_1, \ldots \boldsymbol{x}_\nu, \boldsymbol{x}_{\nu+1}, \ldots \boldsymbol{x}_N) \equiv \Psi(\boldsymbol{x}_1, \ldots \boldsymbol{x}_{\nu+1}, \boldsymbol{x}_\nu, \ldots \boldsymbol{x}_N) \quad \text{if} \quad \boldsymbol{x}_\nu = \boldsymbol{x}_{\nu+1} \, .$$

On the other hand, Ψ is required to change sign on interchanging two particles, and hence the above equation can only hold if Ψ equals zero if the coordinates of any two particles are equal. Thus

$$\rho_2(\boldsymbol{x}, \boldsymbol{x}') = 0 \quad \text{if} \quad \boldsymbol{x}' = \boldsymbol{x} \, .$$

This indicates the occurrence of the so-called Fermi-hole which is absent in bose-particle systems.

The form of the wave function (172) may be cast as a determinant, named after J. C. Slater. In so-called EPRB-experiments (EPRB=Einstein, Podolsky, Rosen [52], Bohm [53]) which were originally devised to test possible correlations between two macroscopically distant fermions in a singlet state, the associated wave function is just a 2×2 determinant. The respective two one-particle states are in this context commonly referred to as "entangled states".

The requirement of antisymmetry, which is equivalent to the Pauli exclusion principle, is a strong subsidiary condition in solving the Schrödinger equation (166). Wave functions associated with fermions constitute only a small subset of the set of functions that satisfy the Schrödinger equation (166).

It should clearly be stated that the antisymmetry of the wave function is definitely not a consequence of our stochastic approach, but rather has to be required as an additional property, as in standard quantum mechanics.

The derivation of the time-dependent Schrödinger equation (164) can again be extended to the case where the particles move in an electromagnetic field. The external potential becomes time-dependent then ($V_{ext.}(\boldsymbol{r}) \to V(\boldsymbol{r}, t)$) and $\widehat{\boldsymbol{p}}_j$ has to replaced with $\widehat{\boldsymbol{P}}_j(\boldsymbol{r}, t) = \widehat{\boldsymbol{p}}_j - e\,\boldsymbol{A}(\boldsymbol{r}_j, t)$.

22. A Borderline Case of Entanglement

We consider a hydrogen molecule whose nuclei are located at $\boldsymbol{R}_A$ and $\boldsymbol{R}_B$, respectively. The Hamiltonian of the two electrons is given by

$$\hat{H} = \sum_{k=1}^{2} \left[\frac{(-i\hbar\,\nabla_k - e\boldsymbol{A}(\boldsymbol{r}_k, t))^2}{2\,m_0} + V(\boldsymbol{r}_k) \right] + \frac{e^2}{4\pi\,\epsilon_0} \sum_{k,l \neq k} \frac{1}{|\boldsymbol{r}_k - \boldsymbol{r}_l|} \qquad (173)$$

where

$$V(\boldsymbol{r}) = -\frac{e^2}{4\pi\,\epsilon_0|\boldsymbol{r}-\boldsymbol{R}_A|} - \frac{e^2}{4\pi\,\epsilon_0|\boldsymbol{r}-\boldsymbol{R}_B|}\,,$$

and ϵ_0 denotes the electric constant.

We first assume that there is no external field ($\boldsymbol{A}(\boldsymbol{r},t) \equiv 0$) and that the 2-electron wave function has for large proton-proton separation, that is when $R_{AB} = |\boldsymbol{R}_A - \boldsymbol{R}_B| \gg$Bohr radius, still the entangled form of a **singlet state** dictated by the Pauli principle

$$\Psi(\boldsymbol{r}_1,\boldsymbol{r}_2) = \frac{1}{\sqrt{2}}\left[\underline{\psi}(\boldsymbol{r}_1\,,\uparrow)\otimes\underline{\psi}(\boldsymbol{r}_2\,,\downarrow) - \underline{\psi}(\boldsymbol{r}_1\,,\downarrow)\otimes\underline{\psi}(\boldsymbol{r}_2\,,\uparrow)\right] \tag{174}$$

where

$$\underline{\psi}(\boldsymbol{r},\sigma) = \left[a_\sigma(R_{AB})\,\varphi_A(\boldsymbol{r}) + b_\sigma(R_{AB})\,\varphi_B(\boldsymbol{r})\right]\underline{\chi}(\sigma) \tag{175}$$

and

$$\sigma = \uparrow(\downarrow)\,;\quad a_\sigma^2 + b_\sigma^2 = 1\,;\quad \varphi_{A/B}(\boldsymbol{r}) = \varphi(\boldsymbol{r}-\boldsymbol{R}_{A/B})$$

with the property

$$\int |\varphi(\boldsymbol{r}-\boldsymbol{R}_{A/B})|^2\,d^3r = 1\,.$$

Here the integrand denotes the electronic 1s-orbital of a single hydrogen atom, and $\rho_{A/B}(\boldsymbol{r},t) = |\varphi(\boldsymbol{r}-\boldsymbol{R}_{A/B})|^2$ is the associated probability density. Furthermore, the unit spinors $\underline{\chi}(\sigma)$ have the property

$$\underline{\chi}^\dagger(\sigma')\,\underline{\chi}(\sigma) = \delta_{\sigma'\sigma}\,.$$

Under the supposition that $R_{AB} = |\boldsymbol{R}_A - \boldsymbol{R}_B|$ is sufficiently large, say 10 cm or even larger, the expectation value $\langle\widehat{H}\rangle$ attains a minimum for either

$$(a_\uparrow \to 1\,, a_\downarrow \to 0) \hookrightarrow (b_\uparrow \to 0\,, b_\downarrow \to 1) \quad \text{“case l”}$$

or

$$(a_\uparrow \to 0\,, a_\downarrow \to 1) \hookrightarrow (b_\uparrow \to 1\,, b_\downarrow \to 0) \quad \text{“case r”}\,.$$

For both cases $\langle\widehat{H}\rangle$ yields the correct value, viz. -2 Ryd, as has to be expected for two hydrogen atoms, each of which possesses the energy -1 Ryd. Since the spin structure does not reflect the symmetry of the potential, one forms a symmetry-adapted linear combination

$$\Psi_i(\boldsymbol{r}_1,\boldsymbol{r}_2) = \frac{1}{\sqrt{2}}\left[\Psi_l(\boldsymbol{r}_1,\boldsymbol{r}_2) + \Psi_r(\boldsymbol{r}_1,\boldsymbol{r}_2)\right],$$

where

$$\Psi_l(\boldsymbol{r}_1,\boldsymbol{r}_2) = \frac{1}{\sqrt{2}}\begin{vmatrix} \varphi_A(\boldsymbol{r}_1)\,\underline{\chi}(\uparrow) & \varphi_A(\boldsymbol{r}_2)\,\underline{\chi}(\uparrow) \\ \varphi_B(\boldsymbol{r}_1)\,\underline{\chi}(\downarrow) & \varphi_B(\boldsymbol{r}_2)\,\underline{\chi}(\downarrow) \end{vmatrix}$$

and

$$\Psi_r(\boldsymbol{r}_1,\boldsymbol{r}_2) = \frac{1}{\sqrt{2}}\begin{vmatrix} \varphi_B(\boldsymbol{r}_1)\,\underline{\chi}(\uparrow) & \varphi_B(\boldsymbol{r}_2)\,\underline{\chi}(\uparrow) \\ \varphi_A(\boldsymbol{r}_1)\,\underline{\chi}(\downarrow) & \varphi_A(\boldsymbol{r}_2)\,\underline{\chi}(\downarrow) \end{vmatrix}\,.$$

The two 2-electron functions are associated with the same energy which hence applies to $\Psi_i(\boldsymbol{r}_1, \boldsymbol{r}_2)$ as well. As a consequence of the symmetry of $\Psi_i(\boldsymbol{r}_1, \boldsymbol{r}_2)$ in $\boldsymbol{r}_1$ and $\boldsymbol{r}_2$ we have

$$\rho(\boldsymbol{r}_1) = \int |\Psi(\boldsymbol{r}_1, \boldsymbol{r}_2)|^2 \, d^3r_2 = \rho_A(\boldsymbol{r}_1) + \rho_B(\boldsymbol{r}_1)$$

and

$$\rho(\boldsymbol{r}_2) = \int |\Psi(\boldsymbol{r}_1, \boldsymbol{r}_2)|^2 \, d^3r_1 = \rho_A(\boldsymbol{r}_2) + \rho_B(\boldsymbol{r}_2) \, .$$

Moreover

$$\int \rho_{A/B}(\boldsymbol{r}_{1/2}) \, d^3r_{1/2} = \frac{1}{2} \quad \text{and}$$

$$\int \rho_{A/B}(\boldsymbol{r}_1) \, d^3r_1 + \int \rho_{A/B}(\boldsymbol{r}_2) \, d^3r_2 = 1 \, . \tag{176}$$

That means that each electron appears in each of the atoms (A and B) with the same probability. This has rather implausible consequences if one exposes, for example, **one** of the atoms (say A) to a Laser puls of frequency ω. Now $\boldsymbol{A}(\boldsymbol{r}, t)$ is no longer zero. The associate perturbation operator has the form

$$V_{perturb}(\boldsymbol{r}_1, \boldsymbol{r}_2, t) = \begin{cases} \frac{ie\hbar}{m_0} \sum_{k=1}^{2} \boldsymbol{A}(\boldsymbol{r}_k, t) \cdot \nabla_k & \text{if } \boldsymbol{r}_1, \boldsymbol{r}_2 \text{ in or near atom A} \\ 0 & \text{else} \end{cases}$$

which promotes the 2-electron system to an excited state

$$\Psi_f(\boldsymbol{r}_1, \boldsymbol{r}_2) = \frac{1}{\sqrt{2}} \left[\Psi_l^{(f)}(\boldsymbol{r}_1, \boldsymbol{r}_2) + \Psi_r^{(f)}(\boldsymbol{r}_1, \boldsymbol{r}_2) \right]$$

where

$$\Psi_l^{(f)}(\boldsymbol{r}_1, \boldsymbol{r}_2) = \frac{1}{\sqrt{2}} \begin{vmatrix} \varphi_A^{(f)}(\boldsymbol{r}_1) \, \underline{\chi}(\uparrow) & \varphi_A^{(f)}(\boldsymbol{r}_2) \, \underline{\chi}(\uparrow) \\ \varphi_B(\boldsymbol{r}_1) \, \underline{\chi}(\downarrow) & \varphi_B(\boldsymbol{r}_2) \, \underline{\chi}(\downarrow) \end{vmatrix}$$

and

$$\Psi_r^{(f)}(\boldsymbol{r}_1, \boldsymbol{r}_2) = \frac{1}{\sqrt{2}} \begin{vmatrix} \varphi_B^{(f)}(\boldsymbol{r}_1) \, \underline{\chi}(\uparrow) & \varphi_B^{(f)}(\boldsymbol{r}_2) \, \underline{\chi}(\uparrow) \\ \varphi_A(\boldsymbol{r}_1) \, \underline{\chi}(\downarrow) & \varphi_A(\boldsymbol{r}_2) \, \underline{\chi}(\downarrow) \end{vmatrix} .$$

Here $\varphi_{A/B}^{(f)}(\boldsymbol{r})$ describes an outgoing wave which has in principle the asymptotic form

$$\varphi_{A/B}^{(f)}(\boldsymbol{r}) \cong \frac{1}{r_{A/B}} \, e^{ik\,r_{A/B}} \, Y_{10}(\widehat{r}_{A/B}) \, ; \quad r_{AB} \equiv |\boldsymbol{r} - \boldsymbol{R}_{A/B}|$$

with $Y_{10}(\widehat{r}_{A/B})$ denoting the spherical harmonic for $l = 1, m = 0$, and k is given by $\hbar^2 k^2/2\,m_0 = -1\,\text{Ryd} + \hbar\,\omega$. We have assumed linearly polarized Laser light with the quantization axis of $Y_{10}(\widehat{r}_{A/B})$ coinciding with the axis of polarization. Moreover we have disregarded the residual charge left with each atom as part of the electronic charge is emitted.

Although only the illuminated volume of atom A can contribute to the transition matrix element

$$M_{fi} = \int_{atom\ A} \int \Psi_f^*(\boldsymbol{r}_1, \boldsymbol{r}_2)\, V_{perturb}(\boldsymbol{r}_1, \boldsymbol{r}_2)\, \Psi_i(\boldsymbol{r}_1, \boldsymbol{r}_2)\, d^3r_1\, d^3r_2$$

the final state $\Psi_f(\boldsymbol{r}_1, \boldsymbol{r}_2)$ yields a current density

$$\begin{aligned} \boldsymbol{j}(\boldsymbol{r}) &= \frac{\hbar}{i\, m_0} \int [\Psi_f^*(\boldsymbol{r}, \boldsymbol{r}_2)\nabla\Psi_f(\boldsymbol{r}, \boldsymbol{r}_2) - c.c.]\, d^3r_2 \\ &= \frac{\hbar}{i\, m_0} \int [\Psi_f^*(\boldsymbol{r}_1, \boldsymbol{r})\nabla\Psi_f(\boldsymbol{r}_1, \boldsymbol{r}) - c.c.]\, d^3r_1 \\ &= \boldsymbol{j}_A(\boldsymbol{r} - \boldsymbol{R}_A) + \boldsymbol{j}_B(\boldsymbol{r} - \boldsymbol{R}_B) \end{aligned}$$

where $\boldsymbol{j}_{A/B}$ is associated with $\varphi^{(f)}_{A/B}$, and hence $\boldsymbol{j}(\boldsymbol{r})$ contains also a photo emission current coming from the non-illuminated atom B at a distance of 10 cm away from A. Similar considerations apply if one excites the molecule to a bound state which would spontaneously decay back then to the ground-state by emitting fluorescent light. If one repeats the excitation sufficiently often one would obtain as many fluorescence photons coming from the illuminated atom as from the non-illuminated one. There is no experimental evidence that anything like that could ever happen.

We are hence led to conclude that the concept of entanglement (i. e. the Pauli exclusion principle when dealing with fermions) does not apply anymore if the atoms are separated by a macroscopic distance. The reason may be traced back to the definition (4) of the probability density $\rho(\boldsymbol{r})$ as the relative residence time that a particle spends in an elementary volume $\Delta^3 r$ around $\boldsymbol{r}$, provided it is bound in a potential and thus occurs repeatedly in that volume. On pulling the two atoms of a H_2-molecule gradually apart one arrives at a situation where one of the two electrons remains captured near the nucleus of atom A for a while, and accordingly the second electron stays captured near the nucleus of atom B for the same time. The Coulomb repulsion between the two electrons effects a correlated separation of the two electrons into the two regions.[2] If the inter-nuclear distance becomes large compared to the linear dimensions of the atoms, the time spans for tunneling of the "A-electron" (marked by the index "1") into the B-region and vice versa become enormously long compared to the time required to traverse the associated atom. The time T for the photo-excitation process will therefore be many orders of magnitude shorter than the tunneling time. Given this situation, the definition (4) yields

$$\rho(\boldsymbol{r}_{1/2}) = \begin{cases} \rho_{A/B}(\boldsymbol{r}_{1/2}) & \text{for } \boldsymbol{r}_{1/2} \text{ around nucleus A/B} \\ 0 & \text{else} \end{cases}$$

where - different from Eq.(176) - the densities $\rho_{A/B}(\boldsymbol{r}_{1/2})$ now integrate to unity. The two electrons do not appear entangled any more, and only the A-atom will now emit an electron under the exposure of light.

[2] The possibility that both electrons accumulate in one of the atoms can safely be excluded. In such a case the other atom would be left ionized requiring an energy ΔE of about 1Ryd. Within a time Δt that excess energy must disappear again where Δt results from $\Delta E\, \Delta t \approx \hbar$. This yields $\Delta t \approx 5 \cdot 10^{-17}$ s, thus excluding the possibility for one of the electrons to go back to the ionized atom 10 cm away at a speed well below light velocity.

23. Van der Waals Interaction

Chemical bonding occurs when N_{at} atoms get so close to each other that their total number of electrons ($N_e = \sum_{i=1}^{N_{at}} N_{e\,i}$) becomes associated with a new common wave function which is anti-symmetric with respect to the exchange of any two electrons from the entire set of N_e electrons. At larger distances the atoms interact only weakly by so-called van der Waals forces. The associated interaction energy has first been treated by Eisenschitz and London [56]. Further studies (s. e. g. Dzyaloshinskii et al. [57]) are based on the idea that the atoms undergo density fluctuations that give rise to temporary dipole moments. The latter cause interatomic attraction. From our point of view density fluctuations of the electronic probability density cannot occur in the groundstate of any system because it would inevitably cause radiative emission. This follows immediately from our treatment of spontaneous light emission in Section 19. We shall therefore go back to the original idea of the Eisenschitz-London paper.

We consider two atoms **A** and **B** whose electronic densities are spherical. Their nuclei are centered at $\boldsymbol{R}_A$ and $\boldsymbol{R}_B$, respectively. The individual atoms of that pair may in general be different, consist of a sodium and a potassium atom, for example. The interatomic distance is assumed larger than the sum of the atomic radii so that the overlap of the electronic densities may be regarded as zero on the scale of interest. It is this situation which we have analyzed in the previous section for two hydrogen atoms whose two-electron wave function factorizes at this distance into the wave functions of the individual atoms. Accordingly, we have in the present case for the groundstate Ψ_0 of the atomic pair

$$\Psi_0(\boldsymbol{r}_{A1}, \boldsymbol{r}_{A2}, \ldots \boldsymbol{r}_{AN_A}, \boldsymbol{r}_{B1}, \boldsymbol{r}_{B2}, \ldots \boldsymbol{r}_{BN_B}) = \phi_0^A(\boldsymbol{r}_{A1}, \boldsymbol{r}_{A2}, \ldots \boldsymbol{r}_{AN_A})\, \phi_0^B (\boldsymbol{r}_{B1}, \boldsymbol{r}_{B2}, \ldots \boldsymbol{r}_{BN_B})$$

with N_A and N_B denoting the number of electrons of the respective atoms, and ϕ_0^A, ϕ_0^B represent antisymmetric wave functions. They are normalized to unity:

$$\int |\phi_0^A|^2\, d^3r_{A\,1}\; d^3r_{A\,2} \ldots d^3r_{A\,N_A} = 1$$

and

$$\int |\phi_0^B|^2\, d^3r_{B\,1}\; d^3r_{B\,2} \ldots d^3r_{B\,N_B} = 1\,.$$

For simplicity we have dropped the spin coordinates. The two wave functions are solutions to the associated Schrödinger equations

$$\widehat{H}_{A/B}\, \phi_0^{A/B} = E_0^{A/B}\, \phi_0^{A/B}$$

where

$$\widehat{H}_{A/B} = \sum_{i=1}^{N_{A/B}} \left[-\frac{1}{2} \nabla^2_{A/B\,i} - \frac{Z_{A/B}}{|\boldsymbol{r}_{A/B\,i} - \boldsymbol{R}_{A/B}|} \right] + \frac{1}{2} \sum_{\substack{i,j \\ i \neq j}} \frac{1}{|\boldsymbol{r}_{A/B\,i} - \boldsymbol{r}_{A/B\,j}|}$$

with $Z_{A/B}$ denoting the respective atomic number. The indices $i\,,j$ in the second sum run over $N_{A/B}$ coordinates, and hence this sum describes only the electron-electron interaction within the associated atom.

To simplify the notation we have introduced atomic Hartree units, i. e. the Bohr radius r_B as the unit length and 1 Hartree as the unit of energy.

Obviously, Ψ_0 satifies the Schrödinger equation

$$(\widehat{H}_A + \widehat{H}_B)\Psi_0 = (E_0^A + E_0^B)\Psi_0$$

and hence

$$E = \langle\Psi_0|\widehat{H}_A + \widehat{H}_B|\Psi_0\rangle = E_0^A + E_0^B\,. \tag{177}$$

However, the actual Hamiltonian describing the $(N_A + N_B)$-electron system and the two nuclei is given by

$$\begin{aligned}\widehat{H}_{total} = \widehat{H}_A + \widehat{H}_B + \sum_{i=1}^{N_A}\sum_{j=1}^{N_B}\frac{1}{|\boldsymbol{r}_{A\,i} - \boldsymbol{r}_{B\,j}|} - \sum_{j=1}^{N_B}\frac{Z_A}{|\boldsymbol{r}_{B\,j} - \boldsymbol{R}_A|}\\ - \sum_{j=1}^{N_A}\frac{Z_B}{|\boldsymbol{r}_{A\,i} - \boldsymbol{R}_B|} + \frac{Z_A\,Z_B}{|\boldsymbol{R}_A - \boldsymbol{R}_B|}\,.\end{aligned} \tag{178}$$

The electronic densities $\rho_0^A(\boldsymbol{r})$, $\rho_0^B(\boldsymbol{r})$ are defined by

$$\begin{aligned}\rho_0^A(\boldsymbol{r}) \overset{def}{=} \tilde{\rho}_0^A(\boldsymbol{r} - \boldsymbol{R}_A) &= N_A \int |\phi_0^A(\boldsymbol{r}, \boldsymbol{r}_{A2}, \ldots \boldsymbol{r}_{A\,N_A})|^2\, d^3r_{A\,2}\; d^3r_{A\,3}\ldots d^3r_{A\,N_A}\\ \rho_0^B(\boldsymbol{r}) \overset{def}{=} \tilde{\rho}_0^B(\boldsymbol{r} - \boldsymbol{R}_B) &= N_B \int |\phi_0^B(\boldsymbol{r}, \boldsymbol{r}_{B2}, \ldots \boldsymbol{r}_{B\,N_B})|^2\, d^3r_{B\,2}\; d^3r_{B\,3}\ldots d^3r_{B\,N_B}\end{aligned} \tag{179}$$

On forming the expectation value $\langle\Psi_0|\widehat{H}_{total}|\Psi_0\rangle$ the single sum over the $\boldsymbol{r}_B$-coordinates in Eq.(178) yields because of (179)

$$-Z_A \int \frac{\rho_0^B(\boldsymbol{r})}{|\boldsymbol{r} - \boldsymbol{R}_A|}\, d^3r = -\frac{Z_A\,Z_B}{|\boldsymbol{R}_B - \boldsymbol{R}_A|}\,.$$

The latter holds since N_B equals Z_B for neutral atoms, and $\rho_0^B(\boldsymbol{r})$ is assumed spherically symmetric with respect to $\boldsymbol{R}_B$.

Similarly, when dealing with the double-sum in Eq.(178), if one first performs the integration over the $\boldsymbol{r}_B$-coordinates, one obtains as an intermediate result

$$\sum_{i=1}^{N_A}\int \frac{\rho_0^B(\boldsymbol{r})}{|\boldsymbol{r}_{A\,i} - \boldsymbol{r}|}\, d^3r = \sum_{j=1}^{N_A}\frac{Z_B}{|\boldsymbol{r}_{A\,i} - \boldsymbol{R}_B|}$$

with the latter again being a consequence of the spherical symmetry of $\rho_0^B(\boldsymbol{r})$ with respect to $\boldsymbol{R}_B$. Performing the remaining integration of this expression and of the third sum in Eq.(178) over the $\boldsymbol{r}_A$-coordinates, one recognizes that the two expressions cancel. Hence, one has - because of perfect mutual screening - despite the occurrence of the extra sums in expression (178)

$$E = \langle \Psi_0 | \widehat{H}_{total} | \Psi_0 \rangle = E_0^A + E_0^B \tag{180}$$

as before in (177).

That means, at this level of analysis each atom remains unaffected by the presence of the other. However, because of the occurrence of the extra sums in (178) Ψ_0 does not satisfy the Schödinger equation

$$\widehat{H}_{total} \Psi = E_0^{total} \, \Psi \, . \tag{181}$$

As the solution to (181) yields a minimum of the expectation value of $\widehat{H}_{total}$ for the ground-state Ψ_0^{total} one may equivalently state that $\langle \Psi_0 | \widehat{H}_{total} | \Psi_0 \rangle$ can be lowered to the exact total energy E_0^{total}, i. e. to the lowest eigenvalue of $\widehat{H}_{total}$ by appropriately distorting Ψ_0 toward Ψ_0^{total}. This can be done in a simplified way by partially including excited states of the atoms so that one is now dealing with a wave function that slightly departs from the ground-state Ψ_0

$$\Psi_0^{total} = c_0 \, \Psi_0 + c_1 \, \Psi_1 = c_0 \, \phi_0^A \, \phi_0^B + c_1 \, \phi_1^A \, \phi_1^B \tag{182}$$

with c_0 , c_1 denoting real-valued coefficients, and ϕ_1^A stands for some excited state of atom A associated with an energy $E_1^A > E_0^A$. Corresponding definitions apply to atom B. The set of electronic coordinates and their assocation with the respective wave functions remain unchanged. As we assume an only slight departure from the groundstate we have

$$|c_1| \ll 1 \, , \quad \text{and we observe} \quad c_0^2 + c_1^2 = 1 \, . \tag{183}$$

The latter reflects the unaltered norm unity of Ψ.

To simplify the argument, we confine ourselves to just one excited state, a lowest lying state that can be mapped onto a Slater determinant in which the excited orbital possesses a negative parity compared to the corresponding groundstate (s-type) orbital $\psi_0^{A/B}(\boldsymbol{r} - \boldsymbol{R}_{A/B})$ that is replaced in the excitation. We identify the excited one-electron state $\psi_1^{A/B}(\boldsymbol{r} - \boldsymbol{R}_{A/B})$ with a p_z-type orbital whose quantization axis, the z-axis, is chosen along the direction $\boldsymbol{R}_A - \boldsymbol{R}_B$.

Different from $\rho_0^A(\boldsymbol{r})$ the electronic charge density in the excited state

$$\rho_1^A(\boldsymbol{r}) = N_A \int |\phi_1^A(\boldsymbol{r}, \boldsymbol{r}_{A2}, \ldots \boldsymbol{r}_{A\,N_A})|^2 \, d^3 r_{A\,2} \; d^3 r_{A\,3} \ldots d^3 r_{A\,N_A}$$

is not spherical any more, but its centroid still conicides with the nucleus, that is, its electrostatic potential does not contain a dipole-type contribution, merely a short-range quadrupole component that we shall ignore in the following. The same applies to ρ_1^B.

As a result, all considerations above for the groundstate carry over to the present case except for the double-sum in Eq.(178). In forming $|\Psi_0^{total}|^2$ we obtain

$$|\Psi_0^{total}|^2 = c_0^2 |\phi_0^A|^2 |\phi_0^B|^2 + c_1^2 |\phi_1^A|^2 |\phi_1^B|^2 + 2 c_0 \, c_1 \, \phi_0^A \, \phi_1^A \, \phi_0^B \, \phi_1^B \, .$$

It is the occurrence of the cross-term which gives rise to an interatomic potential that has been absent so far.

In forming $\langle\Psi_0^{total}|\widehat{H}_{total}|\Psi_0^{total}\rangle$ one obtains

$$\langle\Psi_0^{total}|\widehat{H}_{total}|\Psi_0^{total}\rangle = c_0^2\,(E_0^A + E_0^B) + c_1^2\,(E_1^A + E_1^B) + \ldots \tag{184}$$

where the dots stand for

$$2c_0c_1 \int \psi_1^A(\boldsymbol{r}_A - \boldsymbol{R}_A)\psi_0^A(\boldsymbol{r}_A - \boldsymbol{R}_A) \times \left[\int \frac{\psi_1^B(\boldsymbol{r}_B - \boldsymbol{R}_B)\psi_0^B(\boldsymbol{r}_B - \boldsymbol{R}_B)}{|\boldsymbol{r}_A - \boldsymbol{r}_B|}\, d^3r_B\right] d^3r_A \tag{185}$$

The numerator in the second integral represents a dipole-type charge density with the dipole axis lying in the $\boldsymbol{R}_A - \boldsymbol{R}_B$-direction. As $|\boldsymbol{r}_A - \boldsymbol{r}_B|$ is assumed to be large compared to the sum of the atomic radii, one may approximate the second integral

$$\int \frac{\psi_1^B(\boldsymbol{r}_B - \boldsymbol{R}_B)\psi_0^B(\boldsymbol{r}_B - \boldsymbol{R}_B)}{|\boldsymbol{r}_A - \boldsymbol{r}_B|}\, d^3r_B \approx \frac{p_B}{|\boldsymbol{r}_A - \boldsymbol{R}_B|^2} \tag{186}$$

where

$$p_B = e'\,|\delta\boldsymbol{r}_B|$$

denotes the dipole moment, and $\pm\, e'$ is the effective charge in the centroid of that charge density above and below the plane across the B-nucleus and perpendicular to the z-axis. The two charges are interconnected by $\delta\boldsymbol{r}_B$.

Similarly, the integral

$$\int \psi_1^A(\boldsymbol{r}_A - \boldsymbol{R}_A)\psi_0^A(\boldsymbol{r}_A - \boldsymbol{R}_A)\,\frac{p_B}{|\boldsymbol{r}_A - \boldsymbol{R}_B|^2}\, d^3r_A$$

may be approximated

$$\int \psi_1^A(\boldsymbol{r}_A - \boldsymbol{R}_A)\psi_0^A(\boldsymbol{r}_A - \boldsymbol{R}_A)\,\frac{p_B}{|\boldsymbol{r}_A - \boldsymbol{R}_B|^2}\, d^3r_A = \frac{2\,p_A\,p_B}{|\boldsymbol{R}_A - \boldsymbol{R}_B|^3}\,.$$

Because of (183) one may set $2\,c_0\,c_1 \approx 2\,c_1$ and replace c_0^2 by $1 - c_1^2$ so that Eq.(184), on employing (185), takes the form

$$E = \langle\Psi_0^{total}|\widehat{H}_{total}|\Psi_0^{total}\rangle = (E_0^A + E_0^B) + c_1^2\,\Delta E_1 + 2\,c_1\frac{2\,p_A\,p_B}{R^3} \tag{187}$$

where

$$\begin{aligned} \Delta E_1 &= (E_1^A + E_1^B) - (E_0^A + E_0^B) \quad \text{or} \\ \Delta E_1 &= E_1^A - E_0^A + E_1^B - E_0^B \quad \text{and} \\ R &\stackrel{def}{=} |\boldsymbol{R}_A - \boldsymbol{R}_B|\,. \end{aligned} \tag{188}$$

The new groundstate is given then by requiring

$$\frac{dE}{dc_1} = 0 \quad \text{that is} \quad c_1\,\Delta E_1 + \frac{2\,p_A\,p_B}{R^3} = 0$$

which yields

$$c_1 = -\frac{1}{\Delta E_1} \frac{2\,p_A\,p_B}{R^3}\,.$$

Inserting this into Eq.(187) one arrives at a total energy that is now different from (180)

$$E = (E_0^A + E_0^B) - \frac{4}{\Delta E_1} \frac{p_A^2\,p_B^2}{R^6}\,. \tag{189}$$

This constitutes van der Waals' well known law on the R-dependence of the attractive potential between neutral atoms whose charge densities do not overlap. By breaking the spherical symmetry of the electronic densities the bi-atomic system lowers its total energy. We shall denote this R-dependent portion by $V_{vdW}(R)$ from now on.

As one would qualitatively expect, the attraction is large for atoms with low lying excitation energies. It is obvious from our derivation that it applies to any combination of orbitals $\psi_0^{A/B}(\boldsymbol{r} - \boldsymbol{R}_{A/B})$ and $\psi_1^{A/B}(\boldsymbol{r} - \boldsymbol{R}_{A/B})$ that lead to a product with dipole character, regardless which of the orbitals refers to the groundstate configuration. Although the groundstate charge density in the bi-atomic system may well be shaped by an incompletely filled p-, d- or f-shell of the respective atoms, it will merely give rise to negligible short range quadrupole (multipole) contributions to the interaction potential.

As indicated, the above derivation can be refined by including more than just one excited pair $\phi_1^A\,\phi_1^B$, that is by allowing for additional pairs which contain higher excited orbitals that also yield dipole-type products with the groundstate orbitals. To this end the ansatz (182) has to be generalized in the form

$$\Psi_0^{total} = c_0\,\Psi_0 + \sum_{(n,m)} c_{n\,m}\,\Psi_{n\,m} = c_0 \phi_0^A\,\phi_0^B + \sum_n \sum_m c_{n\,m}\,\phi_n^A\,\phi_m^B\,; \quad n, m > 0 \tag{190}$$

where

$$c_0^2 + \sum_n \sum_m c_{n\,m}^2 = 1\,.$$

Instead of (189) one then obtains the general expression

$$V_{vdW}(R) = -\left[4 \sum_n \sum_m \frac{p_{n\,A}^2\,p_{m\,B}^2}{E_n^A - E_0^A + E_m^B - E_0^B}\right] \frac{1}{R^6} \tag{191}$$

where

$$V_{vdW}(R) = E(R) - (E_0^A + E_0^B)\,.$$

This result is essentially identical with that derived by Dzyaloshinskii et al.[57], however in an exceedingly involved, completely different way based on quantum field theory and "long-wave electrodynamical fluctuations". The paranthesized expression is commonly denoted by C_6. The expression obtained by Dzyaloshinskii et al. contains a factor of 6 in front of the double-sum instead of 4. Some experimental results seem to favor our smaller factor. (For details see Feibelman [58], p.394.)

If the two atoms are identical (A=B) and if one only considers the contributions for $n = m$ to the above sum Eq.(191) takes the form

$$V_{vdW}^{(n,n)}(R) = -\left[2\sum_n \frac{p_{n\,A}^4}{\Delta E_n}\right]\frac{1}{R^6}\,. \tag{192}$$

The parenthesized expression closely resembles the static polarizability α of the atom which - to first perturbation order - attains the form

$$\alpha = 2\sum_n \frac{p_{n\,A}^2}{\Delta E_n}\,. \tag{193}$$

The similarity of expression (192) and α has led to quote $-\nabla V_{vdw}(\boldsymbol{r})$ occasionally by the name "dispersion force" which is exceedingly misleading. Obviously, only the **static** polarizibility displays a certain affinity with V_{vdW}. Optical dispersion relates to the frequency dependence of the dynamical polarizability.

In determining C_6 from Eq.(191)) one can easily run into numerical inaccuracies because of the large spatial extent of excited orbitals. To get an estimate of C_6 we determine $p_{1\,A}$ from the polarizibility α of the atom under study. In Table 1 we have listed results on $V_{vdW}(R) = -C_6/R^6$ and on α for the noble gases where we have reduced the sums in (192) and (193) to the strongly dominating first summand and shortened the notation accordingly by setting $p_{1\,A} = p_A$. Further, we have introduced

$$p_0 = 1e \cdot 1r_B = 0.848 \cdot 10^{-29}\,C\,m$$

as a unit for the atomic dipole and observed that in familiar units (ϵ_0=vacuum permittivity)

$$\alpha = \frac{2\,\hat{p}_A^2}{\Delta E_1\, 4\pi\,\epsilon_0}\,; \qquad \hat{p}_A = p_A\, p_0\,;$$
$$\frac{p_0^2}{4\pi\,\epsilon_0} = 0.646 \cdot 10^{-23}\,eV\,cm^3 \tag{194}$$

Table 1. Polarizabilities in $[10^{-24}cm^3]$, lowest excitation energies and the vdW-prefactor C_6 in units of $Hartree \cdot r_B^6$ for the noble gases $p_A = \hat{p}_A/p_0$ [Eq.(194)]

	α_{exp}	$\alpha(p_A)$	p_A	$\Delta E_1[eV]$	$C_6^{present}$	C_6^{ref}
He	0.204	0.245	0.8	21.13	1.04	1.29 (1.45)
Ne	0.396	0.47	1.4	16.84	6.22	7.5
Ar	1.645	1.8	2.3	11.83	64.0	71.5
Kr	2.49	2.76	2.63	9.99	128.7	145.5
Xe	4.05	4.57	3.09	8.44	293.5	331.5

The results denoted by C_6^{ref} were obtained by Hult et al.[59], the paranthesized value is due to Kohn et al.[60]. The rather involved calculations of these authors are based on perturbational density funtional theory. The experimental data on ΔE_1 and α were taken from *The Smithonian Physical Tables* [61]. As the theoretical expressions for α and C_6 both contain p_A we have chosen a compromise value for the latter quantity to reconcile the former quantities with least error. First-principles calculations on ΔE_n and $p_{n\,A}$ require

Table 2. C_6 in units of $Hartree \cdot r_B^6$ [Eq.(192), $n = 1$] for the noble gases; $p_A = \hat{p}_A/p_0$ [Eq.(194)]

	p_A	$\Delta E_1[eV]$	$C_6^{present}$	C_6^{ref}
He	0.82	21.13	1.17	1.29 (1.45)
Ne	1.44	16.84	7.00	7.5
Ar	2.37	11.83	72.03	71.5
Kr	2.71	9.99	144.85	145.5
Xe	3.18	8.44	330.33	331.5

considerable numerical effort because excited atomic states are not easily accessible. Their linear dimensions increase quickly with n. For that reason we rather rely on experimental data on ΔE_1 and α. To demonstrate the sensibility with which C_6 responds to errors in p_A due to its appearance in fourth power, we have listed in Table 2 recalculated $C_6^{present}$-values which result from an increase of p_A bei 3% compared to the p_A-values given in Table 1. It is obvious from the above considerations which steps have to be taken to improve on the accuracy of our results. Furthermore, as one can see from Eq.(191), our approach is actually not restricted to atoms since it is merely based on dipole-associated virtual excitations of the interacting systems which may be molecules, clusters or even macroscopic solid objects.

24. Decomposing an Experimental Setup into the Quantum System under Study and a Remainder. Schrödinger's Cat

One of the puzzling credos of the Copenhagen interpretation of quantum mechanics consists in the conviction that an experimental setup for performing measurements on microscopic particles, has to be subdivided "somehow" into the particles under study and a remainder that functions as a classical system. This decomposition is known under the name "Heisenberg-cut". Yet from an unbiased point of view it appears to be self-evident that an experimental setup as a whole represents a many-particle system each part of which is subjected to the same laws of quantum mechanics as the particular portion that constitutes the object under study, an electron in a diffraction chamber, for example. We shall use this example to demonstrate the consistency of this standpoint, but we limit ourselves to considering a system that merely consists of just one specific apparatus plus a particle undergoing diffraction in it. The generalization to the inclusion of the entire environment is obvious from the ensuing considerations.

We assume that the system is made up of N particles, a subset consisting of atomic nuclei which we number by a label α, and N_e electrons, one of which representing the single particle of interest, the "test particle". To keep the notation simple, we limit ourselves to

considering only electrostatic particle interactions of the kind described by the many-body potential (162). If the test particle has left the cathode of the setup it is kept by electrodes, diaphragms and lenses at a macroscopic distance away from all kinds of surfaces it might strike and where it might get captured. Thus, the associated one-particle wave function $\psi_e(\boldsymbol{r},t)$ which describes the electron on its way through the apparatus to the screen or detector, has de facto zero overlap with the wave function of the $N-1$ remaining particles of the apparatus. Still, in standard setups it is intended that the particle hits a secluded portion of material on its way to the monitoring device, a diffracting single crystalline foil of metal, for example. But in the majority of cases the contact time is so short compared to the electronic excitation times of the material that the test electron cannot mingle with the other electrons. Below we shall briefly discuss prominent exceptions.
Similar to the case of the H_2-molecule with macroscopically distant nuclei, one is justified then in assuming a factorization of the total wave function

$$\Psi_N(\boldsymbol{r},\boldsymbol{r}_2,\ldots\boldsymbol{r}_N,t) = \psi_e(\boldsymbol{r},t)\,\Psi_{N-1}(\boldsymbol{r}_2,\boldsymbol{r}_3,\ldots\boldsymbol{r}_N,t) \tag{195}$$

where the spin coordinates have again been suppressed for simplicity. We emphasize that $\Psi_N(\boldsymbol{r},\boldsymbol{r}_2,\ldots\boldsymbol{r}_N,t)$ in the present case constitutes the wave function of $N = N_e + N_n^{app.}$ particles: of the test electron, of the $N_e - 1$ electrons that belong to the apparatus **and in addition** of the N_n^{app} nuclei of the latter. If we insert this wave function into the associated N-particle Schrödinger equation (164) we obtain

$$\Psi_{N-1}\, i\hbar\,\dot{\psi}_e + \psi_e\, i\hbar\,\dot{\Psi}_{N-1} = \psi_e\,\widehat{H}_0^{(N-1)}\,\Psi_{N-1} + \Psi_{N-1}\,\widehat{H}_0^e\,\psi_e + V_{tot}\psi_e\,\Psi_{N-1} \tag{196}$$

where V_{tot} denotes the total (Coulomb) interaction potential between all particles

$$V_{tot} = V_{apparatus}(\boldsymbol{r}_2,\boldsymbol{r}_3,\ldots\boldsymbol{r}_N) + V_e^{N-1}(\boldsymbol{r},\boldsymbol{r}_2\ldots\boldsymbol{r}_N)$$

with V_e^{N-1} referring to the Coulomb interaction of the test particle with all charges of the apparatus

$$V_e^{N-1}(\boldsymbol{r},\boldsymbol{r}_2\ldots\boldsymbol{r}_N) = \sum_{j=2}^{N}\frac{e^2\,Z_j}{4\pi\,\epsilon_0\,|\boldsymbol{r}_j-\boldsymbol{r}|}\,. \tag{197}$$

From Eq.(165) we have

$$\widehat{H}_0 = \sum_{j=1}^{N}\left[\frac{\widehat{\boldsymbol{p}}_j^2}{2\,m_0} + V_{ext.}(\boldsymbol{r}_j,t)\right] = \widehat{H}_0^{N-1} + \underbrace{\left[\frac{\widehat{\boldsymbol{p}}^2}{2\,m_0} + V_{ext.}(\boldsymbol{r},t)\right]}_{=\widehat{H}_0^e}\,.$$

where $V_{ext.}(\boldsymbol{r},t)$ denotes some extra potential set up outside the apparatus. In general it would at least be the gravitational potential, in which case it would be time-independent. In Eq.(197) $|Z_j|$ stands for the number of elementary charges, i. e.

$$Z_j = \begin{cases} -Z_\alpha & \text{if } j \text{ runs over the } \alpha^{th} \text{ nucleus} \\ 1 & \text{if } j \text{ refers to an electron}\,. \end{cases}$$

If one multiplies Eq.(196) by Ψ^*_{N-1} and performs an integration with respect to $\boldsymbol{r}_2, \ldots \boldsymbol{r}_N$ one obtains

$$i\hbar\,\dot{\psi}_e + \psi_e \int \Psi^*_{N-1}\left[i\hbar\,\frac{\partial}{\partial t}\Psi_{N-1} - \left(\widehat{H}_0^{(N-1)} + V_{apparatus}\right)\Psi_{N-1}\right] d^3r_2 \ldots d^3r_N$$
$$= \widehat{H}_0^e\,\psi_e + \widehat{V}_e\,\psi_e \quad (198)$$

where $\widehat{V}_e(\boldsymbol{r}, t)$ represents a one-electron potential defined as

$$\widehat{V}_e(\boldsymbol{r}, t) = \int \Psi^*_{N-1} \sum_{j=2}^{N} \frac{e^2\,Z_j}{4\pi\,\epsilon_0\,|\boldsymbol{r}_j - \boldsymbol{r}|}\,\Psi_{N-1}\,d^3r_2 \ldots d^3r_N\,. \quad (199)$$

Since the bracketed expression under the integral in the above equation vanishes, we arrive at

$$i\hbar\,\frac{\partial}{\partial t}\psi_e(\boldsymbol{r}, t) = [\widehat{H}_0^e + \widehat{V}_e(\boldsymbol{r}, t)]\,\psi_e(\boldsymbol{r}, t)\,. \quad (200)$$

Thus, the wave function of the electron under study obeys, in fact, a one-particle Schrödinger equation.

However, in the above derivation we have ignored the response of the wavefunction Ψ_{N-1} of the apparatus to the presence of the test particle. This is is hidden in $\sum_{j=2}^{N} V_{ext.}(\boldsymbol{r}_j, t)$ being a constituent of $\widehat{H}_0^{N-1}$. Not only the gravitational potential acts on the apparatus but also the "still external" test particle through its Coulomb potential. Inclusion of this effect amounts to adding this extra potential

$$\Delta V_{ext.}(\boldsymbol{r}_2, \boldsymbol{r}_3, \ldots \boldsymbol{r}_N) = \int \psi_e^*(\boldsymbol{r}, t)\,V_e^{N-1}(\boldsymbol{r}, \boldsymbol{r}_2 \ldots \boldsymbol{r}_N, t)\psi_e(\boldsymbol{r}, t)\,d^3r$$

to the unmodified expression for $\widehat{H}_0^{N-1}$ in Eq.(198). With the appearance of $\Delta V_{ext.}$ the wave function of the apparatus changes and becomes a functional of $\psi_e(\boldsymbol{r}, t)$. This, in turn, gives rise to a change of $\widehat{V}_e(\boldsymbol{r}, t)$ as follows from its definition Eq.(199). That means that the test particle feels the potential of the charge distribution which it induces in the apparatus. It should be clearly recognized that this - admittedly small - effect turns the one-particle Eq.(200) into a non-linear partial differential equation. That is to say that at this level of description even the simplest realistic case of a particle travelling in a vacuum chamber leads to a non-linear Schrödinger equation. In the following considerations we ignore this charge induction effect.

There are certain cases in which the contact time of the test particle is not short enough, and hence there is a non-vanishing probability that the particle mingles with those of the target. To get a rough picture of this situation, we describe the wave function instead of Eq.(195) by

$$\Psi_N(\boldsymbol{r}, \boldsymbol{r}_2, \ldots \boldsymbol{r}_N, t) = c_0(t)\,\psi_e(\boldsymbol{r}, t)\,\Psi_{N-1}(\boldsymbol{r}_2, \boldsymbol{r}_3, \ldots \boldsymbol{r}_N, t)$$
$$+ c_1(t)\,\Psi_N^{capt}(\boldsymbol{r}, \boldsymbol{r}_2, \ldots \boldsymbol{r}_N, t) \quad (201)$$

where $c_0(t)$ and $c_1(t)$ are real-valued functions with the property $|c_0(t)|^2 + |c_1(t)|^2 = 1$, in particular

$$c_0(t) = e^{-\frac{t}{2\tau}}$$

and hence

$$|c_0(t)|^2 = e^{-\frac{t}{\tau}}\,; \quad |c_1(t)|^2 = 1 - e^{-\frac{t}{\tau}}\,. \tag{202}$$

Here τ refers to a characteristic interaction time with the target, and $|c_1(t)|^2$ is the probability with which the test electron is captured by the target. Thereby it loses its identity as the "test electron". The latter effect is expressed by the property of Ψ_N^{capt} being antisymmetric with respect to the interchange of **any** two particles out of the set of N_e electrons.

Inserting Ψ_N from Eq.(201) into the Schrödinger equation (164) we obtain

$$\begin{aligned} c_0(t)\left[\Psi_{N-1}\, i\hbar\, \dot{\psi}_e + \psi_e\, i\hbar\, \dot{\Psi}_{N-1}\right] + i\hbar\, \dot{c}_0(t)\, \psi_e \Psi_{N-1} + c_1(t)\, i\hbar\, \dot{\Psi}_N^{capt} \\ + i\hbar\, \dot{c}_1(t)\, \Psi_N^{capt} \\ = c_0(t)\left[\psi_e\, \widehat{H}_0^{(N-1)}\, \Psi_{N-1} + \Psi_{N-1}\, \widehat{H}_0^e\, \psi_e + V_{tot}\psi_e\, \Psi_{N-1}\right] \\ + c_1(t)\, \widehat{H}\, \Psi_N^{capt}\,. \end{aligned} \tag{203}$$

The functions Ψ_{N-1} and Ψ_N^{capt} satisfy the associated time-dependent Schrödinger equations

$$i\hbar\, \dot{\Psi}_{N-1} = \left[\widehat{H}_0^{(N-1)} + V_{apparatus}\right] \Psi_{N-1} \tag{204}$$

and

$$i\hbar\, \dot{\Psi}_N^{capt} = \widehat{H}\, \Psi_N^{capt}\,.$$

If we insert this into Eq.(203), multiply the result in front by Ψ_{N-1}^* and perform an integration over $\boldsymbol{r}_2, \boldsymbol{r}_3, \ldots \boldsymbol{r}_N$, we obtain

$$\begin{aligned} c_0(t)\left[i\hbar\, \frac{\partial}{\partial t} - \widehat{H}_0^e - V_e'(\boldsymbol{r}, t)\right] \psi_e(\boldsymbol{r}, t) + \\ i\hbar\, \dot{c}_1(t) \int \Psi_{N-1}^*(\boldsymbol{r}_2, \ldots \boldsymbol{r}_N, t)\, \Psi_N^{capt}(\boldsymbol{r}, \boldsymbol{r}_2 \ldots \boldsymbol{r}_N, t)\, d^3r_2\, d^3r_3 \ldots d^3r_N = 0\,. \end{aligned} \tag{205}$$

Here we have used $i\hbar\, \frac{\partial}{\partial t}\, c_0(t) = -i\, \frac{\hbar}{2\tau}\, c_0(t)$ and set

$$V_e'(\boldsymbol{r}, t) = V_e(\boldsymbol{r}, t) + i\, \tilde{V}_e \quad \text{where} \quad \frac{\tilde{V}_e}{\hbar} \equiv \frac{1}{2\tau}\,.$$

The imaginary part of $V_e'(\boldsymbol{r}, t)$ is commonly referred to as "optical potential".

According to our classification of the electron under study as either "distinguishable" or "non-distinguishable" the associated total probability density $\rho(\boldsymbol{r}, t)$ splits (almost quantitatively) into the "either- and or-probability"

$$\rho(\boldsymbol{r}, t) = \underbrace{|c_0(t)|^2\, |\psi_e(\boldsymbol{r}, t)|^2}_{=\rho_0(\boldsymbol{r}, t)} + \underbrace{|c_1(t)|^2 \int |\Psi_N^{capt}(\boldsymbol{r}, \boldsymbol{r}_2, \ldots \boldsymbol{r}_N, t)|^2\, d^3r_2, d^3r_3 \ldots d^3r_N}_{\equiv \rho_1(\boldsymbol{r}, t)}$$

which means

$$S_e(\boldsymbol{r}, t) = \int \Psi_{N-1}^*(\boldsymbol{r}_2, \boldsymbol{r}_3, \ldots \boldsymbol{r}_N)\, \Psi_N^{capt}(\boldsymbol{r}, \boldsymbol{r}_2, \boldsymbol{r}_3, \ldots \boldsymbol{r}_N)\, d^3r_2, d^3r_3 \ldots d^3r_N \approx 0\, \forall\, \boldsymbol{r}, t\,.$$

It follows then from Eq.(205) that $\psi_e(\boldsymbol{r},t)$ solves the modified Schrödinger equation

$$\left[i\hbar\frac{\partial}{\partial t} - \widehat{H}_0^e - V_e'(\boldsymbol{r},t)\right]\psi_e(\boldsymbol{r},t) = 0\,, \qquad (206)$$

which describes situations one encounters, for example, in experiments on low energy electron diffraction (LEED) at surfaces of solids.

The mod squared of the actually not completely vanishing overlap

$$S(t) = \int S_e(\boldsymbol{r},t)\,\psi_e^*(\boldsymbol{r},t)\,d^3r$$

determines the transition probability $1/\tau$. Eq.(201) and the resulting Eqs.(204) and (206) are pivotal in describing generic quantum mechanical processes, a subset of which plays the role of "measurements"[3]. For example, when one is dealing with a setup where an electron traverses the legendary double slit diaphragm, defined by the potential (199), the function $\Psi_N^{capt}(\boldsymbol{r},\boldsymbol{r}_2,\ldots\boldsymbol{r}_N,t)$ describes the situation when the electron has been captured by the detector which is a part of the "apparatus". In spirit this in keeping with a statement by Hartle and Gell-Mann [54]: *"In a theory of the whole thing there can be no fundamental division into observer and observed"* Our approach reflects even more directly the standpoint taken by v. Kampen [55]: *"The measuring act is fully described by the Schroedinger equation for object and apparatus together..."*

It is the archetypal combination of a particular setup and "pointer readings" of a detector that enables the experimentalist to determine certain properties of the one-particle quantum system by extracting the sought-for information from the solution to the corresponding Schrödinger (or Pauli) equation that yields $\boldsymbol{j}(\boldsymbol{r}) = \rho(\boldsymbol{r},t)\,\boldsymbol{v}(\boldsymbol{r},t)$ at $\boldsymbol{r}_{detector}$. Eigenvalues of Hermitian operators can only be obtained via this detour, and for fundamental, mostly experimental, reasons, only with limited accuracy.

The paradoxical situation which one runs into if one endows the "observer" (or "measurer") with an unrealistic meaning, is illustrated by Schrödinger's cat example [62]: An alpha-particle emitted from some radioactive material triggers a device that kills a cat in a closed box by releasing a poisonous gas. Of course, the moment of radioactive decay does in no way depend on the particular setup. Our description of this process would be based on Eq.(201) where $\Psi_N(\boldsymbol{r},\boldsymbol{r}_2,\ldots\boldsymbol{r}_N,t)$ on the left-hand side now represents the wave function $\Psi_{gas+cat}(t)$ of the system cat plus gas, $\psi_e(\boldsymbol{r},t)$ has to be replaced by an N-particle wave function ψ_{poison} referring to the only weakly "cat-overlapping" molecules of the poisonous gas set free by the device, and $\Psi_{N-1}(\boldsymbol{r}_2,\boldsymbol{r}_3,\ldots\boldsymbol{r}_N,t)$ has to be identified with the many-particle wave function Ψ_{cat} of the live cat. After the elapse of a time $\approx\tau$ the system's wave function $\Psi_{gas+cat}(t)$ has attained the form $\Psi_{capture}(t)$ where the poisonous molecules are now a part of the cat. It solves the time-dependent Schrödinger equation of the united system. The time-evolution of $\Psi_{capture}(t)$ describes all the atomic (chemical) processes that eventually lead to the cat's death. It is this time-dependent process that is familiar from ab initio calculations on chemical reactions. The latter are completely "self-controlled". There is definitely no "observer-induced" influence. From this point of view it appears to

[3]We side here emphatically with John Bell [36] who pleads in his article *"Against Measurement"* for more common sense in describing what is actually happening: the time evolution of a particular **experiment**.

be rather absurd that orthodox quantum mechanics interprets Eq.(201) with the explained new meaning of the wave functions as a superposition of a "live" and a "dead"-state of the cat, and only on opening the lid of the box by an observer, $\Psi_{gas+cat}$ collapses onto the wave function of a live or dead cat.

25. The Origin of Particle Spin

In 1925 Uhlenbeck and Goudsmit [63] suggested in a widely recognized paper that Pauli's idea [64] of a fourth quantum number in the description of electronic states of atoms might be associated with the rotation of an electron about its own axis thus giving rise to an extra angular moment. From the analysis of atomic spectra it was clear that the magnetic moment generated by such a rotation of the electron as a charged sphere had to exactly equal the Bohr magneton

$$\mu_B = \frac{e\hbar}{2m_0} .$$

There was also experimental evidence that the associated mechanical spin moment $\vec{S}$ - different from the atomic orbital momentum - would not obey the classical law of magneto-mechanical parallelism according to which μ_B should differ from $\vec{S}$ by a factor $\frac{e}{2m_0}$. In actual fact this factor had been found to be $\frac{e}{m_0}$ instead so that

$$|\vec{S}| = \frac{\hbar}{2} ,$$

and hence

$$\mu_B = g \frac{e}{2m_0} \frac{\hbar}{2} . \tag{207}$$

We ignore here and in the following the minute departure of g from 2 due to quantum electrodynamical corrections.

The radius of the rotating electron sphere was identified with the classical electron radius $2.8 \cdot 10^{-13}$ cm. As van der Waerden [65] reports, Lorentz immediately demonstrated to Uhlenbeck and Goudsmit that the electron mass would actually be larger than that of a proton if the magnetic moment of a Bohr magneton would be confined to that sphere. Moreover, the speed at the equator of the rotating sphere would by far exceed the velocity of light. Although these objections definitely disqualified the rotating sphere as a model of electron spin, it is still used, tacitly implied or appears semantically concealed as "intrinsic property" in the analysis of most of the present-day experiments involving spin-orientation or spin flips.

If an "eigen-rotation" cannot explain the occurrence of a mechanical spin moment associated with a gyratory electronic motion, what else can be responsible for it? The following considerations are based on the idea that **"spin" is not a property of the particle** but is rather a property of its quantum mechanical state. Ohanian [66] arrives at the same conclusion summarizing an early analysis by Gordon in 1928. He states:

This means that neither the spin nor the magnetic moment are internal properties of the electron - they have nothing to do with the internal structure of the electron, but only with the structure of its wave field.

However, as for his basic message we definitely disagree with Ohanian. He presents reasons that there is formally a strong similarity between a spin-polarized particle wave and a circularly polarized electromagnetic plane wave. His arguments are valid for the case of a longitudinally polarized free particle wave, but they do not apply to a transversally polarized particle wave.

Our description of particle motion as modified by stochastic vacuum forces makes it particularly suggestive to correlate - similar to the explanation of zero-point motion of oscillators - particle spin with the quivering motion that results from those forces and vanishes as $\hbar$ tends to zero. (This applies, of course, to all the other quantum mechanical ground-state properties as well.) To illustrate this we consider the simplest case of a hydrogen electron exposed to a magnetic field $\boldsymbol{B} = B_Z \, \boldsymbol{e}_z$ in its ground state $\psi_0(\boldsymbol{r})$. In Fig.6 we

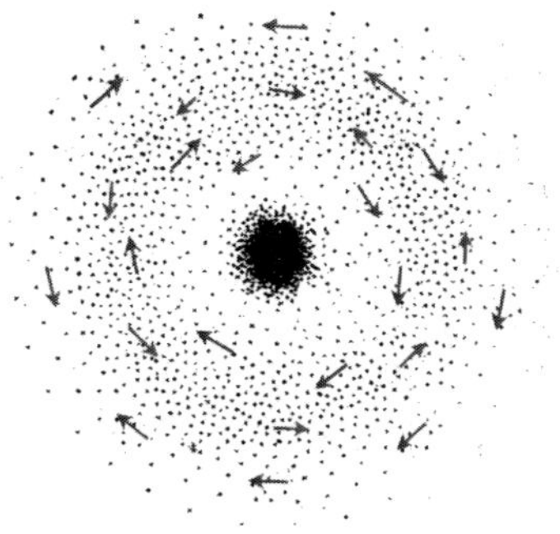

Figure 6. Spin effective components of the quivering motion.

show a schematic distribution of positions that the electron has successively taken at times t_i and equal time intervals $t_{i+1} - t_i = \Delta t$ where Δt is very small compared to T. This time span has been introduced in Section 3 in connection with defining the probability density $\rho(\boldsymbol{r})$. The $z-$axis is thought to run through the atomic center perpendicular to the plotting plane. At each of the points the electron possesses a velocity which we decompose into a radial and a $z-$component, and in a component perpendicular to the $z-$axis. Only the latter are indicated by arrows. For symmetry reasons there will be as many positive as negative radial and $z-$components in the elementary volume around each point. They average out. We subdivide the set of arrows into two subsets associated with left-hand and right-hand circular motion, respectively. One might surmise that in the presence of the magnetic field one of the sets becomes empty and the other set now gives rise to a net circular current so as to minimize the total energy in that electronic state. As a defining property, the energy gain is proportional to B_z and vanishes as $\hbar \rightarrow 0$. As we know from Section 14 the magnetic field causes in the state $\psi_0(\boldsymbol{r})$ a current density $\boldsymbol{j}(\boldsymbol{r}) = \frac{e}{m_0} |\psi_0(\boldsymbol{r})|^2 \, \boldsymbol{A}(\boldsymbol{r})$, but the energy gain from this goes as B_z^2 because of $\Delta E = \int \boldsymbol{j} \cdot \boldsymbol{A} \, d^3 r$ and $\boldsymbol{B} = \nabla \times \boldsymbol{A}$. The omission of the empty subset of arrows does not at all change the distribution of points defining $\psi_0(\boldsymbol{r})$. The energy gain is provided by the vacuum fluctuations in complete analogy to the zero-point energy of oscillators.

We thus arrive at the conclusion that the circular current which occurs on allowing the quivering motion of the electron to become asymmetric does not change the probability density which is characteristic of real-valued solutions to the Schrödinger equation. But it definitely yields a physical effect that has so far been outside our formal framework. In discussing certain properties of solutions to the Dirac equation Schrödinger [67] was led to a similar interpretation of particle spin and coined the irregular particle motion causing it "Zitterbewegung".

The additional spin-dependent interaction with a magnetic field occurs also in complex-valued states $\psi(\boldsymbol{r}) = |\psi(\boldsymbol{r})|e^{i\varphi(\boldsymbol{r})}$. In that case there is an additional set of arrows superimposed on those shown in Fig.1. That set consists of arrows depicting $\boldsymbol{v}(\boldsymbol{r}) = \frac{\hbar}{m_0}\nabla\varphi(\boldsymbol{r})$ at the various points distributed according to $|\psi(\boldsymbol{r})|^2$. Clearly, a linear superposition of those arrows is only possible as long as the velocities are within the non-relativistic regime. Otherwise the superposition is affected by spin-orbit coupling as a result of which $\varphi_\uparrow(\boldsymbol{r})$ and $\varphi_\downarrow(\boldsymbol{r})$ now become different. This point will be taken up again in Section 31.

26. Generalizing One-particle Quantum Mechanics by Including Particle Spin

Several suggestions have already been made to incorporate particle spin into a theory that is akin to the ideas of the present article (s. e. g. Dankel [69], Dohrn et al. [70], Nelson [71]). We believe, however, that our approach offers - in the spirit of a statement by v. Weizsäcker [68],[4] - "something immediately comprehensible".

The points associated with the two subset of arrows in Fig.1 define probability densities $\rho_\uparrow(\boldsymbol{r})$ and $\rho_\downarrow(\boldsymbol{r})$ with $\uparrow$ and $\downarrow$ referring to the respective direction of the spin moment. Both densities integrate to unity

$$\int \rho_{\uparrow(\downarrow)}(\boldsymbol{r})\, d^3r = 1\,. \tag{208}$$

To keep the formalism flexible at the outset we consider a situation where the total probability density is not yet a pure "up" or "down" density

$$\rho(\boldsymbol{r}) = |a|^2\,\rho_\uparrow(\boldsymbol{r}) + |b|^2\,\rho_\downarrow(\boldsymbol{r}) \tag{209}$$

with a and b denoting coefficients whose modulus squares sum up to unity

$$|a|^2 + |b|^2 = 1\,. \tag{210}$$

It is obviously not possible to partition the wave function analogously: $\psi(\boldsymbol{r}) = a\,\psi_\uparrow(\boldsymbol{r}) + b\,\psi_\downarrow(\boldsymbol{r})$ because $\psi^*(\boldsymbol{r})\psi(\boldsymbol{r})$ would contain cross-terms. However, if one introduces a two-component spinor of the form

$$\underline{\psi}(\boldsymbol{r}) = \begin{pmatrix} a\,\psi_\uparrow(\boldsymbol{r}) \\ b\,\psi_\downarrow(\boldsymbol{r}) \end{pmatrix} = a\,\psi_\uparrow(\boldsymbol{r}) \begin{pmatrix} 1 \\ 0 \end{pmatrix} + b\,\psi_\downarrow(\boldsymbol{r}) \begin{pmatrix} 0 \\ 1 \end{pmatrix} \tag{211}$$

and its adjoint $\underline{\psi}^\dagger(\boldsymbol{r}) = \left(a^*\psi_\uparrow^*(\boldsymbol{r}),\, b^*\psi_\downarrow^*(\boldsymbol{r})\right)$ where

$$\int |\psi_{\uparrow(\downarrow)}(\boldsymbol{r})|^2\, d^3r = 1\,, \tag{212}$$

[4] "... What we are dissatisfied with is basically not that the old perceptions have failed but that they could not be substituted by something immediately comprehensible."

one obtains as intended

$$\underline{\psi}^{\dagger}(\boldsymbol{r})\underline{\psi}(\boldsymbol{r}) = |a|^2\,|\psi_{\uparrow}(\boldsymbol{r})|^2 + |b|^2\,|\psi_{\downarrow}(\boldsymbol{r})|^2 = \rho(\boldsymbol{r})$$

and

$$\int \underline{\psi}^{\dagger}(\boldsymbol{r})\underline{\psi}(\boldsymbol{r})\,d^3r = 1\,. \tag{213}$$

We consider the Bohr magneton as known from experiments although it will later turn out to be derivable from Dirac's theory (Section 31). Thus, the energy densities of the interaction with the magnetic field for "up"- and "down"-spin may be cast as

$$-\mu_B B_z\,|a|^2\,\psi_{\uparrow}^{*}(\boldsymbol{r})\,\psi_{\uparrow}(\boldsymbol{r}) \quad \text{and} \quad +\mu_B B_z\,|b|^2\,\psi_{\downarrow}^{*}(\boldsymbol{r})\,\psi_{\downarrow}(\boldsymbol{r})\,.$$

from which the total interaction density results as

$$u_{magn.}(\boldsymbol{r}) = -\underline{\psi}^{\dagger}(\boldsymbol{r})\,\mu_B\underline{\underline{B}}\,\underline{\psi}(\boldsymbol{r}) \tag{214}$$

where we have introduced a matrix

$$\underline{\underline{B}} = \begin{pmatrix} B_z & 0 \\ 0 & -B_z \end{pmatrix}. \tag{215}$$

Likewise, we may cast the non spin-dependent energy density of the electron as

$$\underline{\psi}^{\dagger}(\boldsymbol{r})\,\widehat{H}\,\underline{\psi}(\boldsymbol{r})$$

where

$$\widehat{H} = \widehat{H}_0 + V(\boldsymbol{r}) \quad \text{and} \quad \widehat{H}_0 = \frac{(\widehat{\boldsymbol{p}} - e\boldsymbol{A}(\boldsymbol{r}))^2}{2\,m_0}\,, \tag{216}$$

and with $V(\boldsymbol{r})$ denoting some potential in which the electron moves.

27. The Time-Dependent Non-Relativistic Pauli Equation

The basic two constituents of our approach, viz. $|\psi(\boldsymbol{r},t)|^2$ and $\nabla\varphi(\boldsymbol{r},t)$ remain unaffected by our incorporation of spin. Hence, it is completely in line with the conceptual idea of our approach to assume that the two theorems of Ehrenfest stay unaffected as well. That means, according to Ehrenfest's Second Theorem

$$\langle \boldsymbol{v}\rangle = \frac{d}{dt}\langle \boldsymbol{r}\,\rangle = \int \Big[\dot{\underline{\psi}}^{\dagger}(\boldsymbol{r},t)\,\boldsymbol{r}\,\underline{\psi}(\boldsymbol{r},t) + \underline{\psi}^{\dagger}(\boldsymbol{r},t)\,\boldsymbol{r}\,\dot{\underline{\psi}}(\boldsymbol{r},t)\Big]\,d^3r\,, \tag{217}$$

and we have alternatively from Eq.(96)

$$\langle \boldsymbol{v}\rangle = \frac{1}{m_0}\int \underline{\psi}^{\dagger}(\boldsymbol{r},t)\widehat{\boldsymbol{P}}\,\underline{\psi}(\boldsymbol{r},t)\,, \tag{218}$$

which holds without modification also for the spinors we have introduced. Exploiting the relation

$$[\widehat{H}_0\, \boldsymbol{r} - \boldsymbol{r}\, \widehat{H}_0]\, \underline{\psi}(\boldsymbol{r},t) = -i\,\frac{\hbar}{m_0}\,\widehat{\boldsymbol{P}}\,\underline{\psi}(\boldsymbol{r},t)\,,$$

which follows from simply applying the chain rule, we may combine Eqs.(217) and (218) to obtain

$$\int \left(\left[\widehat{H}_0 + i\hbar\,\frac{\partial}{\partial t}\right] \underline{\psi}^\dagger(\boldsymbol{r},t)\right) \boldsymbol{r}\,\underline{\psi}(\boldsymbol{r},t)\,d^3r$$
$$-\int \underline{\psi}^\dagger(\boldsymbol{r},t)\,\boldsymbol{r}\left[\widehat{H}_0 - i\,\hbar\frac{\partial}{\partial t}\right] \underline{\psi}(\boldsymbol{r},t)\,d^3r = 0\,. \tag{219}$$

This equation holds for any t if $\underline{\psi}(\boldsymbol{r},t)$ satisfies

$$\left[\widehat{H}_0 - i\,\hbar\frac{\partial}{\partial t}\right] \underline{\psi}(\boldsymbol{r},t) = -\underline{\underline{D}}\,\underline{\psi}(\boldsymbol{r}) - F(\boldsymbol{r},t)\,\underline{\psi}(\boldsymbol{r},t)\,, \tag{220}$$

and correspondingly

$$\left[\widehat{H}_0 + i\,\hbar\frac{\partial}{\partial t}\right] \underline{\psi}^\dagger(\boldsymbol{r},t) = -\underline{\psi}^\dagger\,(\boldsymbol{r})\,\underline{\underline{D}} - \underline{\psi}^\dagger(\boldsymbol{r})\,F(\boldsymbol{r},t)\,,$$

where $F(\boldsymbol{r},t)$ is some integrable real-valued function and $\underline{\underline{D}}$ denotes some unitary 2×2-matrix that will be specified later to meet requirements of Ehrenfest's first theorem.

The expectation value of the force exercised on an electron which moves in a potential $V(\boldsymbol{r})$ and simultaneously - through its magnetic moment - feels a force in a spatially varying magnetic field $\boldsymbol{B}(z,t) = B_z(z,t)\,\boldsymbol{e}_z$ may be cast as

$$\langle \boldsymbol{F}\rangle = -\int \underline{\psi}^\dagger(\boldsymbol{r},t)\,\left[\nabla\{V(\boldsymbol{r}) + \mu_B\,\underline{\underline{B}}\}\right]\,\underline{\psi}(\boldsymbol{r},t)\,d^3r - \langle e\dot{\boldsymbol{A}}(\boldsymbol{r},t)\rangle\,. \tag{221}$$

The appearance of the induction-derived force $-\langle e\dot{\boldsymbol{A}}(\boldsymbol{r},t)\rangle$ is a consequence of Eq.(93).

We perform an integration by parts on the first integral and obtain

$$\langle \boldsymbol{F}\rangle = \int \left[\nabla\underline{\psi}^\dagger(\boldsymbol{r},t)\right]\{V(\boldsymbol{r}) + \mu_B\,\underline{\underline{B}}\}\,\underline{\psi}(\boldsymbol{r},t)\,d^3r$$
$$+\int \underline{\psi}^\dagger(\boldsymbol{r},t)\,\{V(\boldsymbol{r}) + \mu_B\,\underline{\underline{B}}\}\,\nabla\underline{\psi}(\boldsymbol{r},t)\,d^3r - \langle e\dot{\boldsymbol{A}}(\boldsymbol{r},t)\rangle\,. \tag{222}$$

From Eq.(96) we have

$$\frac{d}{dt}\,\langle \boldsymbol{p}\rangle = \frac{d}{dt}\int \underline{\psi}^\dagger(\boldsymbol{r},t)\,[-i\hbar\nabla - e\,\boldsymbol{A}(\boldsymbol{r},t)]\,\underline{\psi}(\boldsymbol{r},t)\,d^3r$$

which we rewrite

$$\frac{d}{dt}\,\langle \boldsymbol{p}\rangle = \int \left(-i\hbar\frac{\partial}{\partial t}\underline{\psi}^\dagger\right)\nabla\underline{\psi}\,d^3r + \underbrace{\int \underline{\psi}^\dagger\,\nabla\left(-i\hbar\frac{\partial}{\partial t}\underline{\psi}\right)\,d^3r}_{=-\int \nabla\underline{\psi}^\dagger\left(-i\hbar\frac{\partial}{\partial t}\underline{\psi}\right)d^3r} - \int \underline{\psi}^\dagger\,\underline{\psi}\,e\,\dot{\boldsymbol{A}}\,d^3r\,.$$

On forming $\langle \boldsymbol{F} \rangle - \langle \dot{\boldsymbol{p}} \rangle = 0$ (Ehrenfest's First Theorem) we obtain

$$\int \left(\nabla \underline{\psi}^{\dagger}\right) \left\{V + \mu_B \underline{\underline{B}} - i\hbar \frac{\partial}{\partial t}\right\} \underline{\psi}\, d^3 r$$
$$+ \int \left[\left(+i\hbar \frac{\partial}{\partial t} \underline{\psi}^{\dagger}\right) \nabla \underline{\psi} + \underline{\psi}^{\dagger} \{V + \mu_B \underline{\underline{B}}\} \nabla \underline{\psi}\right] d^3 r = 0\,. \tag{223}$$

If we here eliminate the time-derivatives using Eqs.(220) and identify $\underline{\underline{D}}$ with $\mu_B \underline{\underline{B}}$ this equation takes the form:

$$- \int \left[(\nabla \underline{\psi}^{\dagger})\, \widehat{H}_0\, \underline{\psi} + (\widehat{H}_0\, \underline{\psi}^{\dagger}) \nabla \underline{\psi}\right] d^3 r + \int \left[(\nabla \underline{\psi}^{\dagger}(\boldsymbol{r}, t))\, \underline{\psi}(\boldsymbol{r}, t) + \underline{\psi}^{\dagger}(\boldsymbol{r}, t)\, \nabla \underline{\psi}(\boldsymbol{r}, t)\right]$$
$$\times \{V(\boldsymbol{r}) - F(\boldsymbol{r}, t)\}\, d^3 r = 0\,.$$

Since the first integral vanishes we arrive at

$$\int \nabla \rho(\boldsymbol{r}, t) \{V(\boldsymbol{r}) - F(\boldsymbol{r}, t)\}\, d^3 r = 0 \quad \forall t\,. \tag{224}$$

We first consider the possibility that the expression in curly brackets does not vanish, but the integral does. As we have emphasized in defining probability densities $\rho(\boldsymbol{r}, t)$ and average velocities $\boldsymbol{v}(\boldsymbol{r}, t)$ through Eqs.(4) and (5), non-stationary states require a certain sample-time T of the particle under study to allow its time-derived probability density $\rho(\boldsymbol{r}, t)$ to become quasi-stationary. Hence, if we introduce at $t = t_0$ a small perturbational potential $V(\boldsymbol{r}) \rightarrow V(\boldsymbol{r}) + \delta v(\boldsymbol{r}, t)$ where $t_0 \leq t \ll T$, the probability density $\rho(\boldsymbol{r}, t)$, and thus its gradient remain practically unaffected, but the bracketed expression is now definitely different. We are hence led to conclude that Eq.(224) can only be satisfied if $F(\boldsymbol{r}, t) \equiv V(\boldsymbol{r})$ holds for any time. That means - because of Eq.(220) - that the spinor function $\underline{\psi}(\boldsymbol{r}, t)$ solves

$$\left[\widehat{H}_0 + V(\boldsymbol{r}) + \mu_B \underline{\underline{B}}\right] \underline{\psi}(\boldsymbol{r}, t) = i\,\hbar \frac{\partial}{\partial t} \underline{\psi}(\boldsymbol{r}, t) \tag{225}$$

with $\widehat{H}_0$ as defined in Eq.(216). This constitutes the time-dependent non-relativistic Pauli equation.

28. The Cayley-Klein Parameters and Pauli Spin Matrices

We want to adapt Eq.(225) to a situation where the direction of the magnetic field no longer coincides with the z-axis of the coordinate system. This can be achieved by exploiting a surprising alternative to the standard form of rotating the coordinate system employing orthogonal 3×3 matrices. The idea goes back to Felix Klein (s. e. g. H. Goldstein [72]) and is related to earlier work of Cayley. He considers the rotation of the coordinate system $(x, y, z \rightarrow x', y', z')$ to be performed in three steps described by the Euler angles ϕ, θ and ψ shown in Fig.7. Instead of representing the position vector $\boldsymbol{r}$ by a column matrix he uses a 2×2-matrix $\underline{\underline{P}}(\boldsymbol{r})$ of the form

$$\underline{\underline{P}}(x, y, z) := \begin{pmatrix} z & x - i\,y \\ x + i\,y & -z \end{pmatrix}. \tag{226}$$

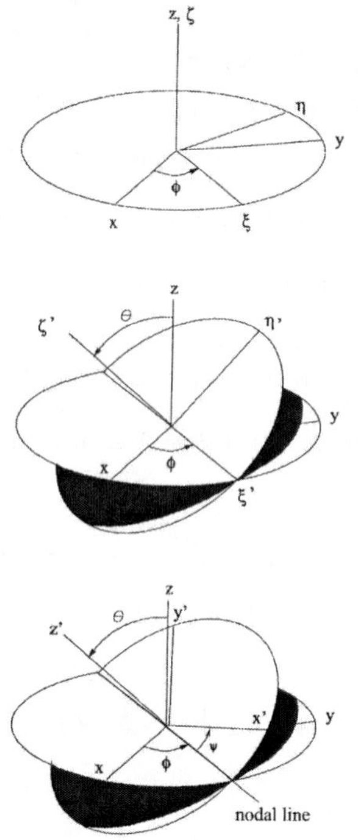

Figure 7. Euler angles.

In place of the standard 3×3-rotation matrix one now has a 2×2-unimodular matrix

$$\underline{\underline{Q}}(\theta, \phi, \psi) = \begin{pmatrix} \alpha & \beta \\ \gamma & \delta \end{pmatrix} \tag{227}$$

whose elements - the so-called Cayley-Klein parameters - are connected with the Euler angles through

$$\begin{aligned} \alpha &= e^{\frac{i}{2}(\psi+\phi)} \cos\frac{\theta}{2} \\ \beta &= i e^{\frac{i}{2}(\psi-\phi)} \sin\frac{\theta}{2} \\ \gamma &= i e^{-\frac{i}{2}(\psi-\phi)} \sin\frac{\theta}{2} \\ \delta &= e^{-\frac{i}{2}(\psi+\phi)} \cos\frac{\theta}{2}\,. \end{aligned} \tag{228}$$

After the three steps of the rotation have been performed the original position vector $\boldsymbol{r} = (x, y, z)$ is now associated with the new coordinates x', y', z' that may be obtained from the transform

$$\underline{\underline{Q}}\,\underline{\underline{P}}\,\underline{\underline{Q}}^{+} = \underline{\underline{P}}'(x', y', z') = \begin{pmatrix} z' & x' - i\,y' \\ x' + i\,y' & -z' \end{pmatrix} \tag{229}$$

where $\underline{\underline{Q}}^{+}$ denotes the adjoint of $\underline{\underline{Q}}$, and we have

$$\underline{\underline{Q}}^{+}\,\underline{\underline{Q}} = \underline{\underline{Q}}\,\underline{\underline{Q}}^{+} = \underline{\underline{1}} \tag{230}$$

The matrix $\underline{\underline{B}}$ had been defined in Eq.(215) as

$$\underline{\underline{B}} = \begin{pmatrix} B_z & 0 \\ 0 & -B_z \end{pmatrix} = B_z \begin{pmatrix} 1 & 0 \\ 0 & -1 \end{pmatrix} . \tag{231}$$

In Klein's representation the point $\boldsymbol{r} = (0, 0, z)$ attains the analogous form

$$\underline{\underline{P}}(\boldsymbol{r}) = z \begin{pmatrix} 1 & 0 \\ 0 & -1 \end{pmatrix} .$$

Hence, $\underline{\underline{B}}$ has to be required to transform under coordinate rotation as $\underline{\underline{P}}$:

$$\underline{\underline{B}}' = \underline{\underline{Q}}\,\underline{\underline{B}}\,\underline{\underline{Q}}^{+} . \tag{232}$$

For a general orientation of the coordinate system with respect to the magnetic field $\underline{\underline{B}}$ has the form analogous to $\underline{\underline{P}}$ in Eq.(226), viz.

$$\underline{\underline{B}} = \begin{pmatrix} B_z & B_x - i\,B_y \\ B_x + i\,B_y & -B_z \end{pmatrix} . \tag{233}$$

This matrix can be decomposed

$$\underline{\underline{B}} = B_x\,\underline{\underline{\sigma}}_x + B_y\,\underline{\underline{\sigma}}_y + B_z\,\underline{\underline{\sigma}}_z\,, \tag{234}$$

where the three matrices on the right-hand side are just the Pauli spin matrices

$$\underline{\underline{\sigma}}_x = \begin{pmatrix} 0 & 1 \\ 1 & 0 \end{pmatrix} \quad \underline{\underline{\sigma}}_y = \begin{pmatrix} 0 & -i \\ i & 0 \end{pmatrix} \quad \underline{\underline{\sigma}}_z = \begin{pmatrix} 1 & 0 \\ 0 & -1 \end{pmatrix} . \tag{235}$$

They are commonly lumped together in the form of a vector

$$\vec{\sigma} = \underline{\underline{\sigma}}_x\,\boldsymbol{e}_x + \underline{\underline{\sigma}}_y\,\boldsymbol{e}_y + \underline{\underline{\sigma}}_z\,\boldsymbol{e}_z\,. \tag{236}$$

The matrix $\underline{\underline{B}}$ in Eq.(234) may therefore be cast as

$$\underline{\underline{B}} = \vec{\sigma}\cdot\boldsymbol{B}\,. \tag{237}$$

The Pauli equation (225) then attains the familiar form

$$\left[\widehat{H}_0 + V(\boldsymbol{r}) + \mu_B\,\vec{\sigma}\cdot\boldsymbol{B}\right]\underline{\psi}(\boldsymbol{r},t) = i\hbar\,\frac{\partial}{\partial t}\,\underline{\psi}(\boldsymbol{r},t)\,. \tag{238}$$

Actually, the spinor in this equation should be marked by a prime because it has changed under the transform as well. We have dropped the prime for simplicity. Since the density of the magnetic interaction energy is, of course, invariant under rotation of the coordinate system

$$u_{magn.}(\boldsymbol{r}) = u'_{magn.}(\boldsymbol{r}'),$$

it can be shown then that the new ψ' is connected with the original ψ through

$$\underline{\psi}' = \underline{\underline{Q}}\,\underline{\psi} \tag{239}$$

and correspondingly

$$\underline{\psi}'^{\dagger} = \left(\underline{\underline{Q}}\,\underline{\psi}\right)^{\dagger} = \underline{\psi}^{\dagger}\,\underline{\underline{Q}}^{+}.$$

This becomes obvious from forming

$$\underline{\psi}'^{\dagger}\,\underline{\underline{B}}'\,\underline{\psi}' \;(=: u'_{mag}(\boldsymbol{r}',t)) = \underline{\psi}^{\dagger}\,\underline{\underline{Q}}^{+}\,\underline{\underline{B}}'\,\underline{\underline{Q}}\,\underline{\psi}.$$

If we insert Eq.(232) on the right-hand side we obtain

$$\underline{\psi}^{\dagger}\,\underline{\underline{Q}}^{+}\,\underline{\underline{B}}'\,\underline{\underline{Q}}\,\underline{\psi} = \underline{\psi}^{\dagger}\,\underbrace{\underline{\underline{Q}}^{+}\,\underline{\underline{Q}}}_{=\underline{\underline{1}}}\,\underline{\underline{B}}\,\underbrace{\underline{\underline{Q}}^{+}\,\underline{\underline{Q}}}_{=\underline{\underline{1}}}\,\underline{\psi} = u_{mag}(\boldsymbol{r},t).$$

Because of

$$\underline{\psi}'^{\dagger}\,\underline{\psi}' = \underline{\psi}^{\dagger}\,\underline{\underline{Q}}^{+}\,\underline{\underline{Q}}\,\underline{\psi} = \underline{\psi}^{\dagger}\,\underline{\psi} = \rho(\boldsymbol{r},t)$$

the probability density is also invariant under rotation of the coordinate system which is consistent with our idea of a spin-defining motion decomposition at the beginning of our considerations. Moreover, if the state of the particle in the original coordinate system has the form

$$\psi_{\uparrow}(\boldsymbol{r})\begin{pmatrix}1\\0\end{pmatrix} \quad \text{or} \quad \psi_{\downarrow}(\boldsymbol{r})\begin{pmatrix}0\\1\end{pmatrix}, \tag{240}$$

it becomes after coordinate rotation

$$\underline{\psi}'_{\uparrow}(\boldsymbol{r}) = \psi_{\uparrow}(\boldsymbol{r})\,\underline{\underline{Q}}(\boldsymbol{r})\begin{pmatrix}1\\0\end{pmatrix} = \psi_{\uparrow}(\boldsymbol{r})\left[\alpha(\boldsymbol{r})\begin{pmatrix}1\\0\end{pmatrix} - \beta^{*}(\boldsymbol{r})\begin{pmatrix}0\\1\end{pmatrix}\right]$$

or

$$\underline{\psi}'_{\downarrow}(\boldsymbol{r}) = \psi_{\downarrow}(\boldsymbol{r})\,\underline{\underline{Q}}(\boldsymbol{r})\begin{pmatrix}0\\1\end{pmatrix} = \psi_{\downarrow}(\boldsymbol{r})\left[\beta(\boldsymbol{r})\begin{pmatrix}0\\1\end{pmatrix} + \alpha^{*}(\boldsymbol{r})\begin{pmatrix}1\\0\end{pmatrix}\right],$$

where $\alpha(\boldsymbol{r})$ and $\beta(\boldsymbol{r})$ are the Cayley-Klein parameters describing the rotation which we have allowed here to be different at different positions $\boldsymbol{r}$.

The spin orientation with respect to the direction of a magnetic field is already uniquely defined by the two angles θ and ϕ. Hence one is at liberty to choose ψ at will without loss of generality. It is convenient to set $\psi = -\pi/2$. We consider the projection of the unit vector $\boldsymbol{e}'_z$ onto the original x/y-plane where it makes an angle φ with the x-axis. This angle and the Euler-angle ϕ are interrelated

$$\phi = \varphi + \frac{\pi}{2}.$$

If one inserts this relation into Eqs.(227) and (228), $\underline{\underline{Q}}$ takes the familiar form

$$\underline{\underline{Q}}(\boldsymbol{r}) = \begin{pmatrix} \exp[\frac{i}{2}\varphi(\boldsymbol{r})] \cos\frac{\theta(\boldsymbol{r})}{2} & \exp[-\frac{i}{2}\varphi(\boldsymbol{r})] \sin\frac{\theta(\boldsymbol{r})}{2} \\ -\exp[\frac{i}{2}\varphi(\boldsymbol{r})] \sin\frac{\theta(\boldsymbol{r})}{2} & \exp[-\frac{i}{2}\varphi(\boldsymbol{r})] \cos\frac{\theta(\boldsymbol{r})}{2} \end{pmatrix}. \tag{241}$$

The above considerations on the magnetic interaction energy starting with the expression (214) carry over to the spin momentum

$$\langle S_z \rangle = \frac{\hbar}{2} \int \left[|a|^2 \, |\psi_\uparrow(\boldsymbol{r})|^2 - |b|^2 \, |\psi_\downarrow(\boldsymbol{r})|^2 \right] d^3r\,. \tag{242}$$

The symbol S_z refers to the effective spin moment in the z-direction with respect to which the functions $\psi_{\uparrow(\downarrow)}(\boldsymbol{r})$ have been defined. In complete analogy to (231) this expression can be compactified by introducing

$$\underline{\underline{S}}_z = \frac{\hbar}{2} \begin{pmatrix} 1 & 0 \\ 0 & -1 \end{pmatrix} \tag{243}$$

so that

$$\langle S_z \rangle = \int \underline{\psi}^+(\boldsymbol{r}) \, \underline{\underline{S}}_z \, \underline{\psi}(\boldsymbol{r}) \, d^3r\,. \tag{244}$$

In case that the functions $\psi_{\uparrow(\downarrow)}(\boldsymbol{r})$ refer to a z'-direction that belongs to a rotated coordinate sytem x', y', z', we have in analogy to Eq.(232)

$$\underline{\underline{S}}_{z'} = \underline{\underline{Q}}\,\underline{\underline{S}}_z\,\underline{\underline{Q}}^+ .$$

If we use the analogous relations pertaining to Eqs.(233) up to (237) we may cast $\underline{\underline{S}}_{z'}$ as

$$\underline{\underline{S}}_{z'} = \frac{\hbar}{2} \left[\hat{\alpha}_x \, \underline{\underline{\sigma}}_x + \hat{\alpha}_y \, \underline{\underline{\sigma}}_y + \hat{\alpha}_z \, \underline{\underline{\sigma}}_z \right] \tag{245}$$

with $\hat{\alpha}_x, \hat{\alpha}_y, \hat{\alpha}_z$ denoting the component of the unit vector $\boldsymbol{e}_{z'}$ in the z'-direction

$$\begin{aligned} \hat{\alpha}_x &= \cos\varphi \, \sin\theta \\ \hat{\alpha}_y &= \sin\varphi \, \sin\theta \\ \hat{\alpha}_z &= \cos\theta\,. \end{aligned}$$

It is convenient to introduce a vector $\vec{S}$ (commonly referred to as “spin operator”), which is analogous to $\vec{\sigma}$, by setting

$$\vec{S} = \frac{\hbar}{2} \vec{\sigma}\,. \tag{246}$$

Eq.(245) may then be cast

$$\underline{\underline{S}}_{z'} = \boldsymbol{e}_{z'} \cdot \vec{S}\,,$$

and hence we have

$$\langle S_{z'}\rangle = \int \underline{\psi}^{+}(\boldsymbol{r})\, \underline{\underline{S}}_{z'}\, \underline{\psi}(\boldsymbol{r})\, d^3r = \boldsymbol{e}_{z'} \cdot \langle \vec{S}\rangle$$

where

$$\langle \vec{S}\rangle = \int \underline{\psi}^{+}(\boldsymbol{r})\, \vec{S}\, \underline{\psi}(\boldsymbol{r})\, d^3r\,. \tag{247}$$

If $\underline{\psi}(\boldsymbol{r},t)$ has the form (240), Eq.(247) yields $\langle \vec{S}\rangle = \pm\frac{\hbar}{2}\,\boldsymbol{e}_z$. On the other hand, if $\underline{\psi}(\boldsymbol{r},t)$ possesses two non-vanishing components, there will always be a coordinate system that is rotated with respect to the present one, in which $\langle \vec{S}\rangle$ becomes $\pm\frac{\hbar}{2}\,\boldsymbol{e}_{z'}$. One only has to turn the pertinent z'-axis in the plane spanned by the original direction of $\langle \vec{S}\rangle$ and the original z-axis until $\boldsymbol{e}_{z'}$ is parallel or anti-parallel to $\langle \vec{S}\rangle$.

29. Spin Precession in a Magnetic Field

So far we have assumed the magnetic field and the spin direction to be collinear. As an example for a non-collinear situation we consider an electron that is bound within an atom where it is initially exposed to a magnetic field along some direction. We omit here discussing the details of its spin alignment due to some minute time dependent perturbations and simply assume that it has eventually attained a stationary spinor state in which its spin momentum points parallel or anti-parallel to the direction of the magnetic field. If one now changes non-adiabatically the direction (and in general inevitably also the magnitude) of the magnetic field, the spin momentum can - without an appropriate external torque - not adjust to the new field direction, and hence the previously existing collinearity no longer obtains. As we shall show by discussing the pertinent solution to the time-dependent Pauli equation (238), the spin momentum now precesses about the new direction of the magnetic field in a completely classical way.

We identify the initial direction of the magnetic field with the z'-axis of a "primed" coordinate system in which the spin-aligned state of the electron has the form

$$\underline{\psi}'(\boldsymbol{r}') = \psi_0'(\boldsymbol{r}')\begin{pmatrix}1\\0\end{pmatrix} \tag{248}$$

where $\psi_0'(\boldsymbol{r}')$ is the energetically lowest lying solution to the Schrödinger equation of the one-particle system under study. We denote this solution by $\psi_0(\boldsymbol{r})$ in the unprimed coordinate system in which the new magnetic field lies along the z-direction and in which the spinor (248) can be cast as

$$\underline{\psi}(\boldsymbol{r}) = \underline{\underline{Q}}^{+}\,\underline{\psi}'(\boldsymbol{r}') = \underbrace{\psi_0(\boldsymbol{r})\,e^{-i\frac{\varphi}{2}}\cos\frac{\theta}{2}\begin{pmatrix}1\\0\end{pmatrix}}_{=\underline{\psi}_{0\uparrow}(\boldsymbol{r})} + \underbrace{\psi_0(\boldsymbol{r})\,e^{i\frac{\varphi}{2}}\sin\frac{\theta}{2}\begin{pmatrix}0\\1\end{pmatrix}}_{=\underline{\psi}_{0\downarrow}(\boldsymbol{r})}, \tag{249}$$

where $\theta, \varphi, \psi (= \ 0)$ are the Euler angles that refer to the interrelation $(x', y', z') \to (x, y, z))$. Hence we have

$$\underline{\psi}(\boldsymbol{r}) = \underline{\psi}_{0\uparrow}(\boldsymbol{r}) + \underline{\psi}_{0\downarrow}(\boldsymbol{r})\,. \tag{250}$$

Note that the unit spinors in Eq.(249) are referenced to the new z-axis!

We now consider the Pauli equation (238) for the time-independent case in the absence of a magnetic field in which case $\underline{\psi}_{0\uparrow(\downarrow)}(\boldsymbol{r})$ are independent degenerate solutions and $\psi_0(\boldsymbol{r})$ satisfies the associated Schrödinger equation

$$\widehat{H}(\boldsymbol{r})\,\psi_0(\boldsymbol{r}) = E_0\,\psi_0(\boldsymbol{r})\,.$$

For $B_z \neq 0$ the two spinors belong to different energies $E_{0\uparrow(\downarrow)} = E_0 \pm \mu_B\,B_z$ and their sum does not satisfy the time-independent Pauli equation any more. However

$$\underline{\psi}(\boldsymbol{r},t) = \underline{\psi}_{0\uparrow}(\boldsymbol{r})\,e^{-\frac{i}{\hbar}E_{0\uparrow}\,t} + \underline{\psi}_{0\downarrow}(\boldsymbol{r})\,e^{-\frac{i}{\hbar}E_{0\downarrow}\,t} \tag{251}$$

solves the time-dependent Pauli equation (225) if we disregard effects of second and higher order in the magnetic field. We now insert the definitions of $\underline{\psi}_{0\uparrow(\downarrow)}(\boldsymbol{r})$ from above and obtain

$$\underline{\psi}(\boldsymbol{r},t) = \psi_0(\boldsymbol{r})\left[e^{-i\frac{(\varphi-\omega_L\,t)}{2}}\,\cos\frac{\theta}{2}\begin{pmatrix}1\\0\end{pmatrix} + e^{i\frac{(\varphi-\omega_L\,t)}{2}}\,\sin\frac{\theta}{2}\begin{pmatrix}0\\1\end{pmatrix}\right]e^{-\frac{i}{\hbar}E_0\,t}. \tag{252}$$

Here we have made use of $E_{0\uparrow(\downarrow)} = E_0 \pm \mu_B\,B_z$ and introduced the frequency ω_L which is defined through

$$E_{0\uparrow} - E_{0\downarrow} = 2\,\mu_B B_z = \hbar\omega_L\,. \tag{253}$$

In complete analogy to Eq.(249) we form

$$\underline{\psi}^{+}(\boldsymbol{r},t) = \underline{\psi}'^{+}(\boldsymbol{r}',t)\,\underline{\underline{Q}}$$

and calculate the expectation value of $\vec{S}$

$$\begin{aligned}\langle\vec{S}\rangle &= \int \underline{\psi}^{+}(\boldsymbol{r},t)\,\vec{S}\,\underline{\psi}(\boldsymbol{r},t)\,d^3r = \int \underline{\psi}'^{+}(\boldsymbol{r}',t)\,\underline{\underline{Q}}\,\vec{S}\,\underline{\underline{Q}}^{+}\,\underline{\psi}'(\boldsymbol{r}',t)\,d^3r' \\ &= \frac{\hbar}{2}\left[\cos(\varphi-\omega_L t)\,\sin\theta\,\boldsymbol{e}_x + \sin(\varphi-\omega_L t)\,\sin\theta\,\boldsymbol{e}_y + \cos\theta\,\boldsymbol{e}_z\right].\end{aligned} \tag{254}$$

Thus, the vector $\langle\vec{S}\rangle$ of the spin momentum moves on a circular cone with an apex angle of $2\,\theta$ about the direction of the magnetic field and its projection onto the x/y-plane rotates at an angular frequency ω_L, the "Larmor frequency", about the z-axis. According to Eq.(253) this frequency is given by

$$\omega_L = \frac{\mu_B B_z}{\hbar/2}\,. \tag{255}$$

The spin precession is completely analogous to that of a classical spinning top which rotates about its symmetry axis at an angular frequency ω and is exposed to the gravitational field of the earth. The precession frequency ω_P is in this case given by

$$\omega_P = \frac{F\,r_s}{L}\,,$$

where L denotes the absolute value of the angular momentum, F is the absolute value of the gravitational force acting on the top's centroid, and r_s is the distance of the centroid from the point of support. In case one has instead of a gravitational field a magnetic field and if the spinning top possesses a magnetic moment μ_B, one has $F\, r_s = \mu_B B_z$. Inserting this into the classical equation for ω_P and setting $L = \hbar/2$ one obtains exactly the expression (255) for the Larmor frequency. If the magneto-mechanical parallelism would also hold for the spin momentum, that is if g in Eq.(207) were equal to one, the magnetic moment would be $\frac{1}{2}\,\mu_B$, and the precession frequency would be smaller by a factor of 2 in striking disagreement with the experiment.

The completely classical behavior of a precessing spin moment in a magnetic field can also be made evident by the following consideration.

Using Eqs.(207) and (254) we may express the spin-derived magnetic moment $\vec{M}_{Spin}$ as

$$\vec{M}_{Spin} = \mu_B\,\left[\cos(\varphi - \omega_L t)\,\sin\theta\,\boldsymbol{e}_x + \sin(\varphi - \omega_L t)\,\sin\theta\,\boldsymbol{e}_y + \cos\theta\,\boldsymbol{e}_z\right] .$$

The time derivative of Eq.(254) can be written

$$\frac{d}{dt} < \vec{S} >= \mu_B\, B_z\,\left[\sin(\varphi - \omega_L t)\,\sin\theta\,\boldsymbol{e}_x - \cos(\varphi - \omega_L t)\,\sin\theta\,\boldsymbol{e}_y\right] , \qquad (256)$$

where we have used $\omega_L = 2\mu\, B_z/\hbar$. We observe that $\boldsymbol{B} = B_z\,\boldsymbol{e}_z$ and

$$\boldsymbol{e}_x \times \boldsymbol{e}_z = -\boldsymbol{e}_y\,; \quad \boldsymbol{e}_y \times \boldsymbol{e}_z = \boldsymbol{e}_x\,; \quad \boldsymbol{e}_z \times \boldsymbol{e}_z = 0\,.$$

Hence, the right-hand side of Eq.(256) can be cast as

$$\mu_B\, B_z\,\left[\sin(\varphi - \omega_L t)\,\sin\theta\,\boldsymbol{e}_x - \cos(\varphi - \omega_L t)\,\sin\theta\,\boldsymbol{e}_y\right] = \vec{M}_{Spin} \times \boldsymbol{B}\,.$$

The result may be written

$$\frac{d}{dt} < \vec{S} >= \vec{M}_{Spin} \times \boldsymbol{B}\,. \qquad (257)$$

This is identical with the classical equation of motion describing the temporal behavior of a spinning top that is acted upon by a torque $\vec{M}_{Spin} \times \boldsymbol{B}$. It corresponds to Ehrenfest's First Theorem, and it is this equation (257) which governs the phenomena encountered in electron and nuclear spin resonance. (S. e. g. Slichter [73].) In applying magnetic resonance techniques one has to supplement Eq.(257) by perturbational terms that cause a change of the precession cone. An equation of this kind was put forward by Bloch [74] in 1945. If the atom is not exposed to a time-dependent perturbation the spin keeps precessing on the cone without changing its apex angle even when the strength of the magnetic field adiabatically increases or decreases. A change of the absolute value of $\boldsymbol{B}$ only changes the Larmor frequency ω_L.

As opposed to the impression that is commonly invited by even the most recent literature, Eq.(257) constitutes a purely quantum mechanical result and is in no ways "semi-classical" or "macroscopical". The fact that from our derivation $|\langle\vec{S}\rangle_z|$ may attain any value, seems to contradict the principle of "orientation quantization" according to which $|\langle\vec{S}\rangle_z|$ may only equal $\pm\hbar/2$. Clearly, if $\langle\vec{S}\rangle_z$ is not parallel or anti-parallel to $\boldsymbol{B}$ but rather precesses about the direction of the latter, the electron emits magnetic dipole radiation until its spin is aligned. But this is a weak interaction, and therefore the state of non-alignment may well be regarded as meta-stable in certain experimental situations.

Spin precession in a magnetic field exhibits a peculiar feature that relates to the occurrence of the argument $\frac{\varphi}{2}$ in the exponential functions of Eq.(249). To see that we assume $\underline{\psi}(\boldsymbol{r})$ to represent a wavepacket of a free particle that traverses a homogeneous magnetic field in an orthogonal direction. When the wavepacket enters the magnetic field the spin component perpendicular to the field may point in the x-direction which is also the direction of flight. We then have

$$\varphi = 0 \quad \text{and hence} \quad e^{\pm i\,\frac{\varphi}{2}} = 1\,.$$

During the flight θ stays constant. When the wavepacket leaves the magnetic field after a full precession period we have

$$\varphi = 2\pi \quad \text{which means} \quad e^{\pm i\,\frac{\varphi}{2}} = -1\,.$$

Hence $\underline{\psi}(\boldsymbol{r})$ has changed its sign, or one may just as well say, its phase has been shifted by π. However, as can be seen from Eq.(254), $\langle \vec{S} \rangle$ points in the same direction as at the beginning of the precession. This phase shift is well detectable in double-beam experiments with spin-polarized neutrons (s. e. g. Rauch [75], Werner et al. [76]).

30. A Theory of the Stern-Gerlach Experiment

"...Phenomena of this kind made physicists despair of finding any consistent space-time picture of what goes on the atomic and subatomic scale...many came to hold not only that it is difficult to find a coherent picture but that it is wrong to look for one..."

John Bell [77]

Quite a few attempts have already been made on a theory of the Stern-Gerlach (SG-) experiment [78]. For a recent rather complete update of the pertinent literature see Home et al.[79]. But a coherent picture of the fundamental mechanism is still missing. Most physicists seem to favor the idea that the electronic state of the atom on entering the magnet constitutes a linear combination of spin states "up" and "down", and the modulus square of the associated coefficients defines the probability of the atom for being either pulled up or down, that is parallel or anti-parallel to the magnetic field gradient. On detection of the atom in the "up"- or "down"-beam the atomic wave function collapses onto the respective component of the linear combination. From our point of view this is unjustifiably associating the process of detection with some mystical influence of "observing", based on pure claim: the atomic beam would behave differently if it would not be detected. By contrast, we believe that the outcome of the experiment is completely determined by the time-dependent Pauli equation and is hence a result of a "quantum mechanics without observer".

Our approach implies a linear combination of spin states as well, that is we describe the electronic 1s-state of the atom that we shall consider below by

$$\underline{\psi}_{atom}(\boldsymbol{r},t) = \psi_{1s}(\boldsymbol{r} - \boldsymbol{v}\,t)\left[a_\uparrow \begin{pmatrix}1\\0\end{pmatrix} + a_\downarrow \begin{pmatrix}0\\1\end{pmatrix}\right]$$

where $\boldsymbol{v}$ denotes the velocity of the atom, and the coefficients $a_\uparrow, a_\downarrow$ have the property $|a_\uparrow|^2 + |a_\downarrow|^2 = 1$. The unit spinors are referenced to the direction of the field gradient $\frac{\partial B_z}{\partial z}\,\boldsymbol{e}_z$. Hence, the expectation value of the force acting on the atom in the SG-magnet is given by Eq.(221) if we neglect the induction derived term and assume electrostatic forces being absent

$$\langle \boldsymbol{F}_{atom} \rangle = \mu_B \int \underline{\psi}^{\dagger}_{atom}(\boldsymbol{r},t)\,\frac{\partial B_z}{\partial z}\,\underline{\psi}_{atom}(\boldsymbol{r},t)\,d^3r\,\boldsymbol{e}_z\,.$$

For simplicity we equate the field gradient to a constant so that $\langle \boldsymbol{F}_{atom} \rangle$ reduces to

$$\langle \boldsymbol{F}_{atom} \rangle = \mu_B\,\frac{\partial B_z}{\partial z}\,[|a_\uparrow|^2 - |a_\downarrow|^2]\,\boldsymbol{e}_z\,.$$

It can obviously attain any value between $-\mu_B\,\frac{\partial B_z}{\partial z}\,\boldsymbol{e}_z$ and $+\mu_B\,\frac{\partial B_z}{\partial z}\,\boldsymbol{e}_z$ depending on the value of the coefficients when the atom enters the magnet. Therefore a splitting into two well separated beams cannot possibly occur as long as there is no particular mechanism which inhibits a random distribution. In the following we shall outline such a possible mechanism.

We assume that the reader is sufficiently familiar with the essential features of the experimental setup. To simplify the line of argument we content ourselves with considering the experiment by Wrede [80] who used a primary beam of hydrogen atoms in a setup that was practically identical with that of Stern and Gerlach. Hydrogen offers the advantage of reducing the spin-orientation problem to that of a single electron. The standpoint we take here is akin to that of Mott and Massey [81] who remark: *"From these arguments we must conclude that it is meaningless to assign to the free electron a magnetic moment. It is a property of the electron that when it is bound in an S state in an atom, the atom has a magnetic moment."*[5]

The hydrogen atoms effuse from some source where they are (almost unavoidably) exposed to the terrestrial magnetic field or at least to the weak fringe field of the SG-magnet. That field causes a weak Zeeman-splitting of the spin up and spin down level of the electronic 1s-state. Because of the weakness of the splitting the two Zeeman-levels are at the temperature of the source equally occupied, that is, 50% of the effusing atoms have their electronic spins oriented parallel to the weak external field, the spins of the remaining 50% atoms are anti-parallel. As the atoms approach the SG-magnet they feel in a co-moving coordinate system a magnetic field whose field strength increases continuously and will in general change its direction. We assume for simplicity that the spin orientation is transverse and that the atom moves along the x-axis of a laboratory-fixed coordinate system so that changes of the spin orientation will only take place in the y/z-plane parallel to the respective plane of the co-moving coordinate system. As soon as the field direction in the co-moving coordinate system departs by a small angle $\delta\theta$ from the original direction of $\vec{B} = B_z\,\boldsymbol{e}_z$ at the onset of the atom's trajectory, a small y-component $\vec{B}_y = B_z\,\sin\delta\theta\,\boldsymbol{e}_y$ of the field appears as a consequence of which the magnetic moment of the atom experiences a torque $-\mu_B\,B_z\,\sin\delta\theta\,\boldsymbol{e}_\varphi$, where $\boldsymbol{e}_\varphi$ denotes the unit vector in the direction of increasing azimuth

[5] However, we want to modify this debatable statement by saying that also free electrons display a magnetic moment when they are exposed to a magnetic field where their motion perpendicular to the field becomes confined to a circular area of a certain diameter.

angle φ in the x/y-plane. This torque causes a change $\dot{\boldsymbol{L}}$ of the spin angular momentum

$$\dot{\boldsymbol{L}} = -\frac{\hbar}{2} \sin \delta\theta \, \omega_L \, \boldsymbol{e}_\varphi \,,$$

where we have used $2\mu_B \, B_z = \hbar \, \omega_L$ (Eq.(253)). Hence, the spin momentum starts processing about the new direction of the magnetic field. We ignore the slight tilt of the co-moving new x/y-plane perpendicular the new field direction.

We envisage a short time span for which we assume the changes of θ to be small so that

$$\sin \delta\theta \approx \delta\theta = \dot{\theta} \, t \,, \tag{258}$$

where $t = 0$ coincides with the beginning of the rotation of the field. The following considerations exploit the typical experimental condition that the precession frequency ω_L is some orders of magnitude larger than the speed of the field rotation. (In the terrestrial magnetic field of magnitude $\approx 5{\cdot}10^{-5}\,T$ the precession frequency of the electronic spin is about $10^6\,s^{-1}$. At an atomic speed of $10^5\,\mathrm{cm\,s^{-1}}$, a distance of about 10 cm and a maximum rotation angle of $\pi/2$ one has $\dot{\theta} \approx 10^4\,\mathrm{s^{-1}}$.) As will become apparent from the following calculations we may limit ourselves to a short time span comprising only few precession periods during which the magnetic field rotates only by a small angle ($\theta \ll 2\pi$) so that one is justified in assuming $\dot{\theta}$ to be constant:

$$\dot{\theta} = const.$$

The unit vector $\boldsymbol{e}_\varphi$ may be decomposed

$$\boldsymbol{e}_\varphi = -\boldsymbol{e}_x \, \sin \varphi + \boldsymbol{e}_y \, \cos \varphi \,. \tag{259}$$

At $t = 0$ we have $\varphi(t = 0) = -\frac{\pi}{2}$, that is $\boldsymbol{e}_\varphi = \boldsymbol{e}_x$. Thus, it is advisable to replace φ with $\varphi + \frac{\pi}{2}$, but we omit denoting the new azimuth angle differently. Hence we have $\varphi = 0$ for $t = 0$, and we obtain instead of Eq.(259)

$$\boldsymbol{e}_\varphi = \boldsymbol{e}_x \, \cos \varphi + \boldsymbol{e}_y \, \sin \varphi \,.$$

The spin precession that now occurs is anti-clockwise

$$\dot{\varphi} = -\omega_L \quad \text{that is} \quad \varphi = -\omega_L \, t \,.$$

Thus

$$\dot{\boldsymbol{L}} = \frac{\hbar}{2} \dot{\theta} \, \omega_L \, \left[\boldsymbol{e}_x \, t \, \cos \omega_L \, t - \boldsymbol{e}_y \, t \, \sin \omega_L \, t \right] \,.$$

This results in a change of the angular momentum after one precession period $T = 2\pi/\omega_L$

$$\Delta \boldsymbol{L} = \frac{\hbar}{2} \dot{\theta} \left[\omega_L \, \boldsymbol{e}_x \int_0^T t \, \cos \omega_L \, t \, dt - \boldsymbol{e}_y \, \omega_L \int_0^T t \, \sin \omega_L \, t \, dt \right] .$$

Hence, using

$$\int_0^{2\pi} \xi \, \sin \xi \, d\xi = -2\pi \quad \text{and} \quad \int_0^{2\pi} \xi \, \cos \xi \, d\xi = 0 \,,$$

we may $\Delta\boldsymbol{L}(T)$ cast as

$$\Delta\boldsymbol{L}(T) = \frac{\hbar}{2}\underbrace{\dot{\theta}\,T}_{\equiv\Delta\theta}\,\boldsymbol{e}_y$$

that is in the spirit of our approximation (258)

$$\Delta\boldsymbol{L}(T) = \frac{\hbar}{2}\sin\Delta\theta\,\boldsymbol{e}_y\,.$$

The y-component of the magnetic field which equaled zero at the beginning of the rotation is now given by $\vec{B}_y = B_z \sin\Delta\theta\,\boldsymbol{e}_y$. That means: after one precession period T the magnetic field and the atomic spin angular momentum have turned by the same angle $\Delta\theta$. The spin orientation follows the magnetic field - within the present approximation - without slip, that is adiabatically. (This is similar to the physics of a spinning artillery shell whose spin axis follows the course of the shell's bending trajectory leaving only a small precession angle.) Thus, the atoms enter the SG-magnet (almost) fully oriented with respect to the SG-magnetic field. This applies to the atoms with anti-parallel spin orientation accordingly. Hence, the two beams leaving the SG-magnet reflect merely the two kinds of atoms associated with the two Zeeman levels before they leave the reservoir.

It is worth mentioning that Leu [82] carried out Stern-Gerlach-type experiments using beams of Na-, K-, Zn-, Cd- and Tl-atoms instead of Ag-atoms. The Zn- and Cd-atoms possess two s-valence electrons which results in a zero net spin momentum of the atoms and consequently one does not observe a beam splitting in the Stern-Gerlach magnet. On the other hand, Tl-atoms possess a 6p-valence electron that is subjected to spin-orbit coupling. This gives rise to a Landé factor $g = \frac{2}{3}$ as a result of which the effective magnetic moment is for $M_j = \frac{1}{2}$ given by

$$\mu_{eff} = \mu_B\, g\, M_j = \frac{1}{3}\,\mu_B\,.$$

This is, in fact, confirmed by the experiments.

If one were dealing with atoms that possess a total angular momenta $J = (l \pm \frac{1}{2})\,\hbar$ associated with $2l+2$ different magnetic quantum numbers M_j, one would have $2l+2$ different states in the initial weak field and therefore as many different sorts of atoms entering the Stern-Gerlach magnet where they are deflected according to their magmetic moment. That means one would have $2l+2$ different beams instead of 2.

Our explanation of the SG-experiment is much in the spirit of Stern's conjecture that the spin of an atom responds adiabatically to the directional change of the magnetic field in which it has originally been aligned. In cooperation with Phipps [83] he devised an experiment where one of the beams at the exit of a first SG-magnet was focused into a linear set of three successive magnets whose weaker, essentially homogeneous fields pointed in three different directions perpendicular to the atomic trajectory. The difference between these directions was 120Â°. If the spin of the selected beam was pointing up after leaving the first SG-magnet and assuming that the spin would adiabatically adjust to the local magnetic field on its passage through the three magnets, it was thus to be expected that it would be finally back to its previous "up"-orientation. To test this the beam was sent into a second SG-magnet identically oriented as the first. There was only one beam coming out of this magnet indicating that the spin was pointing again in the same direction as on entering the three “turn magnets”. In other words: even after a turn of 360ö no slip between spin orientation and the direction of the magnetic field had occurred. We mention here only in passing that our result on the Phipps-Stern experiment agrees with that of Rosen and Zener [84] published already in 1932. Different from our more summary analysis these authors attempt to stay close to explicitly solving the time-dependent Pauli equation.

Surprisingly, the interpretation of the SG-experiment as demonstrating a coherent splitting of the de Broglie-wave of the incoming atom into two beams has become the most popular view on which a host of considerations on “measurement” is based. Papers on the so-called “Humpty-Dumpty-problem” (s. e. g. Englert et al. [85]) deal explicitly with a possible reconstruction of the original single wave by appropriately merging the two coherent beams at a spot reached later. We believe that such thought experiments are without substance. As we have clearly demonstrated, the SG-magnet does not cause a splitting of the incoming matter wave. The SG-situation is distinctly different from that in neutron spin-flip experiments by Rauch and coworkers [86] where a transverse spin polarized beam of neutrons hits a plate of a Si single crystal such that each matter wave packet splits up into two widely separated beams of packets due to dynamical diffraction within the crystal. This diffraction process is spin-independent. The two beams are coherently merged then by dynamical diffraction at a second Si-plate.

Many authors give the impression as if there were not a shadow of doubt that Stern-Gerlach experiments with charged free particles (like electrons) are just as feasible as with spin-carrying neutral atoms. Bohr had very early pointed out (s. Wheeler and Zurek [87]) that such experiments could not possibly succeed because “the Lorentz force would inevitably blur any Stern-Gerlach pattern”. Nevertheless, the literature on EPRB- (Einstein-Podolsky-Rosen-Bohm) correlation with pairs of fermions in a singlet state (s. e. g. Einstein et al. [52], Bohm [53]) abounds with allusions to “measuring separately the $x/y/z$-spin components” of the particles by means of Stern-Gerlach magnets. (S. e. g. Wigner [92].) Even when one were dealing with neutral fermions what kind of mechanism should yield such information on those spin **components**? How would the time evolution of the respective solution to the time-dependent Pauli equation look like in this case?

31. The Time-Dependent Dirac Equation

In trying to extend the theory to relativistic systems we retain the following two fundamental assumptions that characterize the non-relativistic quantum mechanics we have been dealing with so far:

1. The universal existence of stochastic forces that necessitate an ensemble description of the one-particle system under study. The fundamental constituents of this approach are: $\rho(\boldsymbol{r},t)$ for the occurrence of the particle at $\boldsymbol{r}$ and time t and $\boldsymbol{p}(\boldsymbol{r},t)$ for the associated ensemble average of the particle momentum

2. Lumping together the two real-valued functions $\rho(\boldsymbol{r},t)$ und $\boldsymbol{p}(\boldsymbol{r},t)$ in the form of a complex-valued function $\psi(\boldsymbol{r},t)$

3.

$$\psi(\boldsymbol{r},t) = \sqrt{\rho(\boldsymbol{r},t)}\; e^{i\,\varphi(\boldsymbol{r},t)} \tag{260}$$

where

$$\boldsymbol{p}(\boldsymbol{r},t) = \hbar\,\nabla\varphi(\boldsymbol{r},t)\,. \tag{261}$$

From $\psi(\boldsymbol{r},t) = |\psi(\boldsymbol{r},t)|\, e^{i\,\varphi(\boldsymbol{r},t)}$ one then obtains the momentum current density

$$\boldsymbol{j}_p(\boldsymbol{r},t) = \rho(\boldsymbol{r},t)\,\boldsymbol{p}(\boldsymbol{r},t) = \frac{1}{2}\left[\psi^*(\boldsymbol{r},t)\,\hat{\boldsymbol{p}}\,\psi(\boldsymbol{r},t) - \psi(\boldsymbol{r},t)\,\hat{\boldsymbol{p}}\,\psi^*(\boldsymbol{r},t)\right] \tag{262}$$

where $\hat{\boldsymbol{p}} \overset{\text{def}}{=} -i\,\hbar\,\nabla.$

Eq.(261) implies that $\boldsymbol{p}(\boldsymbol{r},t)$ is irrotational, that is, the stochastic forces do not cause friction.

From Eq.(262) follows for the expectation value of the particle momentum

$$< \boldsymbol{p}(t) >= \int \boldsymbol{j}_p(\boldsymbol{r},t)\,d^3r = \int \psi^*(\boldsymbol{r},t)\,\hat{\boldsymbol{p}}\,\psi(\boldsymbol{r},t)\,d^3r\,. \tag{263}$$

If one replaces $\psi(\boldsymbol{r},t)$ with its Fourier integral

$$\psi(\boldsymbol{r},t) = (2\pi)^{-\frac{3}{2}} \int C(\boldsymbol{k},t)\, e^{i\,\boldsymbol{k}\cdot\boldsymbol{r}}\, d^3k\,,$$

one obtains on insertion into Eq.(263)

$$< \boldsymbol{p}(t) >= \int \psi^*(\boldsymbol{r},t)\,\hat{\boldsymbol{p}}\,\psi(\boldsymbol{r},t)\,d^3r = \int C^*(\boldsymbol{k},t)\,\hbar\,\boldsymbol{k}\;C(\boldsymbol{k},t)\,d^3k\,, \tag{264}$$

and analogously

$$\int \psi^*(\boldsymbol{r},t)\frac{\hat{\boldsymbol{p}}^2}{2\,m_0}\,\psi(\boldsymbol{r},t)\,d^3r = \int C^*(\boldsymbol{k},t)\,\frac{\hbar^2\,\boldsymbol{k}^2}{2\,m_0}\;C(\boldsymbol{k},t)\,d^3k\,. \tag{265}$$

Newton's modified second law (23) which we have derived for the non-relativistic case, contains an additional "quantum force" $\boldsymbol{F}_{QP} = -\nabla V_{QP}$ whose expectation value equals zero. As a result one arrives at Ehrenfest's two theorems.

$$< \boldsymbol{v} >= \frac{d}{dt} < \boldsymbol{r} >=< \nabla_{\boldsymbol{p}} E(\boldsymbol{p}) > \tag{266}$$

and

$$\frac{d}{dt} < \boldsymbol{p} >=< \boldsymbol{F} >=< -\nabla V >\,. \tag{267}$$

The salient point here is that these two equations apply to the non-relativistic case and we require them to remain unaffected in the relativistic case if the particle is assumed - as before - to perform a dissipationless motion under stochastic extra forces.

Conversely, one can derive the time-dependent Schrödinger equation just by starting from Eqs.(266) and(267) and going along the same line of argument used in our derivation of the time-dependent Pauli equation in Section 27. In the following we shall refer to the latter. However, instead of

$$E(\boldsymbol{p}) = \frac{\boldsymbol{p}^2}{2m_0} + m_0\, c^2 + V(\boldsymbol{r})$$

we now have

$$E(\boldsymbol{p}) = \underbrace{\sqrt{\boldsymbol{p}^2\, c^2 + m_0^2\, c^4}}_{=E_{kin}+m_0\, c^2} + V(\boldsymbol{r}) \tag{268}$$

with c denoting the velocity of light in vacuo.

Hence $< \nabla_{\boldsymbol{p}} E(\boldsymbol{p}) > = < \boldsymbol{v} >$ in Eq.(266) has to be dealt with differently in the relativistic case. Following Dirac [88] we construct a Fourier-transform $\underline{\underline{H}}_0(\boldsymbol{k})$ that corresponds to the sought-for energy-operator $\hat{H}_{rel.}$ just as $\hbar^2\, k^2/2\, m_0$ in Eq.(264) relates to the expression $\frac{\hat{\boldsymbol{p}}^2}{2\, m_0}$. If one rewrites $E_{kin}(\boldsymbol{k}) + m_0\, c^2$ in Eq.(268) in the form

$$E_{kin}(\boldsymbol{k}) + m_0\, c^2 = \hbar\, c\, \sqrt{\sum_{\mu=0}^{3} k_\mu^2} \quad \text{where} \quad p_\mu = \hbar\, k_\mu \text{ and } \quad k_0 = \frac{m_0\, c}{\hbar}$$

and replaces the right-hand side with a 4×4-matrix $\underline{\underline{H}}_0(\boldsymbol{k})$ defined by

$$\underline{\underline{H}}_0(\boldsymbol{k}) = \hbar\, c \sum_{\mu=0}^{3} \underline{\underline{\alpha}}_\mu\, k_\mu\,,$$

with $\underline{\underline{\alpha}}_\mu$ denoting constant dimensionless 4×4 matrices, the Fourier transform $\underline{\underline{H}}_0(\boldsymbol{k})$ must obviously possess the property

$$\begin{aligned} \underline{\underline{H}}_0^2(\boldsymbol{k}) &= \hbar^2\, c^2 \sum_{\mu=0}^{3} \sum_{\mu'=0}^{3} k_\mu\, k_{\mu'}\, \delta_{\mu\mu'}\, \underline{\underline{1}} \\ &= \frac{\hbar^2\, c^2}{2} \sum_{\mu=0}^{3} \sum_{\mu'=0}^{3} k_\mu\, k_{\mu'} \left[\underline{\underline{\alpha}}_\mu\, \underline{\underline{\alpha}}_{\mu'} + \underline{\underline{\alpha}}_{\mu'}\, \underline{\underline{\alpha}}_\mu\right]. \end{aligned}$$

That means that the matrices $\underline{\underline{\alpha}}_\mu$ have to comply with the requirement

$$\frac{1}{2}\left[\underline{\underline{\alpha}}_\mu\, \underline{\underline{\alpha}}_{\mu'} + \underline{\underline{\alpha}}_{\mu'}\, \underline{\underline{\alpha}}_\mu\right] = \delta_{\mu\mu'}\, \underline{\underline{1}}\,.$$

As can be verified by just performing the multiplications, the matrices $\underline{\underline{\alpha}}_\mu$ meet this requirement if they have the form

$$\underline{\underline{\alpha}}_0 = \begin{pmatrix} \underline{\underline{1}} & \underline{\underline{0}} \\ \underline{\underline{0}} & -\underline{\underline{1}} \end{pmatrix} \quad \text{and} \quad \underline{\underline{\alpha}}_\mu = \begin{pmatrix} \underline{\underline{0}} & \underline{\underline{\sigma}}_\mu \\ \underline{\underline{\sigma}}_\mu & \underline{\underline{0}} \end{pmatrix} \quad \text{for } \mu = 1, 2, 3\,.$$

Here $\underline{\underline{\sigma}}_\mu$ denotes 2×2-matrices that are identical with the Pauli matrices (235). Similar to the latter one can lump the 4×4-matrices $\underline{\underline{\alpha}}_\mu$ together by forming a vector $\underline{\alpha}$ so that $\underline{\underline{H}}_0(\boldsymbol{k})$ may be cast as

$$\underline{\underline{H}}_0(\boldsymbol{k}) = c\, \underline{\alpha} \cdot \hbar \boldsymbol{k} + \underline{\underline{\alpha}}_0\, m_0\, c^2\,. \tag{269}$$

The feasibility of the above line of thought requires a consistent extension of the hitherto discussed spinor function to a bispinor function

$$\underline{\psi}(\boldsymbol{r},t) = \begin{pmatrix} \psi^1_\uparrow(\boldsymbol{r},t) \\ \psi^1_\downarrow(\boldsymbol{r},t) \\ \psi^2_\uparrow(\boldsymbol{r},t) \\ \psi^2_\downarrow(\boldsymbol{r},t) \end{pmatrix}$$

where

$$\psi^{(j)}_{\uparrow(\downarrow)}(\boldsymbol{r},t) = |\psi^{(j)}_{\uparrow(\downarrow)}(\boldsymbol{r},t)|\, e^{i\varphi^{(j)}_{\uparrow(\downarrow)}(\boldsymbol{r},t)}\,, \quad j = 1,2\,.$$

The associated phases $\varphi^{(j)}_{\uparrow(\downarrow)}(\boldsymbol{r},t)$ represent as in Eq.(261) potentials of ensemble averages of momenta which means

$$\boldsymbol{p}(\boldsymbol{r},t) = \sum_{\substack{j=1,2 \\ (\uparrow,\downarrow)}} \frac{|\psi^{(j)}_{\uparrow(\downarrow)}(\boldsymbol{r},t)|^2}{\rho(\boldsymbol{r},t)} \boldsymbol{p}^{(j)}_{\uparrow(\downarrow)}(\boldsymbol{r},t)$$

$$\text{where} \quad \boldsymbol{p}^{(j)}_{\uparrow(\downarrow)}(\boldsymbol{r},t) = \hbar\, \nabla\, \varphi^{(j)}_{\uparrow(\downarrow)}(\boldsymbol{r},t)\,.$$

The quantities $\boldsymbol{p}^{(j)}_{\uparrow(\downarrow)}$ are now different for "spin up" and "spin down" if the particle in question moves in a spatially varying potential. Only in the strictly non-relativistic case the spin generating component of the quivering motion and the orbital motion remain unaffected on superposition. In this case we have $\psi_{\uparrow(\downarrow)}(\boldsymbol{r},t) = |\psi_{\uparrow(\downarrow)}(\boldsymbol{r},t)|\, e^{i\,\varphi(\boldsymbol{r},t)}$.

If one performs a Fourier transform one obtains in complete analogy to Eq.(264) also in the relativistic case

$$< \boldsymbol{p}(t) >= \int \underline{C}^\dagger(\boldsymbol{k},t)\, \hbar\, \boldsymbol{k}\, \underline{C}(\boldsymbol{k},t)\, d^3k\,.$$

Correspondingly one gets

$$\int \underline{C}^\dagger(\boldsymbol{k},t)\, \underline{\underline{H}}_0(\boldsymbol{k})\, \underline{C}(\boldsymbol{k},t)\, d^3k = \int \underline{\psi}^\dagger(\boldsymbol{r},t)\, \underbrace{[c\, \underline{\alpha} \cdot \hat{\boldsymbol{p}} + \underline{\underline{\alpha}}_0\, m_0\, c^2]}_{\stackrel{\text{def}}{=} \underline{\underline{\hat{H}}}_{Dirac}}\, \underline{\psi}(\boldsymbol{r},t)\, d^3r\,.$$

We now form $< \boldsymbol{v} >$ according to

$$< \boldsymbol{v} >=< \nabla_{\boldsymbol{p}} E(\boldsymbol{p}) >= \int \underline{C}^\dagger(\boldsymbol{k},t)\, [\hbar^{-1}\, \nabla_{\boldsymbol{k}} \underline{\underline{H}}_0(\boldsymbol{k})]\, \underline{C}(\boldsymbol{k},t)\, d^3k\,.$$

If we substitute $\underline{C}(\boldsymbol{k})$ by its Fourier transform we obtain

$$< \boldsymbol{v} >= \int \underline{\psi}^\dagger(\boldsymbol{r},t)\, c\, \underline{\alpha}\, \underline{\psi}(\boldsymbol{r},t)\, d^3r\,.$$

Exploiting the identity

$$\underline{\underline{\hat{H}}}_{Dirac}\, \boldsymbol{r} - \boldsymbol{r}\, \underline{\underline{\hat{H}}}_{Dirac} = -i\, c\, \hbar\, \underline{\alpha}\,,$$

and going through the same set of arguments as with deriving the Pauli equation, we arrive at the time-dependent Dirac equation

$$[\underline{\underline{\hat{H}}}_{Dirac} + V(\boldsymbol{r})]\,\underline{\psi}(\boldsymbol{r},t) = i\hbar\,\frac{\partial}{\partial t}\,\underline{\psi}(\boldsymbol{r},t)\,.$$

The derivation can be extended by including electromagnetic fields, again in complete analogy to the derivation of the Pauli equation.

32. Spatial Particle Correlation Beyond the Limit of Entanglement. Spooky Action at a Distance

As discussed in Section 22 the electrons of two hydrogen atoms will respond independently to local perturbations once the inter-atomic distance has become macroscopically large. The electronic wave function factorizes then and becomes the product of two one-particle wave functions. One would therefore expect two free fermions that have moved sufficiently far away in opposite directions with their spins being transverse and anti-parallel, to display the same features. If they were still described by an anti-symmetric wave function the particle properties would remain non-locally intertwined in that each of the particles would appear at distant detectors with only half of the total probability. Therefore a realistic description can only be ensured by a product of two one-particle wave functions, wavepackets moving in opposite directions, one for spin up and the other one for spin down or vice versa, thc choice randomly distributed among the pairs generated in succession. Consequently, there will be a complete loss of the "common-cause"-spin correlation of the particles when they hit differently oriented spin detectors. The latter scatter the incoming fermion depending on the angle which the fermion's spin direction encloses with the scattering plane. To be as concrete as possible we refer in this section to the fundamental experiment by Lamehi-Rachti and Mittig [89] who were able to generate pairs of protons of about 8 MeV with spins paired anti-parallel and moving apart such that the proton's velocities in the center of mass system have the same absolute value but opposite directions. The spin orientation was analyzed by letting each of the protons impinge on a device akin to a Mott detector familiar from polarized electron detection. The incoming proton is scattered at some carbon atom of a carbon foil. Each Mott-type detector is associated with two particle detectors whose axes point to the scattering center and enclose an angle $\pm\alpha$ with the flight direction of the incoming proton. Together with that direction these axes form the scattering plane. The differential cross section of the carbon scatterer for a proton with spin up perpendicular to the scattering plane is given by

$$\sigma(\alpha,\beta) = (|f(\alpha)|^2 + |g(\alpha)|^2)\,[1 - S(\alpha)\sin\beta] \tag{270}$$

where $S(\alpha)$ represents the Sherman function for carbon/proton scattering, β stands for the azimuthal angle in the plane perpendicular to the proton flight direction and $f(\alpha)$ and $g(\alpha)$ denote the scattering and spin-flip amplitude. The latter is associated with spin-orbit coupling [6] which determines also the magnitude of $S(\alpha)$. In view of the objective of this

[6] Although spin-orbit coupling represents a constituent of the relativistic Pauli approximation to the Dirac equation, we consider it here as given.

article, we wish to emphasize at this point that Eq.(270) is a consequence of solving the relativistic Pauli equation, and the experimentally verifiable results that will be discussed below, are another objective consequence which is definitely not affected by the process of particle detection (the "measurement").

In accordance with the notation familiar from EPRB-experiments we denote the Mott-type analyzer at the end of the left proton track by A and that at the end of the right track by B. Furthermore, the detector on the right side of the scattering plane will be characterized by a "+"-sign, that on the left side by a "-"-sign. The two particle detectors of each Mott-type analyzer are located at $\beta = \mp\pi/2$, and α was set $\approx 50°$. Hence the difference between the respective differential cross sections (the "left-right asymmetry") is given by $\Delta\sigma = \sigma^+ - \sigma^- = (|f|^2 + |g|^2)\, 2S$. If the spin of the incoming proton encloses an angle Δ with the normal of the scattering plane, the sin-factor in Eq.(270) becomes $\sin(\Delta \mp \pi/2) = \mp \cos\Delta$ with $\beta = \mp\pi/2$ denoting the positions of the two particle detectors as before. Thus one has $\Delta\sigma = (|f|^2 + |g|^2)\, 2S \cos\Delta$. In order to capture the general case, we introduce an orthogonal Cartesian coordinate system whose x/y-plane is spanned by the two proton tracks before they enter the Mott-type analyzers. The axis of alignment of the proton spins encloses in general an angle φ with the z-axis thus introduced. The pertinent orientation angles of the scattering planes with respect to that z-axis are denoted by θ and ϕ for the normals of the A and B-plane, respectively. That means: $\Delta_A = \theta - \varphi$ and $\Delta_B = \phi - \varphi$. To make contact to the familiar notation, we define a quantity $P^{\pm}_{A(B)}$ through

$$\frac{\sigma^{\pm}_{A(B)}}{2\,[|f|^2 + |g|^2]} = P^{\pm}_{A(B)} = S\,[\tfrac{1}{2S} \pm \tfrac{1}{2}\cos\Delta_{A(B)}] \tag{271}$$

which has the property

$$P^+_{A(B)} + P^-_{A(B)} = 1\,.$$

Obviously, $P^{\pm}_{A(B)}$ is proportional to the count rate of the respective detector, and $\frac{1}{S}\,(P^+_{A(B)} - P^-_{A(B)}) = \cos\Delta_{A(B)}$ describes the degree of spin orientation of the incoming proton with respect to the normal of the associated scattering plane. If the spin of the proton impinging on the analyzer at A is parallel to that normal, that is perpendicular to the associated scattering plane, we have $\Delta_A = 0$ and hence $\frac{1}{S}\,(P^+_A - P^-_A) = 1$. The joint probability of finding the proton pair with one of the protons at A and orientation angle $\Delta_A = \theta - \varphi$ and the other proton at B with orientation angle $\Delta_B = \phi - \varphi - \pi$ is given by

$$P_{joint} = (P^+_A - P^-_A)\,(P^+_B - P^-_B) = P^{++} + P^{--} - P^{+-} - P^{-+} \tag{272}$$

where $P^{\pm\pm} = P^{\pm}_A\, P^{\pm}_B$ and $P^{\pm\mp} = P^{\pm}_A\, P^{\mp}_B$. Because of the definition (271) we have

$$P^{++} + P^{--} + P^{+-} + P^{-+} = 1\,.$$

To make sure that the count rates refer definitely to proton pairs, the counts associated with $P^{\pm\pm}$ and $P^{\pm\mp}$ are filtered by coincidence electronics.

Since for principal reasons one has in general $S < 0$ (in the case under study $S \approx 0.7$), $P^+ - P^- = S\cos\Delta$ can never become unity even when the particle enters the analyzer with its spin perpendicular to the scattering plane, that is when $\Delta = 0$. It is therefore suggestive to introduce an S-independent joint count rate $\hat{P}_{joint} = \frac{1}{S^2} P_{joint}$ which, on combining Eqs.(271) and (272), takes the form

$$\hat{P}_{joint}(\theta, \phi, \varphi) = -\cos(\theta - \varphi)\, \cos(\phi - \varphi)\,. \tag{273}$$

In practice the experiments have been carried out with the scattering plane of the B-analyzer lying in the x/y-plane, which means $\phi = 0$. Since all proton pairs are prepared 100% polarized, that is with their spins aligned parallel and antiparallel with respect to the z-axis, we have also $\varphi = 0$ so that Eq.(273) simplifies to

$$\hat{P}_{joint} = -\cos\theta\,, \tag{274}$$

and this is in agreement with the experimental results.

We emphasize again that this equation has been obtained by assuming a factorization of the two-proton wave function which means that the motion of the "A"-proton is controlled only by the potentials specifying the "A"-analyzer. There is no influence of the potentials that belong to "'B". Analogous statements apply to the "B"-proton. Hence, for each pair of protons there is no correlation between their respective "A" and "B"- scattering processes. However, it has been the objective of the experiments, as the authors expressly state, to demonstrate that there is such a correlation. Yet in order to prove that point, the experiments should have allowed a preparation of proton pairs with an axis of spin alignment that encloses an angle φ with the z-axis as originally assumed above. According to the established terminology that angle has to be regarded as a "hidden variable". The values of φ associated with the various pairs should have random character. One can form then a new expression from $\hat{P}_{joint}(\theta, \phi, \varphi)$ by averaging over φ:

$$\hat{P}_{av}(\theta, \phi) = \int_{-\frac{\pi}{2}}^{\frac{\pi}{2}} \rho(\varphi)\, \hat{P}_{joint}(\theta, \phi, \varphi)\, d\varphi\,. \tag{275}$$

where $\rho(\varphi)$ denotes a weight function normalized to unity. Clearly, in the experiment the averaging occurs automatically and unavoidably.

If one assumes a uniform distribution of φ over the interval π, that is $\rho(\varphi) = \frac{1}{\pi}$, and inserts here $\hat{P}_{joint}(\theta, \phi, \varphi)$ from Eq.(273), one obtains

$$\hat{P}_{av}(\theta, \phi) = -\tfrac{1}{2}\cos(\theta - \phi)\,, \tag{276}$$

where

$$\cos(\theta - \varphi)\, \cos(\phi - \varphi) = \tfrac{1}{2}\cos(\theta - \phi) + \tfrac{1}{2}\cos(\theta + \phi - 2\varphi)$$

has been used. Hence, for $\phi = 0$ as specified in the experiment, Eq.(276) yields

$$\hat{P}_{av}(\theta, \phi) = -\tfrac{1}{2}\cos\theta$$

which differs from (274) by a factor of $\frac{1}{2}$.

At this point it is instructive to contemplate the change that would occur if there would be a non-local correlation between the two analyzers in the following sense:

If the "B"-proton has been specified by the "B"-analyzer as polarized perpendicular to the associated scattering plane, that is if $\varphi = \phi$, and if this property is by some "spooky action at a distance" transferred to the "A"-proton, φ attains the same value for the "A"-proton. The measurement on the "A"-proton would then become "contextual": it would depend on the result obtained for the "B"-proton. Consequently, the detection rate (273) would take the form

$$\hat{P}_{joint} = -\cos(\theta - \phi) = -\cos(\vec{a}, \vec{b}) = -\vec{a} \cdot \vec{b}\,. \tag{277}$$

where we have introduced the quantities $\vec{a}$ and $\vec{b}$ as normal vectors for the "A"- and "B"-scattering plane, respectively, which enclose angles θ and ϕ with the z-axis. For the situation specified by the experiment (viz. $\phi = 0$), this result becomes identical with (274). Thus, a distinction between the two mechanisms is not possible within the given limitations. One might argue that a derivation based on a "spooky-action-at-a-distance"-hypothesis has to be rejected anyway. But the same hypothesis works perfectly for the analogous experiment with pairs of linearly polarized photons where that particular limitation does not exist. (S. Aspect et al. [90].)

By referring to the expectation value

$$\langle \Psi | \vec{\sigma}_A \cdot \vec{a} \otimes \vec{\sigma}_B \cdot \vec{b} | \Psi \rangle = -\cos(\vec{a}, \vec{b})$$

where Ψ denotes the anti-symmetric singlet-state two-proton wave function and $\vec{\sigma}_{A/B}$ the spin operators, Eq.(277) is commonly discussed as "the quantum mechanical prediction" for the experiment in question. Considering all the details of our analysis it is hard to see how this expectation value can have anything to do with the experiment except that it happens to yield the same $-\cos(\vec{a}, \vec{b})$.

We shortly return to the idea pursued by Lamehi-Rachti and Mittig in their paper. In order to exclude the possibility that their result might accidentally coincide with the prediction of a hidden parameter model, they resort to Bell's theorem [91]. It refers to quantities of the type $\hat{P}_{av}(\vec{a}, \vec{b})$ in Eq.(275) which - according to Eq.(276) - becomes equal to $-\frac{1}{2}\cos(\vec{a}, \vec{b})$ if φ is uniformly distributed. In general the weight function $\rho(\varphi)$ will be unknown, and hence a complete lack of correlation between the "A"- and "B"-scattering processes, as implied by our treatment, will not show up simply as a numerical correction factor of the "correlated result". Bell [91] could show that in performing an EPRB-type experiment one is definitely dealing with a non-classical (i. e. non-local) particle correlation if - irrespective of the form of the weight function and irrespective of the kind of hidden variable - the following inequality is violated:

$$|\hat{P}_{av}(\vec{a}, \vec{b}) - \hat{P}_{av}(\vec{a}, \vec{b}')| \leq 2\,|\hat{P}_{av}(\vec{a}', \vec{b}') + \hat{P}_{av}(\vec{a}', \vec{b})|\,,$$

where $\vec{a}, \vec{a}', \vec{b}, \vec{b}'$ denote different analyzer settings. In fact, the authors succeeded in verifying this violation, but it appears to us, because of the limitations discussed above, that this result is absolutely not convincing.

33. Concluding Remarks

In summarizing the essence of quantum mechanics Wigner states in a fundamental article [92] under the headline "What is the state vector?": *"We recognize ....that the state vector is only a shorthand expression of that part of information concerning the past of the system which is relevant for predicting (as far as possible) the future behavior thereof."*

In our view the most impressive success of quantum mechanics in understanding the stability, composition and properties of the building blocks of nature consists in predicting the systematic order in the periodic table, the phenomenon of chemical valency and the ground-state properties of molecules and solids. The state vector of these systems, the ground-state wave function $\Psi(\boldsymbol{r}_1, \boldsymbol{r}_2, \ldots \boldsymbol{r}_N)$, is a function of the particle coordinates $\boldsymbol{r}_1, \boldsymbol{r}_2, \ldots \boldsymbol{r}_N$ in terms of which their Coulomb interaction enters the calculation of the system's total energy. But for every experimentalist there is no doubt that these coordinates are fundamentally inaccessible to measurement, and hence cannot possibly be regarded as "information gained from measurements". As is amply demonstrated by modern ab initio-calculations, the wave function allows one to determine the total energy as a function of nuclear positions, bond angles, vibrational frequencies, lattice constants, elastic moduli, phonon spectra, saturation magnetizations, electric conductivities etc.. These quantities are in the spirit of common-sense true observables whereas the particle coordinates remain definitely hidden parameters. As we have repeatedly explained, this applies to the eigenvalues of hermitian operators as well, thus putting a serious question mark behind "Kochen-Specker"-type [93] and "no-go" theorems (s. e. g.[46]) which are all based on the exasperatingly artificial assumption that "measurements" yield eigenvalues or "probabilities for eigenvalues". Occasionally a certain awareness of this puzzling inconsistency surfaces as in the revealing statement of Wigner's [41] that we have already alluded to in Section 16.

It is deplorable to notice the impropriety with which certain advocates of the Copenhagen school of thought dismiss supporters of Nelson's attempt on developing a "quantum mechanics without observer" as "stranded enthusiasts"(s. Streater [12]), and ironically base their criticism on the old, actually absurd, arguments how indeterminacy enters the theory through measurement and how commuting "observables" correlate with the result of simultaneous measurements. All this has been iterated umpteen times although it is well known to every experimentalist that exactly these "measurements" are inexecutable altogether. With the same insensitivity to reality castigators of the Nelson proponents think it fully justified to equate the physics of photon-correlation experiments with analogous, but actually extremely scarce experiments with massive particles. We believe we have presented ample evidence that quantum mechanics is in detail derivable from classical mechanics plus a modified physical vacuum by allowing the latter to undergo energy fluctuations. Their action on massive particles is calibrated by Planck's constant, and despite their presence the conservation of energy (and with free particles: the conservation of particle momentum) is ensured on average. We hope that the present article can contribute to an unbiased reassessment of present-day quantum mechanics concerning these two questions: 1. Which elements of the old doctrine are obsolete and dispensable? 2. Does "measurement" really play a particular role in quantum mechanics or is its alleged importance simply a misunderstanding?

Appendix: Derivation of the Navier-Stokes Equation

Given an ensemble of N similarly prepared one-particle systems we introduce a transition probability $P^M(\boldsymbol{r}, \vec{\sigma}, t, \Delta t)$ which denotes the probability of a particle being in the elementary volume $d^3\sigma$ around a point $\boldsymbol{r} + \vec{\sigma}$ after a time span Δt if it has been with certainty at point $\boldsymbol{r}$ at time t. This transition probability integrates to unity:

$$\int P^M(\boldsymbol{r}, \vec{\sigma}, t, \Delta t)\, d^3\sigma = 1\,.$$

The superscript "M" stands for "Markov process".

In terms of this transition probability Einstein's law [29] on the mean square displacement of a particle under the action of stochastic forces may be cast as

$$\int \sigma_l\, \sigma_k\, P^M(\boldsymbol{r}, \vec{\sigma}, t, \Delta t)\, d^3\sigma = \delta_{l\,k}\, 2\nu\, \Delta t$$

where $\vec{\sigma} = (\sigma_1, \sigma_2, \sigma_3)$. Here $\nu = \eta / m_0\, n_0$ denotes the "kinematic viscosity" of the embedding system, η represents the common "dynamical viscosity" and n_0 is the particle density of the embedding system.

The Smoluchowski equation [40] describes the temporal change of the probability density at $\boldsymbol{r}$ caused by the motion of the N independent particles under the influence of stochastic forces as a result of which the particles perform transitions from previous positions $\boldsymbol{r} - \vec{\sigma}$ to $\boldsymbol{r}$

$$\rho(\boldsymbol{r}, t + \Delta t) = \int \underbrace{\rho(\boldsymbol{r} - \vec{\sigma}, t) P^M(\boldsymbol{r} - \vec{\sigma}, \vec{\sigma}, t, \Delta t)}_{\equiv G(\boldsymbol{r} - \vec{\sigma}, \vec{\sigma}, t, \Delta t)}\, d^3\sigma\,. \tag{278}$$

One may evaluate the integral on the right-hand side by approximately replacing the integrand $G(\boldsymbol{r} - \vec{\sigma}, \vec{\sigma}, t, \Delta t)$ with a Taylor polynomial of second degree

$$\begin{aligned} G(\boldsymbol{r} - \vec{\sigma}, \vec{\sigma}, t, \Delta t) = G(\boldsymbol{r}, \vec{\sigma}, t, \Delta t) - \sum_{k=1}^{3} \sigma_k \frac{\partial}{\partial x_k} G(\boldsymbol{r}, \vec{\sigma}, t, \Delta t) \\ + \frac{1}{2} \sum_{l,k} \sigma_l\, \sigma_k\, \frac{\partial^2}{\partial x_l\, \partial x_k}\, G(\boldsymbol{r}, \vec{\sigma}, t, \Delta t)\,. \end{aligned}$$

Inserting this expression under the integral of Eq.(278) yields

$$\begin{aligned} \rho(\boldsymbol{r}, t + \Delta t) = \rho(\boldsymbol{r}, t) \quad &- \sum_{k=1}^{3} \frac{\partial}{\partial x_k} [\rho(\boldsymbol{r}, t) \times \underbrace{\int \sigma_k\, P^M(\boldsymbol{r}, \vec{\sigma}, t, \Delta t)\, d^3\sigma]}_{= v_{ck}(\boldsymbol{r}, t)\, \Delta t} \\ &+ \frac{1}{2} \sum_{l,k} \frac{\partial^2}{\partial x_l\, \partial x_k} [\rho(\boldsymbol{r}, t) \underbrace{\int \sigma_l\, \sigma_k\, P^M(\boldsymbol{r}, \vec{\sigma}, t, \Delta t)\, d^3\sigma]}_{= \delta_{lk}\, 2\nu\, \Delta t}\,. \end{aligned}$$

This may be recast as

$$\underset{\text{“Fokker-Planck Equation”}}{\frac{\partial \rho}{\partial t} + \nabla \cdot \boldsymbol{j}_c - \nu \Delta \rho = 0} \qquad \text{where} \quad \boldsymbol{j}_c = \rho\, \boldsymbol{v}_c\,. \tag{279}$$

Apart from obeying the Fokker-Planck equation the system of N particles must satisfy the equation of continuity as well

$$\frac{\partial \rho}{\partial t} + \nabla \cdot \boldsymbol{j} = 0 \quad \text{where} \quad \boldsymbol{j} = \rho\, \boldsymbol{v}\,.$$

On forming the difference of these two equations one obtains

$$\nabla \cdot (\boldsymbol{j}_d + \nu\, \nabla \rho) = 0 \quad \text{where} \quad \boldsymbol{j}_d \equiv \boldsymbol{j} - \boldsymbol{j}_c\,.$$

This is equivalent to

$$\mathbf{j}_d = -\nu\, \nabla \rho \quad \text{“Fick’s Law”}\,.$$

If one writes the diffusion current density $\boldsymbol{j}_d$ in the form:

$$\boldsymbol{j}_d = \rho\, \boldsymbol{u} \quad (\boldsymbol{u} = \text{“osmotic velocity”}),$$

one may recast $\boldsymbol{u}$ as in Eq.(8)

$$\boldsymbol{u}(\boldsymbol{r}, t) = -\nu\, \frac{1}{\rho(\boldsymbol{r}, t)}\, \nabla \rho(\boldsymbol{r}, t)\,.$$

Thus we have $\boldsymbol{v} = \boldsymbol{v}_c + \boldsymbol{u}$ which is just Eq.(6) used in advance in Section 4
. Replacing $\rho(\boldsymbol{r}, t)$ in Eq.(278) with $\rho\, v_{c\,k}$, one obtains

$$\frac{\partial \rho\, v_{c\,k}}{\partial t}|_{scatter} = -\nabla \cdot (\rho\, v_{c\,k}\, \boldsymbol{v}_c) + \nu\, \Delta(\rho\, v_{c\,k})\,,$$

There is an additional (local) change in time of the momentum current density effected by the external force

$$\frac{\partial \rho\, v_{c\,k}}{\partial t}|_{force} = \hat{f}_k(\boldsymbol{r}) \equiv \frac{1}{m_0}\, \rho(\boldsymbol{r})\, F_k(\boldsymbol{r})\,,$$

Invoking the Fokker-Planck equation one can write the sum $\frac{\partial \rho\, v_{c\,k}}{\partial t}|_{force} + \frac{\partial \rho\, v_{c\,k}}{\partial t}|_{scatter}$ in the form

$$\frac{\partial \boldsymbol{v}_c}{\partial t} + (\boldsymbol{v}_c + 2\, \boldsymbol{u}) \cdot \nabla \boldsymbol{v}_c - \nu\, \Delta \boldsymbol{v}_c = \frac{1}{m_0}\, \boldsymbol{F}(\boldsymbol{r})\,. \tag{280}$$

Substituting here $\boldsymbol{v}_c$ by $\boldsymbol{v} - \boldsymbol{u}$ we arrive at our Eq.(7):

$$\frac{\partial}{\partial t}\,(\boldsymbol{v} - \boldsymbol{u}) + [(\boldsymbol{v} + \boldsymbol{u})\nabla(\boldsymbol{v} - \boldsymbol{u})] - \nu\, \Delta(\boldsymbol{v} - \boldsymbol{u}) = \frac{1}{m_0}\, \boldsymbol{F}(\boldsymbol{r})\,.$$

In hydrodynamics $|\boldsymbol{u}|$ is usually neglected compared to $|\boldsymbol{v}|$ and $\frac{1}{m_0}\, \boldsymbol{F}(\boldsymbol{r})$ is given by the internal mass-referenced force $-\frac{1}{\hat{\rho}(\boldsymbol{r},t)}\, \nabla p\, (\boldsymbol{r}, t)$ with $\hat{\rho}(\boldsymbol{r},\, t) \;=\; m_0\, \rho(\boldsymbol{r}, t)$ denoting the massive density and $p\,(\boldsymbol{r}, t)$ the pressure. Hence Eq.(7) takes the familiar Navier-Stokes-form:

$$\underbrace{\hat{\rho}(\boldsymbol{r}, t)\left(\frac{\partial \boldsymbol{v}}{\partial t} + \boldsymbol{v} \cdot \nabla \boldsymbol{v}\right)}_{\frac{d\boldsymbol{v}(\boldsymbol{r},t)}{dt}} - \mu\, \Delta \boldsymbol{v} + \nabla p\, (\boldsymbol{r}, t) = 0 \quad \text{where} \quad \mu = \nu\, \hat{\rho}\,.$$

The derivation of this equation is due to Gebelein [25].

References

[1] D. Wick, *The Infamous Boundary* (Birkhöuser, Boston, 1995).

[2] D. Bohm, *Wholeness and the Implicate Order* (Routledge Kegan Paul, London, 1980), p. 84 .

[3] D. Bohm and J. P. Vigier, *Phys. Rev.* 96, pp. 208-216 (1954).

[4] W. Weizel, *Z. Physik* 134, No. 3, pp. 264-285 (1953), 135, No. 3, pp. 270-273 (1953), 136, No. 5, pp. 582-604 (1954).

[5] E. Nelson, *J. Math. Phys.* 5, pp. 332-343 (1964); *Phys. Rev.* 150, pp. 1079-1085 (1966).

[6] E. Nelson, *Quantum Fluctuations* , (Princeton University Press, 1985).

[7] F. Guerra and L. M. Morato, *Phys. Rev. D* 27, pp. 1774-1786 (1983).

[8] M. Baublitz, *Prog. Theor. Phys.* 80(2), pp. 232-244 (1988).

[9] L. de la Peña and A. H. Cetto, *The quantum dice - An introduction to stochastic electrodynamics* , (Kluwer, Dordrecht 1996).

[10] N. C. Petroni and L. M. Morato, *J. Phys. A: Math. Gen.* 33, pp. 5833-5848 (2000).

[11] T. C. Wallstrom, *Phys. Rev. A* 49, pp. 1613-1617 (1994).

[12] R. F. Streater, *Lost Causes in and beyond Physics* , (Springer-Verlag, Berlin, Heidelberg, 2007).

[13] K. Namsrai, *Nonlocal Quantum Field Theory and Stochastic Quantum Mechanics*, Reidel Publishing, Dordrecht (1986).

[14] L. Fritsche and M. Haugk, *Ann. Phys.* (Leipzig) 12, No.6, pp. 371-402 (2003).

[15] G. Grössing, *Phys. Lett. A* 372, pp. 4556-4563 (2008).

[16] G. Grössing, *Physica A* 388, pp. 811-823 (2009).

[17] J. S. Bell, in *Sixty-Two Years of Uncertainty* , edited by A. I. Miller, (Plenum, New York, 1989), p. 17.

[18] L. Bess, *Prog. Theor. Phys.* 49(6), pp. 1889-1910 (1973).

[19] H. E. Puthoff, *Phys. Rev. D* 35, pp. 3266-3269 (1987).

[20] H. E. Puthoff, *Phys. Rev. A* 40, pp. 4857-4862 (1989).

[21] H. E. Boyer, *Scientific American* (August) 253, No. 2, pp. 70-78 (1985).

[22] F. Calogero, *Phys. Lett. A* 228, pp. 335-346 (1997).

[23] A. Carati and L. Galgani, *Nuovo Cimento B* 114, pp. 489-500 (1999).

[24] L. E. Ballentine, *Rev. Mod. Phys.* 42, pp. 358-381 (1970).

[25] H. Gebelein, *Turbulenz*, p. 75, Springer, Berlin (1935).

[26] W. Pauli,, *Handbuch der Physik*, Bd.XXIV/1, 2. Aufl., Springer, Berlin (1933), p. 126.

[27] M. Born and P. Jordan, *Elementare Quantenmechanik*, p.32, Springer, Berlin (1930).

[28] D. Bohm, *Phys. Rev.* 85(2), pp. 166-179 (1952); 85(2), pp. 180-193 (1952).

[29] A. Einstein, *Ann. d. Physik* 17, pp. 549-560 (1905), *Ann. Phys.* 19, pp. 371-381 (1906).

[30] L. de Broglie, *Une Tentative d'Interprétation Causale et Non-Linéaire de la Mécanique Ondulatoire*, Gauthier-Villars, Paris (1956).

[31] B. Mielnik and G. Tengstrand, Intern. *J. Theor. Phys.* 19, pp. 239-250 (1980).

[32] G. Badurek, H. Rauch, and D. Tuppinger, *Phys. Rev. A* 34, pp. 2600-2608 (1986).

[33] W. Thomson, *Proceedings of the Royal Society of Edinburgh*, Vol. VI, 1867, pp. 94-105.

[34] E. Madelung, *Z. Physik* 40(3-4), pp. 322-326 (1926).

[35] W. Heisenberg, *Z. Physik* 43(3-4), pp. 172-198 (1927).

[36] J. Bell, *Physics World* 3, August, pp. 33-40 (1990).

[37] E. P. Wigner in *Quantum Theory of measurement*, edited by J. A. Wheeler and W. H. Zurek, Princeton University Press, Princeton, New Jersey (1983), p. 267.

[38] P. Ehrenfest, *Z. Physik* 45(7-8), pp. 455-457 (1927).

[39] P. Garbaczewski, arXiv; cond-mat/0703147.

[40] M. V. Smoluchowski, *Ann. d. Physik*, 21, pp. 756-780 (1906) and 48, pp. 1103-1112 (1915); s. also: *Abhandlungen über die Brownische Bewegung und verwandte Erscheinungen* (Akademische Verlagsgesellschaft, Leipzig (1923)).

[41] E. P. Wigner in *Quantum Theory of measurement*, edited by J. A. Wheeler and W. H. Zurek, Princeton University Press, Princeton, New Jersey (1983), p. 313.

[42] R. P. Feynman, *Rev. Mod. Phys.* 20, pp. 367-387 (1948).

[43] J. v. Neumann, *Mathematische Grundlagen der Quantenmechanik*, Dover Publications, New York (1943).

[44] N. Bohr, *Phil. Mag.* 26, 1-25, 476-502, 857-875 (1913).

[45] W. Nagourney, J. Sandberg, and H. Dehmelt, *Phys. Rev. Lett.* 56, pp. 2797-2799 (1986).

[46] N. D. Mermin, *Rev. Mod. Phys.* 65, pp. 803-815 (1993).

[47] E. T. Jaynes and F. Cummings, *Proc. IEEE* 51, 89 (1963).

[48] M. D. Crispy and E. T. Jaynes, *Phys. Rev.* 179, pp. 1253-1261 (1969).

[49] E. T. Jaynes, *Phys. Rev. A* 2, pp. 260-262 (1970).

[50] V. Weisskopf und E. Wigner, *Z. Physik*, 63, 54-73 (1930).

[51] E. Fermi, Atti della Reale Accademia nazionale dei Lincei, Rendiconti. *Classe di scienze fisiche, matematiche e naturali*, Serie 6, Vol.5, p.795 (1927).

[52] A. Einstein, B. Podolsky, and N. Rosen, *Phys. Rev.* 47, pp. 777-780 (1935).

[53] D. Bohm, *Quantum Theory* (Prentice Hall, Englewood Cliffs, N.J. (1951)).

[54] J. B. Hartle and M. Gell-Mann in:*Complexity, Entropy and the Physics of Information*, W. Zurek (ed.) Addison-Wesley, Redwood City, CA (1990).

[55] N. G. van Kampen, *Physica A* 153(1), pp. 97-113 (1988).

[56] E. Eisenschitz und F. London, *Z. Physik* 60, pp. 491-527 (1930).

[57] I. E. Dzyaloshinskii, E. M. Lifshitz and L. P. Pitaevskii, Soviet Physics, *Usp* 4, No.2, pp. 153-176 (1961).

[58] Peter J. Feibelman, *Prog. Surf. Sci.* 12, pp. 287-408 (1982).

[59] E. Hult, H. Rydberg, B. I. Lundqvist, and D. C. Langreth, *Phys. Rev. B* 59, pp. 4708-4713 (1999).

[60] W. Kohn, Y. Meir and D. E. Makarov, *Phys. Rev. Lett.* 80, pp. 4153-4156 (1998).

[61] "*Smithsonian Physical Tables*" 9th ed.

[62] E. Schrödinger, "The present situation in quantum mechanics" in: *Quantum Theory and Measurement*, J. A. Wheeler and W. H. Zurek (Eds.), Princeton University Press, Princeton, N. J. (1983), pp. 152-167.

[63] G. E. Uhlenbeck und S. Goudsmit, *Die Naturwissenschaften*, 13(47), pp. 953-954 (1925), *Nature* 127, pp. 264-265 (1926).

[64] W. Pauli, *Z. Physik* 31, pp. 373-385 (1925).

[65] B. L. van der Waerden in *Theoretical Physics in the Twentieth Century. A Memorial Volume to Wolfgang Pauli*, edited by M. Fierz and V. F. Weisskopf, Interscience Publishers, New York (1960), p. 214.

[66] H. C. Ohanian, *Am. J. Phys.* 54 (6), pp.500-505.

[67] E. Schrödinger, Sitzungsber. Preuß. *Akad. Wiss. Phys.-Math. Kl.* 24, pp. 418-431. (1930)

[68] C. F. v. Weizsäcker in: *Zum Weltbild der Physik* S. Hirzel, Leipzig (1945), p.32.

[69] T. Dankel, Archiv. *Rational Mech. Anal.* 37, pp. 192-221 (1971).

[70] D. Dohrn, F. Guerra, and P. Ruggiero in : *Feynman Path Integrals*, edited by S. Albeverio, *Lecture Notes in Physics* 106, Springer, Heidelberg (1979).

[71] E. Nelson, *Quantum fluctuations*, Princeton University Press, Princeton, New Jersey (1985), p.102.

[72] H. Goldstein *Classical Mechanics*, 2nd edition, Addison-Wesley, Reading, MA (1980), pp.148-158.

[73] C. P. Slichter, *Principles of Magnetic Resonance*, 3rd edition, Springer, Berlin, Heidelberg (1990).

[74] F. Bloch, *Phys. Rev.* 70(7-8), pp. 460-474 (1946).

[75] H. Rauch, *Found. Phys.* 23(1), pp. 7-36 (1993).

[76] S. A. Werner, R. Colella and A. W. Overhauser, and C. F. Eagen, *Phys. Rev. Lett.* 35(16), pp. 1053-1055 (1975)

[77] J. Bell, *J. Phys. Colloques* 42, pp. C2-41-C2-62 (1981).

[78] O. Stern und W. Gerlach, *Z. Physik* 9, pp. 349-352 (1922); W. Gerlach und O. Stern, *Ann. d. Physik* 74, pp. 673-699 (1924).

[79] D. Home, A. K. Pan, Md M. Ali and A. S. Majumdar, *Phys. Rev. A* 75(4), id. 042110 (6 pages) (2007).

[80] E. Wrede, *Z. Physik* 41, pp. 569-575 (1927).

[81] N. V. Mott and H. S. W. Massey in *Quantum theory and measurement*, edited by J. A. Wheeler and W. H. Zurek, Princeton University Press, Princeton, New Jersey (1983), p. 703.

[82] A. Leu, *Z. Physik* 41, pp. 551-562 (1927).

[83] T. E. Phipps und O. Stern, *Z. Physik* 73, pp. 185-191 (1932).

[84] N. Rosen and C. Zener, *Phys. Rev.* 40(4), pp. 502-507 (1932).

[85] B.-G. Englert, J. Schwinger and M. O. Scully, *Found. Phys.* 18(10), pp. 1045-1056 (1988).

[86] G. Badurek, H. Rauch, and D. Tuppinger, *Phys. Rev. A* 34, pp. 2600-2608 (1986).

[87] J. A. Wheeler and W. H. Zurek (Eds.) *Quantum theory and measurement*, Princeton University Press, Princeton, New Jersey (1983), p. 699.

[88] P. A. M. Dirac, *Proc. Roy. Soc. (A),* 117, pp. 610-624 (1928).

[89] M. Lamehi-Rachti and W. Mittig, *Phys. Rev. D* 14, pp. 2543-2555 (1976).

[90] A. Aspect, Ph. Grangier, and G. Roger, *Phys. Rev. Lett.* 49, 91 (1982).

[91] J. S. Bell, *Physics* (N.Y.), 1(3), pp. 195-200 (1965).

[92] E. P. Wigner in *Quantum Theory and measurement*, edited by J. A. Wheeler and W. H. Zurek, Princeton University Press, Princeton, New Jersey (1983), p. 292.

[93] S. Kochen and E. P. Specker, *J. Math. Mech.* 17, pp. 59-87 (1967).

In: Space-Time Geometry and Quantum Events
Editor: Ignazio Licata, pp. 155-173
ISBN: 978-1-63117-455-1

Chapter 4

FRACTAL SPACE-TIME AS AN UNDERLYING STRUCTURE OF THE STANDARD MODEL

Ervin Goldfain
Photonics CoE, Welch Allyn Inc., Skaneateles Falls, NY, US

Abstract

The Standard Model of particle physics (SM) is a theoretical framework that integrates our current knowledge of the subatomic world and its fundamental interactions. A key program built in the structure of the SM is the Renormalization Group (RG), whose function is to preserve self-consistency and describe how parameters of the theory evolve with the energy scale. Despite being overwhelmingly supported by experimental data, the SM has many puzzling aspects, such as the large number of parameters, a triplication of chiral families and the existence of three gauge interactions. In contrast with the majority of mainstream proposals advanced over the years, the basic premise of our study is that a satisfactory resolution of challenges confronting the SM requires further advancing the RG program. In particular, understanding the nonlinear dynamics of RG equations and the unavoidable transition from smooth to fractal dimensionality of space-time are critically important for the success of this endeavor. Here we show how the onset of fractal space-time near or above the electroweak scale is likely to settle at least some of these challenges.

Keywords: Standard Model, Quantum Field Theory, Renormalization Group, Landau-Ginzburg-Wilson model, fractal space-time, scale invariance, continuous dimension

List of Abbreviations Used in the Text

SM = Standard Model, EW = electroweak, EWSB = electroweak symmetry breaking, FP = fixed point, QFT = Quantum Field Theory, EFT = effective field theory, RG = Renormalization Group, IR = infrared, UV = ultraviolet, QG = Quantum Gravity, LGW = Landau-Ginzburg-Wilson, QCD = Quantum Chromodynamics.

1. Introduction

As coherent synthesis of Quantum Mechanics and Special Relativity, Quantum Field Theory (QFT) provides a compelling description of phenomena up to the energy scales probed by present accelerators. Nevertheless, there are plausible reasons to suspect that QFT breaks down at some high energy scale (Λ_{UV}), above which it needs to be replaced by a more fundamental theory. Although there is no consensus among theorists on this issue, the underlying reasons may be summarized as follows:

a) New interactions, hidden symmetries or any other exotic extensions of QFT could likely unfold near Λ_{UV},
b) Reconciling classical gravity with QFT appears to be an insurmountable challenge. A major obstacle is that perturbative quantization of classical gravity cannot be extrapolated at energies close to the Planck scale, $\Lambda_{UV} = O(M_{Pl})$. As a result, the theory is said to be "non-renormalizable", meaning that it lacks any predictive power at scales comparable with Λ_{UV}. A number of non-perturbative models of quantum gravity have been proposed as alternative solutions, but it is presently unclear if they yield a truly consistent integration scheme of General Relativity and QFT [1, 2].
c) The current accelerator technology probes energies moderately above the range defining the Standard Model (SM) of particle physics (μ_{SM} = O (TeV)). The prevalent view is that M_{Pl} is the only genuine scale in QFT and stems from the assumption that no dramatic change in physics develops between μ_{SM} and M_{Pl}. But if this assumption is true, there is at present no compelling explanation for the *mass hierarchy*, stemming from the observation that fermion masses are scattered over thirteen orders of magnitude and are confined around $\mu_{SM} << M_{Pl}$. In addition, quantum corrections applied to the Higgs vacuum can shift the Higgs mass close to M_{Pl}, leading to the so-called "*fine-tuning problem*" [3].
d) The dynamics of QFT may undergo transition to classical behavior as a result of *decoherence* [4] or become *unstable* near or below Λ_{UV}. The instability can arise from unbalanced quantum corrections or from the transition to chaos in nonlinear evolution of interacting fields [5, 6, 39, 41].

Due to its limitations in dealing with phenomena on scales nearing Λ_{UV}, the conventional interpretation of QFT is that it represents an "effective" field theory (EFT), which is likely to be amended by new physics emerging above Λ_{UV}. EFT describes phenomena occurring exclusively on low-energy scales $\mu_{SM} << \Lambda_{UV}$, in the same way the continuum theory of elasticity describes the long wavelength excitations of a crystal [7, 8]. In the case of a crystal, the continuum theory breaks down at the scale of the lattice spacing. Likewise, EFT breaks down as the energy scale is ramped up close to Λ_{UV}.

It is instructive to briefly review at this point two examples of successful EFT's:

1) In the Wilson treatment of critical phenomena using the *Renormalization Group* program (RG) [7, 9, 52], quantum fields present in the theory (Φ_μ) depend on the running scale μ and are separated into two components

$$\Phi_\mu^{(l)} : 0 \le \mu \le \frac{\Lambda_{UV}}{s} \tag{1.1}$$

$$\Phi_\mu^{(s)} : \frac{\Lambda_{UV}}{s} \le \mu \le \Lambda_{UV} \tag{1.2}$$

Here, the parameter "s" is an arbitrary scaling factor ($s > 1$), $\Phi_\mu^{(l)}$ and $\Phi_\mu^{(s)}$ are the long and short wavelength excitations and correspond, respectively, to the light and heavy particles carried by Φ_μ. Starting with an EFT defined at Λ_{UV}, the core idea of Wilson's approach is to integrate out all heavy particles contained in the "momentum shell" (1.1, 1.2) and form a new EFT with the remaining fields below the separation scale Λ_{UV}/s. Since μ is considered a running parameter, iterating this process yields a flow of EFT's from Λ_{UV} toward their low-energy limit. It is customary to refer to this iterative process as a *RG flow*(or *RG trajectory*). A key property of local EFT's is that the low-energy endpoint of the RG flow must describe phenomena that are fully decoupled from physical processes occurring near the high-energy limit Λ_{UV}. This property conveys the basic idea behind the concept of *scale invariance*[7, 9, 52].

2) The second example is the SM itself, a robust EFT that has been in place for more than three decades. It includes the $SU(3)\otimes SU(2)\otimes U(1)$ gauge model of strong and electroweak interactions along with the Higgs mechanism that spontaneously breaks the electroweak $SU(2)\otimes U(1)$ group down to the $U(1)$ group of electrodynamics. The SM has been confirmed countless times in all accelerator experiments, including the latest runs of the Large Hadron Collider (LHC). Despite this convincing body of evidence, the SM is confronted with many unsolved challenges [10-12]. Over the years, this has led to an overflow of theoretical extensions targeting the physics beyond the SM scale ($\mu > \mu_{SM}$). The majority of these proposals center on solving some unsatisfactory aspects of the theory while introducing new unknowns. Experiments are expected to provide guidance in pointing to the correct theory yet, so far, LHC searches show no credible hint for physics beyond the SM up to a center-of-mass energy of $\sqrt{s}$ =8 TeV [13]. These results, albeit entirely preliminary, suggest two possible scenarios, namely:

a) SM fields are either decoupled or ultra-weakly coupled to new dynamic structures emerging in the low or intermediate TeV scale,

b) There is an undiscovered and possibly non-trivial connection between the SM and TeV phenomena.

More importantly, this discussion raises a key question: What should be the principles guiding model-building efforts beyond the SM? In contrast with many mainstream proposals on how to tackle this question, our basic premise is that moving beyond the SM requires further advancing the RG program. As we shall argue below, understanding the *nonlinear dynamics of RG flow equations* and the transition from smooth to *fractal dimensionality* of space-time are essential steps for the success of this endeavor.

The article is organized in the following way: section two surveys the principles of the RG program, with emphasis on phase space trajectories and their fixed points. The idea of dimensional regularization and its implications on the emergence of fractal space-time in QFT form the topic of section three. Section four and five describe the asymptotic approach to scale invariance of RG trajectories and presents a natural solution for the so-called *mass hierarchy problem* of the SM. The emergence of a Higgs-like resonance as Bose-Einstein condensate on fractal space-time is introduced in section six. Concluding remarks are gathered in section seven. To facilitate reading, frequently used text abbreviations are also listed at the end of the article.

This work represents a development of ideas published by the author in [5-6, 14-22, 35, 46]. To make its content fully transparent, we have opted for minimal mathematics but adequate clarity and level of detail. Since many of these ideas are under development, concurrent analysis is needed to confirm, expand or falsify our provisional findings.

Before going into details, it is essential to point out that in our work the terms "scale invariance" and "conformal invariance" are considered identical, although they are not synonymous (a field theory can be scale invariant without being conformal invariant [23, 38]).

2. Renormalization Group Trajectories

We begin by briefly reviewing the conventional construction of Lagrangian field theory. Consider a field theory whose action in D - dimensional space-time is given by

$$S[\Phi,\overline{\Phi}]=\int L[\Phi,\overline{\Phi}]d^{D}x \tag{2.1}$$

$$L[\Phi,\overline{\Phi}]=u_{\alpha}P_{\alpha}[\Phi,\overline{\Phi}] \tag{2.2}$$

The theory is fully specified by three primary inputs, namely: a) the field content of the Lagrangian (2.2),b) the set of symmetry constraints imposed on (2.2) and c) the dimension of space-time (D). The basis functionals $P_{\alpha}[\Phi,\overline{\Phi}]$ with $\alpha=1,2,...N$ represent a sum of local products of fields Φ, their conjugates $\overline{\Phi}$ and/or their derivatives and u_{α} are a set of coupling parameters. Here, coupling parameters mean the coefficients describing *interaction strengths* as well as *particle masses*. The summation convention is assumed throughout.

As alluded to in the first section, the RG program posits that the description of the physics embodied in (2.1) can be done in terms of a family of "*effective actions*", each valid at a certain energy scale μ [24]. A key concept of this program is the RG flow, whose role is to define how effective formulations of the theory evolve with μ . According to this program, parameters u_α run with μ following the general system of non-autonomous equations of order "n" [7, 24]

$$\mu\frac{\partial u_\alpha(\mu)}{\partial\mu} = \beta_\alpha(u_1, u_2, ..., u_\alpha, ..., u_N, \mu) \tag{2.3}$$

Using the standard field-theoretic jargon, $u_\alpha(\infty)$ are called "*bare*" parameters of the theory whereas $u_\alpha(0)$ are referred to as the "*renormalized*" parameters. In this picture, the flow (2.3) describes the evolution of u_α from the ultraviolet region of arbitrarily large energies (UV) to the infrared limit of low energies (IR). If d_α represents the canonical mass dimension of u_α, the substitution

$$g_\alpha = \mu^{-d_\alpha} u_\alpha \tag{2.4}$$

transforms (2.3) into a coupled system of autonomous differential equations of order "n" relating *dimensionless* couplings g_α, that is,

$$\mu\frac{\partial g_\alpha(\mu)}{\partial\mu} = \beta_\alpha(g_1, g_2, ..., g_\alpha, ..., g_N) \tag{2.5}$$

A fixed point (FP) is invariant under (2.5) and corresponds to the stationary condition

$$\beta(g_1^*, g_2^*, ..., g_N^*) = \beta(g^*) = 0 \tag{2.6}$$

By definition, integral curves along the vector field (2.5) are called *RG trajectories* parameterized by the running scale μ. The goal of the RG theory consists in constructing trajectories which do not develop divergences in either UV or IR limits. Each such trajectory defines a possible field theory characterized by the continuous mapping

$$\mu \to g(\mu) \to S_\mu[\Phi(\mu), \overline{\Phi}(\mu)] \tag{2.7}$$

RG trajectories encode in a universal way the natural flow of (2.1) towards FP's. It follows that, since FP's are independent from μ, (2.7) unveils the asymptotic approach of field theories to *scale invariance*. A relativistic QFT that is scale invariant and contains massless particles is necessarily a free theory [25].

The fixed-point structure of quantum field theories is typically difficult to extract analytically. In particular, Feynman diagrams fail in the neighborhood of any strong coupling point ($g^* >> 0$) and the behavior of the theory is ill-defined near such points. By contrast, perturbation methods are applicable near any trivial FP ($g^* = 0$) and (2.5) can be evaluated there using the series expansion

$$\mu \frac{\partial g_\alpha(\mu)}{\partial \mu} = a_\alpha^{ij} g_i g_j + b_\alpha^{ijk} g_i g_j g_k + ... \tag{2.8}$$

It is customary to assume that the perturbative flow (2.8) of typical QFT's evolves toward isolated sets of stable FP's [7, 15]. A typical example is the strong FP of Quantum Chromodynamics (QCD), which characterizes the passage to asymptotic freedom of quarks and gluons in the UV limit. Another example is the IR "conformal" FP describing the approach to scale invariance in the low-energy limit of QFT. But in addition to FP's, RG flowscan display *limit cycles* (isolated closed trajectories) and *chaotic behavior*. If either (2.5) or (2.8) are regarded as nonlinear systems of coupled differential equations, the behavior of these systems is characterized by sensitivity to initial conditions, the loss of stability and the emergence of local or global bifurcations [26-30, 34-35, 38, 56].

At first sight, this observation appears to stand in conflict with one foundational postulate of relativistic field theory, *the cluster decomposition principle*. This principle asserts that local processes in any relativistic field theory must be insensitive to distant environment in space-time or energy scale and guarantees the factorization of the S-matrix of scattering amplitudes [7]. A direct consequence of this principle is that transitions amplitudes measured in the laboratory must be insensitive to the physics of short-distance scales. Rather than contradicting the clustering principle, the nonlinear behavior of the RG flow imparts a new twist to it. As we show below, abandoning the notion that the RG flow typically evolves towards isolated and stable FP's, is likely to bring closure to some of the open challenges facing the SM.

3. Dimensional Regularization and the Onset of Fractal Space-Time

The previous section has touched upon the fundamentals of Wilson's RG program. Closely related to this program in QFT is the concept of *dimensional regularization*, which we now briefly outline.

A well-known difficulty of QFT is that perturbative calculations using momentum integrals do not converge [7, 31]. The root cause is that integrands fall off too slowly at large momenta. Infinities arising from the short-wavelength region of the integrals are called *ultraviolet* (UV) divergences. For massive fields, this type of singular behavior in the UV limit is the only anomaly of such integrals. In the zero-mass limit, further singularities show up at small momenta and are called *infrared* (IR)divergences. The zero-mass limit is relevant to critical behavior in statistical and condensed matter physics, as it relates to phenomena that exhibit unbounded correlation length and scale invariance.

Renormalization is a powerful technique for removing both UV and IR divergences and it consists in a two-step program: *regularization* and *subtraction*. One first controls the divergence present in momentum integrals by inserting a suitable regulator, and then brings in a set of counter-terms to cancel out the divergence. Momentum integrals in QFT have the generic form

$$I = \int_0^\infty d^4q\, F(q) \tag{3.1}$$

Two regularization techniques are frequently employed to manage (3.1), namely "momentum cutoff" and "dimensional regularization". When the momentum cutoff scheme is applied for regularization in the UV region, the upper limit of (3.1) is replaced by a finite cutoff Λ_{UV},

$$I \to I_{\Lambda_{UV}} = \int_0^{\Lambda_{UV}} d^4q\, F(q) \tag{3.2}$$

Explicit calculation of the convergent integral (3.2) amounts to a sum of three polynomial terms

$$I_{\Lambda_{UV}} = A(\Lambda_{UV}) + B + C(1/\Lambda_{UV}) \tag{3.3}$$

Dimensional regularization proceeds instead by shifting the momentum integral (3.1) from a four-dimensional space to a continuous D - dimensional space

$$I \to I_D = \int_0^\infty d^Dq\, F(q) \tag{3.4}$$

Introducing the dimensional parameter $\varepsilon = 4 - D$ leads to

$$I_D \to I_\varepsilon = A'(\varepsilon) + B' + C'(1/\varepsilon) \tag{3.5}$$

Historically, the idea of continuous dimension was introduced by Wilson and Fisher [52] and initially used to compute physical quantities of interest as expansions in powers of ε. Later on, Veltman and 't Hooft have shown how this idea can be incorporated in QFT and developed into a reliable renormalization technique [31].

Regularization techniques are not independent from each other. For example, the connection between dimensional and cutoff regularizations is given by [6, 32]

$$\log \frac{\Lambda_{UV}^2}{\mu^2} = \frac{2}{\varepsilon} - \gamma_E + \log 4\pi + \frac{5}{6} \tag{3.7}$$

We find it convenient to present (3.7) is a slightly different form, that is,

$$\varepsilon \sim \frac{1}{\log(\frac{\Lambda_{UV}^2}{\mu^2})} \tag{3.8}$$

It is apparent from (3.8) that the four-dimensional space-time is recovered in either one of these limits:

a) $\Lambda_{UV} \to \infty$ and $0 < \mu << \Lambda_{UV}$,

b) $\Lambda_{UV} < \infty$ and $\mu \to 0$.

However, both limits are in conflict with our current understanding of the far UV and the far IR boundaries of field theory. Theory and experimental observations alike tell us that the notions of infinite or zero energy are, strictly speaking, meaningless. This is to say that either infinite energies (point-like objects) or zero energy (infinite distance scales) lead to divergences whose removal requires the machinery of the RG program. Indeed, there is always a finite cutoff at both ends of either energy or energy density scale (far UV = Planck scale, far IR = finite radius of the observable Universe or the non-vanishing energy density of the vacuum set by cosmological constant). It follows from these considerations that the limit $\varepsilon \to 0$ works as a highly accurate approximation and realistic models near or beyond the SM scale must account for space-time geometries having continuous dimensionality. Fractal space-time defined by the continuous dimension $D = 4 - \varepsilon$ asymptotically approaches ordinary space-time near or below the SM scale, that is, for $\mu \leq \mu_{SM}$.

4. The Asymptotic Approach to Scale Invariance

Section two has surveyed how RG trajectories describe the asymptotic behavior of field theory and the universal approach to scale invariance. One finds it quite natural to also demand that RG trajectories themselves maintain scale invariance as they evolve toward FP's.

To clarify this point, let us return to the system of differential equations (2.5). Taking advantage of the large numerical disparity between μ and Λ_{UV}, we may use (3.8) to rewrite (2.5) as

$$\frac{\partial g_\alpha(\mu)}{\partial[\log(\frac{\mu}{\Lambda_{UV}})]} \sim \frac{\partial g_\alpha(\mu)}{\partial(\frac{\mu}{\Lambda_{UV}})} \Rightarrow \frac{\partial g_\alpha(\varepsilon)}{\partial \varepsilon} = \beta_\alpha(g_1, g_2, \ldots, g_\alpha, \ldots, g_N) \tag{4.1}$$

or

$$F(g_\alpha, \frac{\partial g_\alpha(\varepsilon)}{\partial \varepsilon}) = \frac{\partial g_\alpha(\varepsilon)}{\partial \varepsilon} - \beta_\alpha(g_1, g_2, ..., g_\alpha, ..., g_N) = 0 \quad (4.2)$$

As briefly alluded in section two, realistic RG flows evolve in the presence of *weak perturbations* which affect stability of their FP's [28]. The minimal way to include the effect of weak perturbations of the linear form $\eta(\varepsilon) \sim \varepsilon$ is to modify (4.2) as follows

$$F(g_\alpha, \frac{\partial g_\alpha}{\partial \varepsilon}) = 0 \Rightarrow F[\eta(\varepsilon), g_\alpha, \frac{\partial g_\alpha}{\partial \varepsilon}] \sim F(\varepsilon, g_\alpha, \frac{\partial g_\alpha}{\partial \varepsilon}) = 0 \quad (4.3)$$

By definition, (4.3) represents a system of *autonomous differential equations* if it remains unchanged under the substitutions $\varepsilon \to \varepsilon + c$ and $g_\alpha(\varepsilon) \to g_\alpha(\varepsilon)$, where c is a fixed but arbitrary parameter and $\alpha = 1, 2, ..., N$. Also by definition, (4.3) represents a system of *scale invariant differential equations* if it remains unchanged under the substitutions $\varepsilon \to \lambda\varepsilon$ and $g_\alpha(\varepsilon) \to \lambda g_\alpha(\lambda\varepsilon)$,where λ is a fixed but arbitrary parameter. It is always possible to transform scale invariant equations into autonomous equations [33]. In short,

$$F(\varepsilon + c, g_\alpha, \frac{\partial g_\alpha}{\partial \varepsilon}) = 0 \Leftrightarrow F[\lambda\varepsilon, \lambda g_\alpha(\lambda\varepsilon), \frac{\partial g_\alpha(\lambda\varepsilon)}{\partial \varepsilon}] = 0 \quad (4.4)$$

It can be shown that, any system of ordinary differential equations(4.3) that is scale invariant and whose associated autonomous equation possesses a FP, has a power-series solution in the basin of attraction of this FP. This power-series solution can be presented as [33]

$$g_\alpha(\varepsilon) = \varepsilon(d_\alpha + e_{\alpha,i}\varepsilon^{h_i}) \sim O(\varepsilon) \quad (4.5)$$

Here, d_α and $e_{\alpha,i}$ are finite numerical coefficients and the index $i = 1, 2, ..., M$, with $M = N \times n$. For $h_i > 0$, all couplings and masses vanish in the four-dimensional space-time limit ($\varepsilon = 0$). This confirms what we concluded in section two, namely that massless particles in a scale-invariant QFT residing at the trivial FP ($g_\alpha^*(0) = 0$) are necessarily free. [25]. But more importantly, what (4.5) also reveals is that set of couplings $g_\alpha(\varepsilon)$ assume non-vanishing values on *fractal space-time*, that is, on a space-time with continuous dimensionality ($D = 4 - \varepsilon < 4$). Building on this finding, next section recovers the pattern of massive SM particles and interaction couplings from the transition to chaos of RG equations (4.3).

In closing this section we note that this conclusion is consistent with the Landau-Ginzburg-Wilson (LGW) theory of critical behavior for the so-called Φ^4 model. This model describes the attributes of many statistical systems and field theories approaching criticality in the IR limit. An essential feature of the LGW theory is that the trivial FP $g_\alpha^* = 0$ is unstable and it cannot characterize critical behavior in less than four dimensions ($D <$ 4). There is another non-trivial fixed point, referred to as the *Wilson-Fisher* FP, defined on fractal space-time of arbitrary dimension $D=4-\varepsilon$. The Wilson-Fisher point accounts for the onset of critical behavior and scale invariance in less than four dimensions [9]. One is necessarily led to conclude from this analysis that the couplings of any interacting QFT arise from the fractional dimension of space-time $\varepsilon \neq 0$, as deviations from *trivial* scale invariance at $\varepsilon = 0$ [5-6].

5. Towards a Solution for the Mass Hierarchy Problem

Let us now return to (4.3) and (4.4). Under the reasonable assumptions that the solution (4.5) drifts toward a limit cycle $\overset{0}{g_\alpha}(\varepsilon)$ which becomes unstable near $\varepsilon = 0$, it can be shown that (4.3) and (4.4) undergo transition to chaos driven by the dimensional parameter $\varepsilon \to 0$ [14-15, 35]. The first stage of this transition is a *Feigenbaum cascade* of period-doubling bifurcations of $\overset{0}{g_\alpha}(\varepsilon)$. Numerous examples of this scenario show that the sequence of critical values ε_n, $n=1,2,...$ driving the transition to chaos in (4.3) and (4.4) satisfies the geometric progression [36-37]

$$\varepsilon_n - \varepsilon_\infty = \varepsilon_n - 0 \sim k_n \overline{\delta}^{-n} \tag{5.1}$$

Here, $n \gg 1$ is the index counting the number of cycles created through the period-doubling cascade, $\overline{\delta}$ is the rate of convergence and k_n is a coefficient that becomes asymptotically independent of n as $n \to \infty$. Period-doubling cycles are characterized by $n = 2^p$, for $p \gg 1$. Substituting (5.1) in (4.5) yields the following ladder-like progression of critical couplings

$$\boxed{g_\alpha^*(p) \sim (\overline{\delta})^{-2^p}} \tag{5.2}$$

It can be shown that (5.2) recovers the full mass and flavor content of the SM, including neutrinos, together with the coupling strengths of gauge interactions [6, 15, 35]. Specifically,

- The *trivial FP* of the RG flow consists of the massless photon (γ) and the massless UV gluon (g).
- The *non-trivial FP* of the RG flow is degenerate and consists of massive quarks (q), massive charged leptons and their neutrinos (l, ν) and massive EW bosons (W, Z).
- *Gauge interactions* develop near the non-trivial FP and include electrodynamics (e), the weak interaction (g_W^*) and the strong interaction (g_s^*).

It is instructive to note that a similar treatment applied to QCD is able to retrieve the spectrum of hadron masses [16, 39]. Along the same lines of thought, it can be also shown that the number of fermion generations follows from the stability analysis of RG trajectories [19]. These findings reinforce the point made earlier about the many unexplored implications of the RG program on the SM physics.

6. Higgs-Like Scalar as Bose-Einstein Condensate on Fractal Space-Time

It is widely accepted that the SM embodies our current knowledge of the strong and EW interactions. SM is a self-contained framework of remarkable predictive power whose fundamental degrees of freedom are the spin one-half quarks and leptons, the spin one gauge bosons and the spin-zero Higgs doublet. Symmetry constraints play a key role in fixing the dynamical structure of SM, which exhibits invariance under the combined $SU(3)_L \times SU(2)_Y \times U(1)_{EM}$ gauge group. Despite being confirmed in many independent tests, SM is an *incomplete* framework as it leaves many basic questions unanswered [55]. The reported narrow resonance seen by the LHC, whose mass is centered on $m = 126\,\text{GeV}$, is strongly consistent with a CP even *Higgs-like* boson [10-11]. However, at the time of writing, no consensus has yet been reached on two important points, namely, a) that the Higgs-like boson is the simplest possible type predicted by the SM and b) that the Higgs mechanism based on the Weinberg-Salam potential is the actual source of EWSB [10, 54].

It was shown in [40] that the transition from order to chaos in classical and quantum systems of gauge and Higgs fields is prone to occur somewhere in the low to mid TeV scale. The inability of the Higgs vacuum to survive not too far above the LHC scale explains away the fine-tuning problem and signals the breakdown of the SM in this region [40]. The likely instability of the vacuum in the low to intermediate TeV scale brings up an intriguing speculation on the nature of the Higgs scalar. In particular, what we interpret as the Higgs scalar may actually be a Bose-Einstein condensate of gauge fields on fractal space of dimension $D = 4 - \varepsilon$. The goal of this section is to elaborate on this idea.

First, recall that in relativistic QFT, pure scalar fields are a peculiar class of operators due to the following reasons [7, 42]:

a) Since canonical mass dimension of fields is linearly dependent on their spin j,

$$d_\Phi = j + 1 \tag{6.1}$$

scalar fields have the minimal mass dimension, $d_\Phi = 1$.

b) Self-interacting scalars carry dimensionless coupling parameters in four-dimensional space-time ($d_{u_\Phi} = 0$). As it is known, dimensionless couplings ensure consistency of quantum field theory, in particular compliance with conformal symmetry [42].

c) Scalar fields do not carry any gauge charges (electric, weak hypercharge or color) and are free from chirality.
d) The mapping theorem states that non-abelian gauge fields are indistinguishable from scalars in the infrared limit of field theory [43].

Secondly, fractal space-time has the ability to confine quantum fields in a similar manner with the phenomenon of Anderson localization in condensed matter physics [18, 44, 45].

It follows from these considerations that scalars are the simplest embodiment of quantum fields. They are the most likely to form a *Higgs-like* condensate of gauge bosons on space-time s endowed with low level fractionality ($\varepsilon \ll 1$), that is,

$$\Phi_C = \frac{1}{4}[(W^+ + W^- + Z^0 + \gamma + g) + (W^+ + W^- + Z^0 + \gamma + g)] \tag{6.2}$$

As further explained in the Appendix, a remarkable feature of (6.2) is that it is a weakly coupled cluster of gauge fields having *zero topological charge*. Compliance with this requirement motivates the duplicate construction of (6.2), which contains individual *WW*, *ZZ*, photon and gluon doublets. Stated differently, (6.2) is the only inclusive combination of gauge field doublets that is free from all gauge and topological charges. Table 1 shows a comparative display of properties carried by the SM Higgs and the Higgs-like condensate.

Table 1. SM Higgs doublet versus the Higgs-like condensate.

Scalar field	Original form	Composition	Mass (GeV)	Weak hypercharge	Electric charge	Color	Topological charge
SM Higgs	$\begin{pmatrix} \varphi^+ \\ \varphi^0 \end{pmatrix}$	none	~ 126	$\begin{pmatrix} +1 \\ +1 \end{pmatrix}$	$\begin{pmatrix} +1 \\ 0 \end{pmatrix}$	0	0
Higgs-like condensate	Φ_C	(6.2)	~ 126	0	0	0	0

7. Concluding Remarks

There is a vast body of proposed extensions of the SM offering solutions to its open questions or ideas on how to pursue model-building beyond its boundaries. Most proposals add new layers of complexity to the structure of the SM under the tacit assumption that these must come into play at large energy scales. The widespread belief is that new physics is prone to show up in the form of hidden particles or extended symmetry groups (examples include supersymmetric partners, sterile neutrinos, axions, Kaluza-Klein particles, WIMP's, dark photons and so on [55]).

By contrast, ideas developed here hint that an unexplored range of phenomena emerge from the nontrivial topology of space-time as the dimensional parameter $\varepsilon = 4 - D$ deviates slightly from zero. Our study shows that nonlinear dynamics of RG equations, along with the unavoidable transition to fractal space-time above the EW scale, can settle some of the puzzles surrounding the SM. In particular, the low fractality texture of space-time naturally explains the mass hierarchy problem and suggests the emergence of the Higgs-like resonance as condensate of gauge bosons with a vanishing topological charge. In addition, as detailed in [5, 6, 51], the onset of space-time of low level fractality clarifies the fermion chirality and the violation of CP symmetry in weak interactions, the gauge hierarchy and cosmological constant problems as well as the possible content of non-baryonic dark matter. On this last point and by analogy with the Higgs-like structure (6.2), dark matter may surface as low-energy condensates of gauge bosons on fractal space-time that are likely to quickly annihilate into lepton-antilepton or quark-antiquark pairs [6, 53].

The concept of fractal space-time helps also set the stage for a unified understanding of symmetries that operate within QFT. To this end, recall section four where autonomous equations were shown to be isomorphic with scale invariant equations (relation 4.4). This observation unveils a tantalizing connection between local symmetries of QFT and the dimensional parameter ε. Local infinitesimal transformation in ordinary space-time ($D=4$), including translations, rotations and boosts, may be viewed as local scale transformations characterized by infinitesimal changes of dimension, $\varepsilon \to \varepsilon'$. The same applies to infinitesimal gauge transformations. It follows that *all symmetry groups* of QFT, including the Poincaré and gauge groups, may be deeply related to scale invariance on fractal space-time and the concept of continuous dimension [5-6, 20-21].Furthermore, this connection brings to the fore two important insights:

a) The concept of *arbitrary spin* may be seen as a topological manifestation of fractal space-time.
b) Classical gravitation defined in ordinary space-time is equivalent to field theory on fractal space-time. At least in principle, this observation opens up an unforeseen path to the long-sought unification of General Relativity and the physics of subatomic scales [5-6, 20-21].

Needless to say, our tentative findings need to be further scrutinized and, most importantly, confronted with the experiment. Unfortunately, as of today, many aspects of the weak and strong interactions still elude us and the accelerator data do not provide clear guidance on where to focus the theory next.

Although intriguing, follow up research is required to consolidate or falsify the body of ideas discussed above. For example, to be compelling, the postulated Higgs-like condensate (6.2) has to duplicate all production/decay cross sections and branching ratios predicted by the SM. It must also be consistent with preservation of unitarity in scattering of polarized *WW* bosons. One must also explain why the chaotic behavior of the RG flow is not directly observable in field theories describing separate gauge sectors of the SM, namely *U(1)*, *SU(2)* and *SU(3)* [56].

It is our hope that our work will inspire further developments on the subtle connection between the dynamics of the SM and the fractal structure of space-time above the EW scale.

Appendix: Conserved Topological Charge of Field Theory on Fractal Space-Time

It is known that scale invariance in classical field theory amounts to the condition [23]

$$\partial_\nu S^\nu = 0 \tag{A.1}$$

where the scale current is related to the energy-momentum tensor of the theory via

$$S^\nu = x_\sigma T^{\nu\sigma} \tag{A.2}$$

(A.1) and (A.2) imply that the trace of this tensor vanishes, that is,

$$\partial_\nu S^\nu = T_\nu^\nu = 0 \tag{A.3}$$

The space-time continuum of both classical and quantum physics represents a smooth four-dimensional manifold. Its geometrical properties are fully specified by the metric tensor, which also determines the geodesics of classical particles and light beams. Consider now an ordinary space-time with metric $\eta_{\sigma\nu} = diag(-,+,...,+)$ and constrain all coordinates $\nu = 0,1,2,3$ to be slightly dependent on ε as in

$$x^\nu \to (x^\nu)^{1-\varepsilon} \tag{A.4}$$

with $\varepsilon << 1$. This space-time has low level fractionality and it can be referred to as a *minimal fractal manifold* (MFM). The measure of the MFM generalizes the familiar definition of classical and quantum physics and is given by [50]

$$d\rho(x) = d^4x\, \text{v}(x,\varepsilon) \tag{A.5}$$

$$\text{v}(x,\varepsilon) = \frac{(|x^0||x^1||x^2||x^3|)^{-4\varepsilon}}{(\Gamma(1-\varepsilon))^4} \tag{A.6}$$

Likewise, the ordinary differential operator on the MFM is upgraded to

$$\text{D}_\nu = \frac{1}{\sqrt{\text{v}(x,\varepsilon)}}\partial_\nu[\sqrt{\text{v}(x,\varepsilon)}\cdot] \tag{A.7}$$

Let us now introduce on the MFM a classical scalar field $\varphi(x)$ with Lagrangian L and assume that the potential part of the Lagrangian is a polynomial of order r having the form $V(\varphi) \sim \lambda_0 \varphi^r$. The energy-momentum tensor of the field can be presented as [50]

$$\overline{T_{\sigma\nu}} = \mathrm{v}(x,\varepsilon) T_{\sigma\nu} \tag{A.8}$$

where $\sigma = 0,1,2,3$ and

$$T_{\sigma\nu} = \eta_{\sigma\nu} L + \mathrm{D}_\sigma \varphi \mathrm{D}_\nu \varphi \tag{A.9}$$

The continuity equation reads

$$\partial_\nu \overline{T_\sigma^\nu} = \lambda_0 \varphi^r (\frac{r}{2} - 1) \mathrm{v}^{-r/2} \partial_\sigma \mathrm{v} \tag{A.10}$$

For $r \neq 2$, the energy-momentum tensor is not conserved due to the non-vanishing right-hand term. Adding to $\overline{T_\sigma^\nu}$ a contribution whose four-divergence amounts to the right-hand term of (A.10),

$$\partial_\nu t_\sigma^\nu = -\lambda_0 \varphi^r (\frac{r}{2} - 1) \mathrm{v}^{-r/2} \partial_\sigma \mathrm{v} \tag{A.11}$$

turns (A.10) into

$$\partial_\nu (\overline{T_\sigma^\nu} + t_\sigma^\nu) = \partial_\nu \theta_\sigma^\nu = 0 \tag{A.12}$$

Unlike $\overline{T_\sigma^\nu}$, the newly defined energy-momentum tensor θ_σ^ν is conserving. Its component t_σ^ν is clearly linked to the fractal geometry of the MFM via (A.6). As with any other constant currents, there are also conserved charges which arise from integrating the temporal component of θ_σ^ν over the spatial domain. Since these charges follow from the fractal geometry of space-time, they are intrinsically *topological* in nature and characterized by

$$\boxed{\tau_\sigma \sim \int \theta_\sigma^0 d^3x} \tag{A.13}$$

It is instructive to note that (A.13) falls in line with the concept of *space-time polarization* induced by fractal topology [47-49].

Due to the manifestly neutral attributes of the scalar field embodied in (6.2), it is natural to assume that the Higgs-like condensate Φ_C is the sole state with *zero topological charge* that includes all gauge boson flavors. A logical corollary of this assumption is that, in general, particle-antiparticle pairs of gauge bosons or fermions carry *non-zero* topological charges. In light of this interpretation, conservation of the topological charge (A.13) requires that, in all decay or production channels of Φ_C shown in Tab. 2, a fraction of this charge becomes an "effective" polarization of the space-time manifold. This mechanism explains, at least in principle, why the Higgs-like condensate Φ_C couples to both SM gauge and fermion operators without carrying any gauge charge.

Table 2. Decay and production channels of the Higgs-like condensate

Decay	$\Phi_c \to bb$	$\Phi_c \to \tau\tau$	$\Phi_c \to WW(l\nu l\nu)$	$\Phi_c \to \gamma\gamma$	$\Phi_c \to ZZ(4l)$
Production	$gg \to tb \to \Phi_c$	$W, Z \to \Phi_c$	$gg \to tt \to \Phi_c$	–	–

The entries of this table are, respectively, l = lepton, t = top quark, b = bottom quark, ν = neutrino, τ = tau lepton, γ = photon, g = gluon.

References

[1] M. Shaposhnikov, "Asymptotic safety of gravity and the Higgs-boson mass", *Theoretical and Mathematical Physics*, 170, 2, 229-38 (2012).

[2] H. Nicolai, "*Quantum Gravity: the view from particle physics*", e-print http://arxiv.org/abs/1301.5481 (2013).

[3] M. Shifman, "Reflections and Impressionistic Portrait at the Conference *Frontiers Beyond the Standard Model*, FTPI. Oct. 2012", e-print http://arxiv.org/pdf/1211.0004v1.pdf (2012).

[4] M. Schlosshauer, "*Decoherence and the Quantum-to-Classical Transition*", Springer-Verlag (2007).

[5] E. Goldfain, "Non-equilibrium Theory, Fractional Dynamics and Physics of the Terascale Sector" in *New Developments in the Standard Model*, Nova Science Publishers, 41-74 (2012).

[6] E. Goldfain, "Fractional Field Theory and High-Energy Physics: New Developments" in *Horizons in World Physics*, 279, Nova Science Publishers, 69-92 (2013).

[7] A. Duncan, "*Conceptual Framework of Quantum Field Theory*", Oxford University Press, (2012).

[8] A. Dobado, A. Gomez-Nicola, A. L. Maroto and J. R. Pelaez, "*Effective lagrangians for the standard model*", Springer-Berlin, (1997).

[9] J. Zinn-Justin, "*Renormalization Group: an introduction*", e-print http://www-math.unice.fr/~patras/CargeseConference/ACQFT09_JZinnJustin.pdf, (2009).

[10] M. Strassler, "Looking beyond the Standard Model", *The Higgs Symposium*, e-print http://higgs.ph.ed.ac.uk/sites/default/files/Strassler_Looking%20Beyond%20SM.pdf (2013).

[11] J. Ellis, "The Higgs Boson and Beyond", *The Higgs Symposium*, e-print http://higgs.ph.ed.ac.uk/workshops/higgs-symposium (2013).
[12] S. J. Brodsky, G. de Teramond and M. Karliner, "Puzzles in Hadronic Physics and Novel Quantum Chromodynamics Phenomenology", *Annual Review of Nuclear and Particle Science* 62, 1-35 (2012).
[13] The ATLAS collaboration, "*Search for resonances decaying into top-quark pairs using fully hadronic decays in pp collisions with ATLAS at sqrt(s) = 7 TeV*", e-print http://arxiv.org/abs/1211.2202 (2013).
[14] E. Goldfain, "Bifurcations and pattern formation in particle physics: An introductory study", *Euro Physics Lett.* 82, 11001(2008).
[15] E. Goldfain, "Feigenbaum Attractor and the Generation Structure of Particle Physics", *Int. J. Bifurcation Chaos* 18, 891(2008).
[16] E. Goldfain, "Chaos in Quantum Chromodynamics and the Hadron Spectrum", *Electron J. Theor. Phys.* 7, 23, 75-84 (2010).
[17] E. Goldfain, "On the asymptotic transition to complexity in quantum chromodynamics", *Communications in Nonlinear Science and Numerical Simulation*, 14, 1431-38 (2009).
[18] E. Goldfain, "Derivation of gauge boson masses from the dynamics of Levy flows", *Nonlinear Phenomena in Complex Systems* 8, 4, 366-72 (2005).
[19] E. Goldfain, "Stability of renormalization group trajectories and the fermion flavor problem", *Comm. Nonlin. Science and Numer. Simul.* 13, 9, 1845-50 (2008).
[20] E. Goldfain, "Local scale invariance, Cantorianspace-time and unified field theory", *Chaos, Solitons and Fractals* 23, 3, 701-10 (2005).
[21] E. Goldfain, "Fractional dynamics and the TeV regime of field theory", *Comm. Nonlin. Science and Numer. Simul.*, 13, 3, 666-76, (2008).
[22] E. Goldfain, "Dynamic Instability of the Standard Model and the Fine-Tuning Problem", *Prespacetime Journal*, 12, 12, 1175-81 (2012).
[23] B. Grinstein, J. F. Fortin and A. Stergiou, "Scale Invariance without Conformal Invariance in Relativistic Quantum Field Theory", e-print http://planck12.fuw.edu.pl/talks/grinstein.pdf (2012).
[24] M. Reuter and F. Saueressig, "Quantum Einstein gravity" *New J. Phys.* 14, 055022 (2012).
[25] S. Weinberg, "Minimal Fields of Canonical Dimensionality are Free", *Phys. Rev. D* 86, 105015 (2012).
[26] A. Morozov and A. Niemi, "Can renormalization group flow end in a Big Mess", *Nucl. Phys.* B, 666, 3, 311-36 (2003).
[27] M. A. Luty, J. Polchinski and R. Rattazi, "The *a*-theorem and the asymptotics of 4D quantum field theory", *Journal of High Energy Physics*, 152 (2013).
[28] K. Michael Martini, "*Limit cycles in the Renormalization Group*", e-print http://guava.physics.uiuc.edu/~nigel/courses/563/Essays_2012/PDF/Martini.pdf (2012).
[29] H.W. Hammer and L. Platter, "Efimov physics from a renormalization group perspective", *Phil. Trans. R. Soc.* A, 369 (2011).
[30] A. LeClair, "Renormalization group limit cycles and field theories for elliptic S-matrices", *J. Stat. Mech.*, 08004 (2004).

[31] H. Kleinert and V. Schulte-Frohlinde, "*Critical Properties of Φ^4 theories*", World Scientific, Singapore, 1-489 (2001).
[32] M. Pawlowski and R. Raczka, "*A Higgs-free model for fundamental interactions*", e-print http://arxiv.org/pdf/hep-ph/9503269.pdf (1995).
[33] M. Visser and N. Yunes, "Power Laws, Scale Invariance and the Generalized Frobenius Series", *Intl. Journal of Mod. Phys.* A, 18, 20, 3433-68 (2003).
[34] D. S. Glazek and K. G. Wilson, "Limit cycles in quantum theories", *Phys. Rev. Lett.* 89, 230401 (2002).
[35] E. Goldfain, "Chaotic dynamics of the Renormalization Group flow and Standard Model parameters", *J. Nonlin. Sci* 3, 170-80, (2007).
[36] N. A. Magnitskii and S. V. Sidorov, "New Methods for Chaotic Dynamics", *World Scientific Publishing, Series in Nonlinear Science A*, 58 (2006).
[37] N. A. Magnitskii, "Universal theory of dynamical chaos in nonlinear dissipative systems of differential equations", *Comm. Nonlin. Sci. Numer. Simul.*, 13, 416-33, (2008).
[38] Y. Nakayama, "*A lecture note on scale invariance versus conformal invariance*", e-print http://arxiv.org/pdf/1302.0884v1.pdf (2013).
[39] L Muñoz, C Fernández-Ramírez, A Relaño and J Retamosa, "Chaos in Hadrons", *J. Phys.: Conf. Ser.* 381, 1, 012031 (2012).
[40] E. Goldfain,"Dynamic Instability of the Standard Model & the Fine-Tuning Problem", *Prespacetime Journal*, 12, 12, 1175-81, (2012).
[41] T. S. Biró, S. G. Matinyan and B. Müller, "Chaos in gauge field theory", World Scientific publishing, *Lectures Notes in Physics* 56 (1995).
[42] J. F. Donoghue, E. Golowich and B. R. Holstein, "*Dynamics of the Standard Model*", Cambridge University Press, (1994).
[43] M. Frasca, "*Mapping theorem and Green functions in Yang-Mills theory*", e-print http://pos.sissa.it/archive/conferences/117/039/FacesQCD_039.pdf (2010).
[44] J. P. Chen, "*Bose-Einstein condensation on fractal spaces*", e-print http://math.arizona.edu/~mathphys/school_2012/JoeChen.pdf (2012).
[45] E. Akkermans, "*Statistical Mechanics and Quantum Fields on Fractals*", e-print http://arxiv.org/abs/1210.6763(2012).
[46] E. Goldfain, "Complex Dynamics and the Future of Particle Physics", *Nonlin.Sci. Lett. A* 1, 1, 39-42, (2010).
[47] V. Tarasov, "Electromagnetic Fields on Fractals", *Mod. Phys. Lett.* A, 21, 1587 (2006).
[48] M. Ostoja-Starzewski, "Electromagnetism on anisotropic fractal media", *Z. Angew. Math. Phys.* (2012).
[49] A. S. Balankin, B. Mena, J. Patiño, D. Morales, "Electromagnetic fields in fractal continua", *Phys. Lett. A*, 377, 10-11, 783-88, (2013).
[50] G. Calcagni and G. Nardelli, "*Symmetries and propagator in multi-fractional scalar field theory*", e-print http://eprintweb.org/s/article/arxiv/1210.2754 (2012).
[51] E. Goldfain, "Dynamics of neutrino oscillations and the cosmological constant problem", in *Hadron Models and New Energy Issues*, InfoLearn Quest USA, ISBN 978-1-59973-042-4, 168-75 (2007).
[52] L. P. Kadanoff, "*Statistical Physics: Statics, Dynamics and Renormalization*", World Scientific Publishing Co. (1999).

[53] D. Hooper and T. A. Slatyer, "*Two Emission Mechanisms in the Fermi Bubbles: A Possible Signal for Annihilating Dark Matter*", e-print http://arxiv.org/abs/1302.6589 (2013).

[54] J. Ellis, V. Sanz and T. You, "*Associated Production Evidence against Higgs Impostors and Anomalous Couplings*", e-print http://arxiv.org/pdf/1303.0208v1.pdf (2013).

[55] J. L. Hewett and H. Weerts, "*Fundamental Physics at the Intensity Frontier*", e-print http://arxiv.org/abs/1205.2671 (2012).

[56] A. Morozov, "*Integrability in non-perturbative QFT*", e-print http://arxiv.org/abs/1303.2578 (2013).

In: Space-Time Geometry and Quantum Events
Editor: Ignazio Licata, pp. 175-196
ISBN: 978-1-63117-455-1

Chapter 5

RELATIVITY OF SCALES, FRACTAL SPACE AND QUANTUM POTENTIALS

Laurent Nottale*
CNRS, LUTH, Paris Observatory and Paris-Diderot University,
Meudon CEDEX, France

Abstract

The principles and methods of the theory of the relativity of scales are briefly reminded in this paper, then they are applied to two unsolved problems in physics, high temperature superconductivity and turbulence. We emphasize in particular the concept of quantum-type potentials, since in many situations the effect of the fractality of space – or of the underlying medium – amounts to the addition of such a potential energy to the classical equations of motion. Various equivalent representations of these equations – geodesic, quantum, fluid-mechanical, stochastic, diffusion – are given, as well as several forms of generalized quantum potentials. We give two examples of applications of these concepts to physics: (i) in solid-state physics, we show that the attractive potential energy that binds electrons together into Cooper pairs in high critical temperature superconducting cuprate materials can be understood in terms of a quantum potential that originates from the local fluctuations of the density of charges induced by dopants; (ii) in fluid mechanics, we elaborate de Montera's suggestion of applying the scale relativity theory in velocity space in order to get a new statistical description of turbulence based on a macroscopic Schrödinger-type equation.

1. Introduction

The theory of scale relativity and fractal space-time accounts for a possibly nondifferentiable geometry of the space-time continuum, basing itself on an extension of the principle of relativity to scale transformations of the reference system.

This theory has been initially built with the goal of re-founding quantum mechanics on prime principles [31, 32, 34]. The success of this enterprise [41, 46] has been completed by obtaining new results: in particular, a generalisation of standard quantum mechanics at

*Email: laurent.nottale@obspm.fr

high energy to new forms of scale laws [33, 34, 46], and the discovery of the possibility of macroscopic quantum-type behavior under certain conditions [36].

This new "macroquantum" mechanics no longer rests on the microscopic Planck constant $\hbar$. The parameter which replaces $\hbar$ is specific of the system under consideration, emerges from self-organization of this system and can now be macroscopic or mesoscopic. This theory is specifically adapted to the description of multi-scale systems able to spontaneous self-organization and structuration. Two priviledged domains of applications are therefore astrophysics [34, 35, 36, 37, 39, 46] and biophysics [4, 28, 29, 38, 42, 44, 46].

In this contribution, we develop some aspects of the theory which may be relevant to its explicit use to solving self-organization problems. A special emphasis is made of the concept of macroquantum potential energy. Scale relativity methods are relevant because they provide new mathematical tools to deal with scale-dependent fractal systems, like equations in scale space and scale-dependent derivatives in physical space.

2. Brief Reminder of the Theory

The theory of scale relativity consists of introducing in an explicit way the scale ε of measurement (or of observation) in the physical description. These scale variables can be identified, in a theoretical framework, to the differential elements dx or dt, and, in an experimental or observational framework, to the resolution of the measurement apparatus.

In its framework, the coordinates can be explicit functions of these variables, $X = X(dX)$ (we omit the indices for simplicity of the writing, but the coordinates are in general vectors while the resolution variables are tensors [46, Chap. 3.6]). In case of divergence of these functions toward small scales, they are fractal coordinates. The various quantities which describe the system under consideration become themselves fractal functions, $F = F[X(dX), dX]$. In the simplified case when the fractality of the system is but a consequence of that of space, there is no proper dependence of F in function of dX, and we have merely $F = F[X(dX)]$.

The description of such an explicitly scale dependent system in terms of differential equations needs three levels instead of two. Usually, one makes a transformation of coordinates $X \to X + dX$, then one looks for the effect of this infinitesimal transformation on the system properties, $F \to F + dF$. This leads to write differential equations in terms of space-time coordinates.

But in the new situation, since the coordinates are now scale dependent, one should first state the laws of scale transformation, $\varepsilon \to \varepsilon'$, then their consequences on the coordinates, $X(\varepsilon) \to X'(\varepsilon')$ and finally on the various physical quantities $F[X(\varepsilon)] \to F'[X'(\varepsilon')]$. One of the main methods of the scale relativity theory just consists of describing these scale transformations by differential equations acting in scale space (i.e., the space of the scale variables $\{\varepsilon\}$). In other words, one considers infinitesimal scale transformations, $\ln(\varepsilon/\lambda) \to \ln(\varepsilon/\lambda) + d\ln(\varepsilon/\lambda)$, rather than the discrete iterated transformations that have been most often used in the study of fractal objects [5, 22, 23].

The motion equations in scale relativity are therefore obtained in the framework of a double partial differential calculus acting both in space-time (positions and instants) and in scale space (resolutions), basing oneself on the constraints imposed by the double principle of relativity, of motion and of scale.

2.1. Laws of Scale Transformation

The simplest possible scale differential equation which determines the length of a fractal curve (i.e., a fractal coordinate) $\mathcal{L}$ reads

$$\frac{\partial \mathcal{L}}{\partial \ln \varepsilon} = a + b\mathcal{L}, \tag{1}$$

where $\partial/\partial \ln \varepsilon$ is the dilation operator [33, 46]. Its solution combines a self-similar fractal power-law behavior and a scale-independent contribution:

$$\mathcal{L}(\varepsilon) = \mathcal{L}_0 \left\{ 1 + \left(\frac{\lambda}{\varepsilon}\right)^{\tau_F} \right\}, \tag{2}$$

where λ is an integration constant and where $\tau_F = -b = D_F - 1$. One easily verifies that the fractal part of this expression agrees with the principle of relativity applied to scales. Indeed, under a transformation $\varepsilon \to \varepsilon'$, it transforms as $\mathcal{L} = \mathcal{L}_0(\varepsilon'/\varepsilon)^{\tau_F}$ and therefore it depends only on the ratio between scales and not on the individual scales themselves.

This result indicates that, in a general way, fractal functions are the sum of a differentiable part and of a non-differentiable (fractal) part, and that a spontaneous transition is expected to occur between these two behaviors.

On the basis of this elementary solution, many generalisations have been obtained, in particular:

– log-periodic corrections to power laws:

$$\mathcal{L}(\varepsilon) = a\,\varepsilon^{\nu}\,[1 + b\cos(\omega \ln \varepsilon)]. \tag{3}$$

which is a solution of a second-order differential wave equation in scales.

– laws of "scale dynamics" showing a "scale acceleration":

$$\tau_F = \frac{1}{G} \ln\left(\frac{\lambda_0}{\varepsilon}\right), \quad \ln\left(\frac{\mathcal{L}}{\mathcal{L}_0}\right) = \frac{1}{2G} \ln^2\left(\frac{\lambda_0}{\varepsilon}\right). \tag{4}$$

This law may be the manifestation of a constant "scale force", which describes the difference with the free self-similar case (in analogy with Newton's dynamics of motion). In this case the fractal dimension is no longer constant, but varies in a linear way in terms of the logarithm of resolution. Many manifestations of such a behavior have been identified in human and physical geography [12, 49].

– laws of special scale relativity [33]:

$$\ln \frac{\mathcal{L}(\varepsilon)}{\mathcal{L}_0} = \frac{\tau_0 \ln(\lambda_0/\varepsilon)}{\sqrt{1 - \ln^2(\lambda_0/\varepsilon)/\ln^2(\lambda_0/\lambda_H)}}, \tag{5}$$

$$\tau_F(\varepsilon) = \frac{\tau_0}{\sqrt{1 - \ln^2(\lambda_0/\varepsilon)/\ln^2(\lambda_0/\lambda_H)}}. \tag{6}$$

This case will not be applied in the present contribution, but we recall it here because it is one of the most profound manifestations of scale relativity. Here the length (i.e. the fractal coordinate) and the 'djinn' $\tau_F(\varepsilon) = D_F(\varepsilon) - 1$, which characterizes the fractal dimension now considered as a variable, have become the components of a vector in scale space. In this new law of scale transformation, a limiting scale appears, λ_H, which is impassable and invariant under dilations and contractions, independently of the reference scale λ_0. We have identified this invariant scale to the Planck length $l_{\mathbb{P}} = \sqrt{\hbar G/c^3}$ toward small scales, and to the cosmic length $\mathbb{L} = 1/\sqrt{\Lambda}$ (where Λ is the cosmological constant) toward large scales [33, 34, 46].

Many other scale laws can be constructed as expressions of Euler-Lagrange equations in scale space, which give the general form expected for them [46, Chap. 4], in analogy with the well known construction of the laws of motion as Euler-Lagrange equations in standard space-time.

2.2. Laws of Motion in Fractal Space

The laws of motion are obtained in the scale relativity theory by writing the fundamental equation of dynamics (which reduces to a geodesic equation in the absence of an exterior field) in a fractal space. The non-differentiability and the fractality of coordinates implies at least three consequences [34, 46]:

(1) The number of possible paths is infinite. The description therefore naturally becomes non-deterministic and probabilistic. These virtual paths are identified to the geodesics of the fractal space. The ensemble of these paths constitutes a fluid of geodesics, which is therefore characterized by a velocity field.

(2) Each of these paths is itself fractal. The velocity field is therefore a fractal function, explicitly dependent on resolutions and divergent when the scale interval tends to zero (this divergence is the manifestation of non-differentibility).

(3) Moreover, the non-differentiability also implies a two-valuedness of this fractal function, $(V_+,\ V_-)$. Indeed, two definitions of the velocity field now exist, which are no longer invariant under a transformation $|dt| \to -|dt|$ in the nondifferentiable case.

These three properties of motion in a fractal space lead to describing the geodesic velocity field in terms of a complex fractal function $\widetilde{\mathcal{V}} = (V_+ + V_-)/2 - i(V_+ - V_-)/2$. The $(+)$ and $(-)$ velocity fields can themselves be decomposed in terms of a differentiable part $v_\pm$ and of a fractal (divergent) fluctuation of zero mean $w_\pm$, i.e., $V_\pm = v_\pm + w_\pm$ and therefore the same is true of the full complex velocity field, $\widetilde{\mathcal{V}} = \mathcal{V}(x, y, z, t) + \mathcal{W}(x, y, z, t, dt)$.

Jumping to elementary displacements along these geodesics, this reads $dX_\pm = d_\pm x + d\xi_\pm$, with (in the case of a critical fractal dimension $D_F = 2$ for the geodesics)

$$d_\pm x = v_\pm\ dt, \quad d\xi_\pm = \zeta_\pm \sqrt{2\mathcal{D}}\, |dt|^{1/2}. \tag{7}$$

This case is particularly relevant since it corresponds to a Markov-like situation of loss of information from one point to the following, without correlation nor anti-correlation. Here $\zeta_\pm$ represents a dimensionless stochastic variable such that $<\zeta_\pm>= 0$ and $<\zeta_\pm^2>= 1$. The parameter $\mathcal{D}$ characterizes the amplitude of fractal fluctuations.

These various effects can be combined in terms of a total derivative operator [34] :

$$\frac{\widehat{d}}{dt} = \frac{\partial}{\partial t} + \mathcal{V}.\nabla - i\mathcal{D}\Delta. \tag{8}$$

The fundamental equation of dynamics becomes, when it is written in terms of this operator

$$m \frac{\widehat{d}}{dt} \mathcal{V} = -\nabla\phi. \tag{9}$$

In the absence of an exterior field ϕ, this is a geodesic equation (i.e., a free inertial Galilean-type equation).

The next step consists of making a change of variable in which one connects the velocity field $\mathcal{V} = V - iU$ to the function $\psi = e^{iS/S_0}$ (where S is the action, now complex because the velocity field is complex) according to the relation

$$m\mathcal{V} = -i\, S_0\, \nabla \ln \psi. \tag{10}$$

This equation is but the standard relation between momentum and action $P = \nabla S$, that provides a new expression (now exact) for the principle of correspondance. The parameter S_0 is a constant for the system considered (it identifies to the Planck constant $\hbar$ in standard quantum mechanics). Thanks to this change of variable, the equation of motion can be integrated under the form of a Schrödinger equation [34, 46] generalized to a constant different from $\hbar$,

$$\mathcal{D}^2 \Delta\psi + i\mathcal{D}\frac{\partial}{\partial t}\psi - \frac{\phi}{2m}\psi = 0, \tag{11}$$

where the two parameter introduced hereabove, S_0 and $\mathcal{D}$, are linked by the relation:

$$S_0 = 2m\mathcal{D}. \tag{12}$$

In the case of standard quantum mechanics, $S_0 = \hbar$, so that $\mathcal{D}$ is a generalisation of the Compton length (up to the constant c) and Eq. (12) is a generalisation of the Compton relation

$$\lambda_C = \frac{2\mathcal{D}}{c} = \frac{\hbar}{mc}. \tag{13}$$

We obtain the same result by using the full velocity field including the fractal fluctuations of zero mean [46]. This implies the possible existence of fractal solutions for quantum mechanical equations [7].

By setting finally $\psi = \sqrt{P} \times e^{i\theta}$, with $V = 2\mathcal{D}\nabla\theta$, one can show (see [41, 46] and next section) that $P = |\psi|^2$ gives the number density of virtual geodesics. This function becomes naturally a density of probability or a density of matter (or of radiation) according to the various conditions of an actual experiment (one particle, many particles or a radiation flow). The function ψ, being solution of the Schrödinger equation and subjected to the Born postulate and to the Compton relation, owns therefore all the properties of a wave function.

Reversely, the density ρ and the velocity field V of a fluid in potential motion can be combined in terms of a complex function $\psi = \sqrt{\rho} \times e^{i\theta}$ which may become a wave function solution of a Schrödinger equation under some conditions, in particular in the presence of a quantum-type potential (see next section).

3. Multiple Representations

After this brief summary of the theory (see more details in [46]), let us now elabotate some of its aspects. One of them is the multiplicity of equivalent representations of the same equations. Usually, classical deterministic equations, quantum equations, stochastic equations, fluid mechanics equations, etc.. correspond to different systems and even to different physical laws. But in the scale relativity framework, they are unified as being different representations of the same fundamental equation (the geodesic equation of relativity), subjected to various changes of variable. This is a particularly useful tool for applications in sciences where one makes often use of diffusion equations of the Fokker-Planck type or of fluid mechanics equations.

3.1. Geodesic Representation

The first representation, which can be considered as the root representation, is the geodesic one. The two-valuedness of the velocity field is expressed in this case in terms of the complex velocity field $\mathcal{V} = V - iU$. It implements what makes the essence of the principle of relativity, i.e., the equation of motion must express the fact that any motion should disappear in the proper system of coordinates:

$$\mathcal{V} = 0. \tag{14}$$

By deriving this equation with respect to time, it takes the form of a free inertial equation devoid of any force:

$$\frac{\widehat{d}}{dt}\mathcal{V} = 0, \tag{15}$$

where the 'covariant' derivative operator $\widehat{d}/dt$ includes the terms which account for the effects of the geometry of space – more generally of space-time in the 'relativistic' case (see [46] and references therein). In the case of a fractal space, it reads as we have seen

$$\frac{\widehat{d}}{\partial t} = \frac{\partial}{\partial t} + \mathcal{V}.\nabla - i\mathcal{D}\Delta. \tag{16}$$

3.2. Quantum-Type Representation

We have recalled in the previous section how a wave function ψ can be introduced from the velocity field of geodesics:

$$\mathcal{V} = -2i\mathcal{D}\,\nabla \ln \psi. \tag{17}$$

This mean that the doubling of the velocity field issued from non-diifferentiability is expressed in this case in terms of the modulus and the phase of this wave function. This allows integration of the equation of motion in the form of a Schrödinger equation,

$$\mathcal{D}^2\Delta\psi + i\,\mathcal{D}\frac{\partial}{\partial t}\psi - \frac{\phi}{2m}\psi = 0. \tag{18}$$

By making explicit the modulus and the phase of the wave function, $\psi = \sqrt{P} \times e^{i\theta}$, where the phase is related to the classical velocity field by the relation $V = 2\mathcal{D}\,\nabla\theta$, one can give this equation the form of hydrodynamics equations including a quantum potential. Moreover, it has been recently shown that this transformation is reversible, i.e., by adding a quantum-like potential energy to a classical fluid, it becomes described by a Schrödinger equation and therefore acquires some quantum-type properties [43, 46].

3.3. Fluid Representation with Macroquantum Potential

It is also possible, as we shall now prove it, to go directly from the geodesic representation to the fluid representation with quantum-type potential without writing explicitly the Schrödinger equation.

To this purpose, let us express the complex velocity field in terms of the classical (real) velocity field V and of the number density of geodesics P_N, which is equivalent, as we have seen hereabove, to a probability density P:

$$\mathcal{V} = V - i\mathcal{D}\nabla \ln P. \tag{19}$$

The quantum covariant derivative operator thus reads

$$\frac{\widehat{d}}{\partial t} = \frac{\partial}{\partial t} + V.\nabla - i\mathcal{D}\left(\nabla \ln P.\nabla + \Delta\right). \tag{20}$$

The fundamental equation of dynamics becomes (introducing also an exterior scalar potential ϕ):

$$\left(\frac{\partial}{\partial t} + V.\nabla - i\mathcal{D}\left(\nabla \ln P.\nabla + \Delta\right)\right)\left(V - i\mathcal{D}\nabla \ln P\right) = -\frac{\nabla\phi}{m}. \tag{21}$$

The imaginary part of this equation,

$$\mathcal{D}\left\{(\nabla \ln P.\nabla + \Delta)V + \left(\frac{\partial}{\partial t} + V.\nabla\right)\nabla \ln P\right\} = 0, \tag{22}$$

takes, after some calculations, the following form

$$\nabla\left\{\frac{1}{P}\left(\frac{\partial}{\partial t} + \text{div}(PV)\right)\right\} = 0, \tag{23}$$

and it can finally be integrated in terms of a continuity equation:

$$\frac{\partial P}{\partial t} + \text{div}(PV) = 0. \tag{24}$$

The real part,

$$\left(\frac{\partial}{\partial t} + V.\nabla\right)V = -\frac{\nabla\phi}{m} + \mathcal{D}^2\left(\nabla \ln P.\nabla + \Delta\right)\nabla \ln P, \tag{25}$$

takes the form of an Euler equation,

$$m\left(\frac{\partial}{\partial t} + V.\nabla\right)V = -\nabla(\phi + Q), \tag{26}$$

where Q is an additional potential energy that depends on the probability density P:

$$Q = -2m\mathcal{D}^2 \frac{\Delta\sqrt{P}}{\sqrt{P}}. \tag{27}$$

It is remarkable that we have obtained this result directly, without passing through a quantum-type representation using a wave function nor through a Schrödinger equation.

The additional "fractal" potential is obtained here as a mere manifestation of the fractal geometry of space, in analogy with Newton's potential emerging as a manifestation of the curved geometry of space-time in Einstein's relativistic theory of gravitation. We have suggested ([46] and references therein) that this geometric energy could contribute to the effects which have been attributed in astrophysics to a missing "dark matter" (knowing that all attempts to directly observe this missing mass have so far failed). Another suggestion, relevant to biology, is that such a potential energy could play an important role in the self-organisation and in the morphogenesis of living systems [40, 42].

3.4. Coupled Two-Fluids

Another equivalent possible representation consists of separating the real and imaginary parts of the complex velocity field,

$$\mathcal{V} = V - iU. \tag{28}$$

One obtains in this case a system of equations that describe the velocity fields of two fluids strongly coupled together,

$$\left(\frac{\partial}{\partial t} + V.\nabla\right) V = (U.\nabla + \mathcal{D}\Delta)U - \nabla\left(\frac{\phi}{m}\right), \tag{29}$$

$$\left(\frac{\partial}{\partial t} + V.\nabla\right) U = -(U.\nabla + \mathcal{D}\Delta)V. \tag{30}$$

This representation may be useful in, e.g., numerical simulations of scale relativity / quantum processes [16].

3.5. Diffusion-Type Representation

The fundamental two-valuedness which is a consequence of nondifferentiability has been initially described in terms of two mean velocity fields v_+ and v_-, which transform one into the other by the reflexion $|dt| \leftrightarrow -|dt|$. It is therefore possible to write the equations of motion directly in terms of these two velocity fields. The representation obtained in this way implements the diffusion-like property of a fractal space and is therefore particularly interesting for applications to diffusive systems. Indeed, one obtains a standard Fokker-Planck equation for the velocity v_+, in similarity with a classical stochastic process :

$$\frac{\partial P}{\partial t} + \text{div}(P v_+) = \mathcal{D}\Delta P, \tag{31}$$

where the parameter $\mathcal{D}$ plays the role of a diffusion coefficient. On the contrary, the equation obtained for the velocity field v_- does not correspond to any classical process:

$$\frac{\partial P}{\partial t} + \mathrm{div}(P v_-) = -\mathcal{D}\Delta P. \tag{32}$$

In the scale relativity approach, this equation is *derived* from the geodesic equation on the basis of nondifferentiability, and it is therefore not a classical equation. It cannot be set as a founding equation in the framework of a classical diffusion process, contrary to Nelson's proposal [27], since it becomes self-contradictory with the backward Kolmogorov equation generated by such a classical process [15, 58, 36] [46, p. 384].

3.6. A New Form of Quantum-Type Potential

However, one may remark that in the previous representation the variables are not fully independent, since it involves three quantities P, v_+ and v_- instead of two expected from the velocity doubling. Therefore it should be possible to obtain a system of equations involving only the probability density P and one of the velocity fields, for example v_+. To this purpose, one remarks that v_- is given in terms of these two quantities by the relation:

$$v_- = v_+ - 2\mathcal{D}\nabla \ln P. \tag{33}$$

We also recall that

$$V = v_+ - \mathcal{D}\nabla \ln P. \tag{34}$$

The energy equation now reads

$$E = \frac{1}{2} m\, V^2 + Q + \phi = \frac{1}{2} m \,(v_+ - \mathcal{D}\nabla \ln P)^2 + Q + \phi, \tag{35}$$

where the macroquantum potential can be written

$$Q = -2m\mathcal{D}^2 \, \frac{\Delta\sqrt{P}}{\sqrt{P}} = -m\mathcal{D}^2 \left\{ \Delta \ln P + \frac{1}{2} (\nabla \ln P)^2 \right\}. \tag{36}$$

One of the terms of this "fractal potential" is therefore compensated while another term appears, so that we obtain:

$$E = \frac{1}{2} m\, v_+^2 + \phi - m\,\mathcal{D}\, v_+ . \nabla \ln P - m\mathcal{D}^2 \Delta \ln P. \tag{37}$$

We finally obtain a new representation in terms of a Fokker-Planck equation, which contains the diffusive term $\mathcal{D}\Delta P$ in addition to the continuity equation obtained in the case of the fluid representation (V, P), and an energy equation which includes a new form of quantum potential:

$$\frac{\partial P}{\partial t} + \mathrm{div}(P v_+) = \mathcal{D}\Delta P, \tag{38}$$

$$E = \frac{1}{2} m\, v_+^2 + \phi + Q_+, \tag{39}$$

where the new quantum-type potential reads

$$Q_+ = -m\,\mathcal{D}\,(v_+.\nabla \ln P + \mathcal{D}\Delta \ln P). \tag{40}$$

It now depends not only on the probability density P, but also on the velocity field v_+.

This derivation is once again reversible. This means that a classical diffusive system described by a standard Fokker-Planck equation which would be subjected to such a generalized quantum-type potential would be spontaneously transformed into a quantum-like system described by a Schrödinger equation (18) acting on a wave function $\psi = \sqrt{P} \times e^{i\,\theta}$ where $V = 2\mathcal{D}\nabla\theta$. Thanks to Eq. (34), this wave function is defined in terms of P and v_+ as

$$\psi = \sqrt{P}^{\,1-i} \times e^{i\,\theta_+}, \tag{41}$$

where $v_+ = 2\mathcal{D}\,\nabla\theta_+$.

Such a system, although it is initially diffusive, would therefore acquire some quantum-type properties, but evidently not all of them: the behaviors of coherence, unseparability, indistinguishability or entanglement are specific of a combination of quantum laws and elementarity [59] and cannot be recovered in such a context.

This is nevertheless a remarkable result, which means that a partial reversal of diffusion and a transformation of a classical diffusive system into a quantum-type self-organized one should be possible by applying a quantum-like force to this system. This is possible in an actual experiment consisting of a retro-active loop involving continuous measurements, not only of the density [43] but also of the velocity field v_+, followed by a real time application on the system of a classical force $F_{Q+} = -\nabla Q_+$ simulating the new macroquantum force [47].

One may also wonder whether living systems, which already work in terms of such a feedback loop (involving sensors, then cognitive processes, then actuators) could have naturally included such kinds of quantum-like potentials in their operation through the selection / evolution process, simply because it provides an enormous evolutionary advantage due to its self-organization and morphogenesis negentropic capabilities [42] [46, Chap. 14].

3.7. Quantum Potential Reversal

One of the recently obtained results which may be particularly relevant to the understanding of living systems concerns the reversal of the quantum-type potentiel. What happens when the potential energy keeps its same form, as given by $\Delta\sqrt{P}/\sqrt{P}$ for a given distribution $P(x, y, z)$, while its sign is reversed ? In other words, to what kind of process does the equation

$$\left(\frac{\partial}{\partial t} + V.\nabla\right) V = -\frac{\nabla\phi}{m} - 2\mathcal{D}^2\,\nabla\frac{\Delta\sqrt{P}}{\sqrt{P}}, \tag{42}$$

correspond ?

We have shown [42, 46] that such an Euler equation, when it is combined with a continuity equation, can no longer be integrated under the form of a generalized Schrödinger equation. This process is therefore no longer self-organizing. On the contrary, this is a classical diffusive process, characterized by an entropy increase proportional to time.

Indeed, let us start from a Fokker-Planck equation

$$\frac{\partial P}{\partial t} + \mathrm{div}(Pv) = D\Delta P, \tag{43}$$

which describes a classical diffusion process with diffusion coefficient D. Then make the change of variable

$$V = v - D\nabla \ln P. \tag{44}$$

One finds after some calculations that V and P are now solutions of a continuity equation

$$\frac{\partial P}{\partial t} + \mathrm{div}(Pv) = 0, \tag{45}$$

and of an Euler equation which reads

$$\left(\frac{\partial}{\partial t} + V.\nabla\right) V = -2D^2\, \nabla \frac{\Delta\sqrt{P}}{\sqrt{P}}. \tag{46}$$

In other words, we have obtained a fluid dynamical description of a standard diffusion process in terms of a "diffusion potential" which is exactly the reverse of the macroquantum potential.

4. Application to High-Temperature Superconductivity

4.1. Ginzburg-Landau Non-Linear Schrödinger Equation

The phenomenon of superconductivity is one of the most fascinating of physics. It lies at the heart of a large part of modern physics. Indeed, besides its proper interest for the understanding of condensed mater, it has been used as model for the construction of the electroweak theory through the Higgs field and of other theories in particle physics and other sciences.

Moreover, superconductivity (SC) has led physicists to deep insights about the nature of matter. It has shown that the ancient view of matter as something "solid", in other words "material", was incorrect. The question: "is it possible to walk through walls" is now asked in a different way. Nowadays we know that it is not a property of matter by itself which provides it qualities such as solidity or ability to be crossed, but its interactions.

A first relation of SC with the scale relativity approach can be found in its phenomenological Ginzburg-Landau equation. Indeed, one can recover such a non-linear Schrödinger equation simply by adding a quantum-like potential energy to a standard fluid including a pressure term [43].

Consider indeed an Euler equation with a pressure term and a quantum potential term:

$$\left(\frac{\partial}{\partial t} + V\cdot\nabla\right) V = -\nabla\phi - \frac{\nabla p}{\rho} + 2\mathcal{D}^2\, \nabla\left(\frac{\Delta\sqrt{\rho}}{\sqrt{\rho}}\right). \tag{47}$$

When $\nabla p/\rho = \nabla w$ is itself a gradient, which is the case of an isentropic fluid, and, more generally, of every cases when there is a state equation which links p and ρ, its combination with the continuity equation can be still integrated in terms of a Schrödinger-type equation [36],

$$\mathcal{D}^2 \Delta\psi + i\mathcal{D}\frac{\partial}{\partial t}\psi - \frac{\phi + w}{2}\psi = 0. \tag{48}$$

In the sound approximation, the link between pressure and density writes $p - p_0 = c_s^2(\rho - \rho_0)$, where c_s is the sound speed in the fluid, so that $\nabla p/\rho = c_s^2 \nabla \ln \rho$. Moreover, when $\rho - \rho_0 \ll \rho_0$, one may use the additional approximation $c_s^2 \nabla \ln \rho \approx (c_s^2/\rho_0)\nabla\rho$, and the equation obtained takes the form of the Ginzburg-Landau equation of superconductivity [19],

$$\mathcal{D}^2 \Delta\psi + i\mathcal{D}\frac{\partial}{\partial t}\psi - \beta\,|\psi|^2\,\psi = \frac{1}{2}\,\phi\,\psi, \tag{49}$$

with $\beta = c_s^2/2\rho_0$. In the highly compressible case, the dominant pressure term is rather of the form $p \propto \rho^2$, so that $p/\rho \propto \rho = |\psi|^2$, and one still obtains a non-linear Schrödinger equation of the same kind [30].

Laboratory experiments aiming at implementing this transformation of a classical fluid into a macroscopic quantum-type fluid are presently under development [45, 47].

4.2. A Quantum Potential as Origin of Cooper Pairs in HTS

Another important question concerning SC is that of the microscopic theory which gives rise to such a macroscopic phenomenological behavior.

In superconducting materials, the bounding of electrons in Cooper pairs transforms the electronic gas from a fermionic to a bosonic quantum fluid. The interaction of this fluid with the atoms of the SC material becomes so small that the conducting electrons do not "see" any longer the material. The SC electrons become almost free, all resistance is abolished and one passes from simple conduction to superconduction.

In normal superconductors, the pairing of electrons is a result of their interaction with phonons (see, e.g., [9]). But since 1985, a new form of superconductivity has been discovered which has been named "high temperature superconductivity" (HTS) because the critical temperature, which was of the order of a few kelvins for normal SC, has reached up to 135 K. However, though it has been shown that HTS is still due to the formation of Cooper pairs, the origin of the force that pairs the electrons can no longer be phonons and still remains unknown. Actually, it can be proved that any attractive force between the electrons, as small it could be, would produce their Cooper pairing [18].

Therefore the problem of HTS can be traced back to that of identifying the force that links the electrons. We suggest here that this force actually derives from a standard quantum potential, created by the density fluctuations of charges induced by the dopants.

Most HTS are copper oxide compounds in which superconductivity arises when they are doped either by extra charges but more often by 'holes' (positive charge carrier). Moreover, a systematic electronic inhomogeneity has been reported at the microscopic level, in particular in compounds like $Bi_2Sr_2CaCu_2O_{8+x}$ [51], the local density of states (LDOS) showing 'hills' and 'valleys' of size $\sim$ 30 Angstroms, strongly correlated with the SC gap. Actually, the minima of LDOS modulations preferentially occur at the dopant defects [21]. The regions with sharp coherence peaks, usually associated with strong superconductivity, are found to occur between the dopant defect clusters, near which the SC coherence peaks are suppressed.

Basing ourselves on these observations, we suggest that, at least in this type of compound, the electrons can be traped in the quantum potential well created by these electronic modulations.

Let us give here a summary of this new proposal. We denote by ψ_n the wave function of doping charges which have diffused from the initial site of dopant defects, and by ψ_s the wave function of the fraction of carriers which will be tied in Cooper pairs (only 19-23 % of the total doping induced charge joins the superfluid near optimum doping [55]).

We set $\psi_n = \psi_s + \psi_d$, where ψ_d is the wave function of the fraction of charges which do not participate in the superconductivity.

The doping induced charges constitutes a quantum fluid which is expected to be the solution of a Schrödinger equation (here of standard QM, i.e., written in terms of the microscopic Planck's constant $\hbar$)

$$\frac{\hbar^2}{2m}\Delta\psi_n + i\hbar\frac{\partial\psi_n}{\partial t} = \phi\,\psi_n, \tag{50}$$

where ϕ is a possible external scalar potential, and where we have neglected the magnetic effects as a first step.

Let us separate the two contributions ψ_s and ψ_d in this equation. We obtain:

$$\frac{\hbar^2}{2m}\Delta\psi_s + i\hbar\frac{\partial\psi_s}{\partial t} - \phi\,\psi_s = -\frac{\hbar^2}{2m}\Delta\psi_d - i\hbar\frac{\partial\psi_d}{\partial t} + \phi\,\psi_d. \tag{51}$$

We can now introduce explicitly the probability densities n and the phases θ of the wave functions $\psi_s = \sqrt{n_s} \times e^{i\theta_s}$ and $\psi_d = \sqrt{n_d} \times e^{i\theta_d}$. The velocity fields of the (s) and (d) quantum fluids are given by $V_s = (\hbar/m)\nabla\theta_s$ and $V_d = (\hbar/m)\nabla\theta_d$. As we have seen hereabove, a Schrödinger equation can be put into the form of fluid mechanics-like equations, its imaginary part becoming a continuity equation and the derivative of its real part becoming a Euler equation with quantum potential. Therefore the above equation can be written as:

$$\frac{\partial V_s}{\partial t} + V_s.\nabla V_s = -\frac{\nabla\phi}{m} - \frac{\nabla Q_s}{m} - \left(\frac{\partial V_d}{\partial t} + V_d.\nabla V_d + \frac{\nabla Q_d}{m}\right) \tag{52}$$

$$\frac{\partial n_s}{\partial t} + \mathrm{div}(n_s V_s) = -\frac{\partial n_d}{\partial t} - \mathrm{div}(n_d V_d). \tag{53}$$

But the (d) part of the quantum fluid, which is not involved in the superconductivity, remains essentially static, so that $V_d = 0$ and $\partial n_d/\partial t = 0$. Therefore we obtain for the quantum fluid (s) a new system of fluid equations:

$$\frac{\partial V_s}{\partial t} + V_s.\nabla V_s = -\frac{\nabla\phi}{m} - \frac{\nabla Q_s}{m} - \frac{\nabla Q_d}{m}, \tag{54}$$

$$\frac{\partial n_s}{\partial t} + \mathrm{div}(n_s V_s) = 0, \tag{55}$$

which can be re-integrated under the form of a Schrödinger equation, thanks to the reversibility of this transformation [43]

$$\frac{\hbar^2}{2m}\Delta\psi_s + i\hbar\frac{\partial\psi_s}{\partial t} - (\phi + Q_d)\psi_s = 0. \tag{56}$$

It therefore describes the motion of electrons (s), represented by their wave function ψ_s, in a potential well given by the exterior potential ϕ, but also by an interior quantum potential Q_d which just depends on the local fluctuations of the density n_d of charges,

$$Q_d = -\frac{\hbar^2}{2m}\frac{\Delta\sqrt{n_d}}{\sqrt{n_d}}. \tag{57}$$

Even if in its details this rough model is probably incomplete, we hope this proposal, according to which the quantum potential created by the dopants provides the attractive force needed to link electrons into Cooper pairs, to be globally correct, at least for some of the existing HT superconductors.

Many (up to now) poorly understood features of cuprate HTS can be explained by this model (L. Nottale, in preparation). For example, the quantum potential well involves bound states in which two electrons can be trapped with zero total spin and momentum. One can show that the optimal configuration for obtaining bound states is with 4 dopant defects (oxygen atoms), which bring 8 additional charges. One therefore expects a ratio $n_s/n_n = 2/(8+2) = 0.2$ at optimal doping. This is precisely the observed value [55], for which, to our knowledge, no explanation existed up to now.

The characteristic size of LDOS wells of $\sim$ 30 Angstroms is also easily recovered in this context: the optimal doping being $p = 0.155 = 1/6.5$, the 8 to 10 charges present in the potential well correspond to a surface $(8-10)\times 6.5 = (52-65) = (7.2-8.1)^2$ in units of $d_{\mathrm{CuO}} = 3.9$ Angstroms, i.e. 28-32 Angstroms as observed experimentally.

In this context, the high critical temperature superconductivity would be a geometric multiscale effect. In normal SC, the various elements which permit the superconductivity, Cooper pairing of electrons, formation of a quantum bosonic fluid and coherence of this fluid are simultaneous. In HTS, under the quantum potential hypothesis suggested here, these elements would be partly disconnected and related to different structures at different scales (in relation to the connectivity of the potential wells), achieving a multi-scale fractal structure [13].

If confirmed, this would be a nice application of the concept of quantum potentials [8], here in the context of standard microscopic quantum mechanics.

5. Application to Turbulence

In a recent work, L. de Montera has suggested an original application of the scale relativity theory to the yet unsolved problem of turbulence in fluid mechanics [10]. He has remarked that the Kolmogorov scaling of velocity increments in a Lagrangian description (where one follows an element of fluid, for example thanks to a seeded micro particle [20]),

$$\delta v \propto |\delta t|^{1/2}, \tag{58}$$

was exactly similar to the fractal fluctuation Eq. (7) which is at the basis of the scale relativity description.

The difference is that coordinates remain differentiable, while in this new context velocity becomes non-differentiable, so that accelerations $a = \delta v/\delta t \propto |\delta t|^{-1/2}$ become scale-divergent. Although this power law divergence is clearly limited by the dissipative Kolmogorov small scale, it is nevertheless fairly supported by experimental data, since acceleration of up to 1500 times the acceleration of gravity have been measured in turbulent flows [20, 57]).

De Montera's suggestion therefore amounts to apply the scale relativity method after an additional order of differentiation of the equations. The need for such a shift has already been remarked in the framework of classical stochastic models of turbulence [6, 54].

Let us consider here some possible implications of this new proposal.

The necessary conditions which underlie the construction of the scale relativity covariant derivative are very clearly fulfilled for turbulence (now in velocity space):
(1) The chaotic motion of fluid particles implies an infinity of possible paths.
(2) Each of the paths (realisations of which are achieved by test particles of size $< 100\ \mu$m in a lagrangian approach, [57]) are of fractal dimension $D_F = 2$ in velocity space, at least in the Kolmogorov (K41) regime (Eq. 58) [17, 14].
(3) The two-valuedness of acceleration is manifest in turbulence data. As remarked by Falkovich et al. [11], the usual statistical tools of description of turbulence (correlation function, second order structure function, etc...) are reversible, while turbulence, being a dissipative process, is fundamentally irreversible. The two-valuedness of derivative is just a way to account for the symmetry breaking under the time scale reflexion $\delta t \to -\delta t$. Among the various ways to describe this doubling [48, 50], one of them is particularly adapted to comparison with turbulence data. It consists of remarking that the calculation of a derivative involves a Taylor expansion

$$\frac{dX}{dt} = \frac{X(t+dt) - X(t)}{dt} = \frac{(X(t) + X'(t)dt + \frac{1}{2}X''(t)dt^2 + ...) - X(t)}{dt}, \tag{59}$$

so that one obtains

$$\frac{dX}{dt} = X'(t) + \frac{1}{2}X''(t)\, dt + ... \tag{60}$$

For a standard non fractal function, the contribution $\frac{1}{2}X''(t)dt$ and all the following terms of higher order vanish when $dt \to 0$, so that one recovers the usual result $dX/dt = X'(t)$. But for a fractal function such that its second derivative is scale divergent as $X''(t) \propto 1/dt$, the second order term can no longer be neglected and must contribute to the definition of the derivative [46, Sec. 3.1]. Therefore one may write

$$\frac{d_+X}{dt} = X'(t) + \frac{1}{2}X''(t)\, |dt|, \quad \frac{d_-X}{dt} = X'(t) - \frac{1}{2}X''(t)\, |dt|, \tag{61}$$

then

$$\frac{\widehat{d}X}{dt} = \frac{d_+ + d_-}{2dt}X - i\,\frac{d_+ - d_-}{2dt}X = X'(t) - i\,\frac{1}{2}X''(t)\, |dt|. \tag{62}$$

Lagrangian measurements of turbulence data by N. Mordant [24, 25] confirm this expectation. One finds that the acceleration $a = v'$ and its increments $da = v''dt$ are indeed of the same numerical order: in these data, the dispersions are respectively $\sigma_a = 280\,\mathrm{m/s}^2$ versus $\sigma_{da} = 224\,\mathrm{m/s}^2$.

This result is fully supported by the Sawford-Pope stochastic model of turbulence [52, 54]. Indeed, in this model, the acceleration is given by a Langevin equation including a K41 stochastic fluctuation:

$$\frac{dv}{dt} = -\frac{v}{T_L} + (C_0 \varepsilon)^{1/2} dW, \tag{63}$$

where C_0 is the Kolmogorov constant associated with the second-order Lagrangian structure function and ε the mean dissipated kinetic energy per unit mass. The derivative of acceleration is given by a stochastic differential equation:

$$da = -C\,a\,dt - D\,v\,dt + B\,dW. \tag{64}$$

It therefore includes a Langevin-like damping term $-C\,a$, an harmonic oscillator-like term $-D\,v$ (in v-space) and a fluctuating white noise term $B\,dW$ with $dW \sim \sqrt{dt}$.

The coefficients are given, in terms of the velocity variance σ_v^2, the acceleration variance σ_a^2 and the Lagrangian integral time scale T_L, by:

$$C = \frac{\sigma_a^2}{\sigma_v^2} T_L, \quad D = \frac{\sigma_a^2}{\sigma_v^2}, \quad B = \frac{\sigma_a^2}{\sigma_v} \sqrt{2T_L}. \tag{65}$$

For the same experimental data [24, 25], which are performed with a resolution time scale $\tau_{\min} = 1/6500$ s, the obtained numerical values are $\sigma_v = 0.98$ m/s, $\sigma_a = 280$ m/s^2 and $T_L = 0.0224$ s. This yields respective expected contributions to the standard deviation of da: $D\sigma_v\tau_{\min} = 12$ m/s^2, $C\sigma_a\tau_{\min} = 79$ m/s^2 and $B\sqrt{\tau_{\min}} = 210$ m/s^2, so that the expected standard deviation of da is $\sigma_{da} = 225\,\mathrm{m/s}^2$, in very good agreement with the observed one.

This fundamental result, according to which $\sigma_{da} \approx \sigma_a$ (the ratio being 0.8 in the Mordant experimental data) fully supports the acceleration two-valuedness on an experimental basis (see also [50]).

(4) The dynamics is Newtonian: the equation of dynamics in velocity space is the time derivative of the Navier-Stokes equation, i.e.,

$$\frac{da}{dt} = \dot{F}. \tag{66}$$

Langevin-type friction terms may occur in this equation but they do not change the nature of the dynamics. They will simply add a non-linear contribution in the final Schrödinger equation.

(5) The range of scales is large enough for a K41 regime to be established: in von Karman laboratory fully developed turbulence experiments, the ratio between the small dissipative scale and the large (energy injection) scale is of the order of 10^3 and a K41 regime is actually observed [24].

The application of the scale relativity method is therefore fully supported experimentally in this case. Velocity increments dV can be decomposed into two terms, a classical differentiable part $dv = a\,dt$ and a fractal (scale dependent) fluctuation $d\xi_v$, each of them being two-valued:

$$d_+V = a_+dt + \zeta_+\sqrt{2\mathcal{D}_v\,dt}, \quad d_-V = a_-dt + \zeta_-\sqrt{2\mathcal{D}_v\,dt}, \tag{67}$$

where $<\zeta>= 0$ and $<\zeta^2>= 1$. One recognizes here the K41 scaling of velocity increments as $dt^{1/2}$.

Following the same derivation as in the x-space case, with $\{x, v, a\}$ replaced by $\{v, a, \dot{a}\}$, one introduces a complex acceleration field $\mathcal{A} = A_V - iA_U = (a_+ + a_-)/2 - i\,(a_+ + a_-)/2$ and a total 'covariant' derivative

$$\frac{\widehat{d}}{dt} = \frac{\partial}{dt} + \mathcal{A}.\nabla_v - i\,\mathcal{D}_v\,\Delta_v \tag{68}$$

and then write a 'super-dynamics' equation

$$\frac{\widehat{d}}{dt}\mathcal{A} = \dot{F}. \tag{69}$$

A wave function ψ_v acting in velocity space can be constructed from the acceleration field,

$$\mathcal{A} = -2i\,\mathcal{D}_v\nabla_v \ln\psi_v \tag{70}$$

and the super-dynamics equation can then be integrated under the form of a Schrödinger equation including possible non-linear terms (NLT)

$$\mathcal{D}_v^2\,\Delta_v\psi_v + i\,\mathcal{D}_v\frac{\partial\psi_v}{\partial t} = \frac{\phi}{2}\,\psi_v + NLT, \tag{71}$$

where ϕ_v is a potential (in v-space) from which the force $\dot{F}$ or part of this force derives.

By coming back to a fluid representation – but now in terms of the fluid of potential paths – using as variables $P_v(v) = |\psi_v|^2$ and $a(v)$ (which derives from the phase of the wave function), this equation becomes equivalent to the combination of a Navier-Stokes-like equation written in velocity space and a continuity equation,

$$\frac{da}{dt} = \dot{F} + 2\mathcal{D}_v^2\,\nabla_v\left(\frac{\Delta_v\sqrt{P_v}}{\sqrt{P_v}}\right), \tag{72}$$

$$\frac{\partial P_v}{\partial t} + \text{div}_v(P_v a) = 0. \tag{73}$$

Therefore we have recovered the same equation from which we started (time derivative of Navier-Stokes equation) but a new term has energed, namely, a quantum-type force which is the gradient of a quantum-type potential in v-space.

This approach is particularly well adapted to a generalization of the Sawford-Pope stochastic models of description of turbulence, since it is fundamentally based on the same differential stochastic equations. In particular, the constant that generalizes the Planck constant in scale relativity is now easily identified with the fundamental turbulence parameter of the K41 theory. Indeed, the value of $\mathcal{D}_v$ is directly given, in the K41 regime, by the parameter which commands the whole process, the energy dissipation rate by unit of mass, ε,

$$2\mathcal{D}_v = C_0\, \varepsilon, \tag{74}$$

where estimations of the Kolmogorov numerical constant C_0 vary from 4 to 9. Concerning the two small scale (dissipative) and large scale (energy injection) transitions, one could include them in a scale varying $\mathcal{D}_v$, but a better solution consists of keeping $\mathcal{D}_v$ constant, then to include the transitions subsequently to the whole process in a global way.

The only difference with the Sawford-Pope model is therefore the account for new second order terms in differential equations and for the two-valuedness of velocity issued for irreversibility. It implies the appearance of new terms in each of the stochastic differential equations, namely, a mean 'classical' acceleration $a_+ = A_V + a_U$ in the acceleration equation and a macroquantum force $F_Q = -\nabla Q$ in the equation for the derivative of acceleration.

In the case of isotropic turbulence, we expect the three coordinates in v-space to be independant, and therefore the 3D wave function to be separable into three one-dimensional functions. But, in one dimension, solutions of the Schrödinger equation can be reduced to real functions, so that, within this approximation, $A_V = 0$, and we are left with a new contribution to the acceleration that reads $a_+ = a_U = \mathcal{D}_v \nabla \ln P_v$. This is confirmed in the case where the external potential can be approximated by an harmonic oscillator potential, as in Sawford's stochastic model, since it is known that the harmonic oscillator wave functions, which are expressed in terms of Hermite polynomials, are real.

Finally the completed stochastic differential equations read:

$$\frac{dv}{dt} = -\frac{v}{T_L} + \mathcal{D}_v \nabla_v \ln P_v + \sqrt{C_0\, \varepsilon}\, \frac{dW_v}{dt}, \tag{75}$$

$$\frac{da}{dt} = -C\, a - D\, v + 2\mathcal{D}_v^2\, \nabla_v \left(\frac{\Delta_v \sqrt{P_v}}{\sqrt{P_v}} \right) + B\, \frac{dW_a}{dt}. \tag{76}$$

A full analysis of this system of equations will be published in a forthcoming work, as well as a comparison with experimental data of the theoretical predictions that can be derived from it (de Montera, Lehner and Nottale, in preparation). Many unexplained features which are characteristic of turbulence can be accounted for in this model.

Let us give a simple example of such an ability. One of the main open questions concerning turbulence is that of the origin of the extremely large tails exhibited by the probability density function (PDF) of the acceleration of a Lagrangian particle. Recent developments of new experimental techniques have allowed to certify this highly non Gaussian behavior up to more than 50 times the standard deviation of the acceleration PDF [26]. Several phenomenological models reproduce these tails [1, 2, 3, 53, 57], but, up to our knowledge, they have never been derived directly from the fluid dynamics equations.

The scale relativity new contribution to the acceleration a is given, in function of the PDF of the velocity, by the function

$$A(v) = \mathcal{D}_v\, \frac{\nabla_v P_v}{P_v}. \tag{77}$$

Let us call $V(a) = A^{-1}(a)$ the inverse function of $a = A(v)$. This inverse function may have multiple parts, defined on the various monotonic parts of $A(v)$. We denote $V_k(a)$

these parts. The resulting probability distribution of acceleration is given by the inversion formula:

$$P_a(a) = \sum_k \frac{1}{|A'[V_k(a)]|} P_v[V_k(a)], \tag{78}$$

where the sum is done on each of the monotonic parts of the inverse function $V = A^{-1}$.

It is clear from Eq.(77) that very large accelerations, which will create large tails of their PDF, may come from the minimas of the PDF of velocity. In many cases, one expects divergent values of the acceleration to occur when $P_v(v) \to 0$. Moreover, in the macroscopic quantum-type approach advocated here, the velocity PDF is the square of the modulus of a wave function, $P_v(v) = |\psi_v(v)|^2$. This wave function may in general be positive and negative, and when it crosses $v = 0$, it will usually behave locally as $\psi_v(v) = k\,v$, and therefore the PDF of velocity will generally behave as

$$P_v(v) = \left(\frac{v}{v_0}\right)^2 \tag{79}$$

in the minima. From this behavior of P_v in its minima, one obtains $A(v) = 2\mathcal{D}_v/v$, then $V(a) = 2\mathcal{D}_v/a$ and therefore one gets an asymptotic PDF of acceleration which large tails varying as a^{-4} under this approximation:

$$P_a(a) = \frac{(2\mathcal{D}_v)^3}{v_0^2} \frac{1}{a^4}. \tag{80}$$

Such a predicted a^{-4} behavior (which is effectively obtained, for example, from an harmonic oscillator solution) is already a good approximation of the observed tails, as can be seen from the fact that the fourth moment $a^4 P_a(a)$ is verified in the experimental data to remain finite [26]. Although still rough, this adequation is nevertheless a very encouraging result in favor of de Montera's proposal. A more detailed account and refinement of this result and its comparison with experimental data will be given in a forthcoming publication.

6. Conclusion

The theory of scale relativity, thanks to its account, at a profound level, of the fractal geometry of a system, is particularly adapted to the construction and development of a theoretical description of self-organized systems. It supports the use of statistical and probabilistic tools, but it also suggests to go beyond ordinary probabilities, since the description tool becomes a quantum-like (macroscopic) wave function, which is the solution of a generalized Schrödinger equation. This involves a probability density such that $P = |\psi|^2$, but also phases, which are built from the velocity field of potential paths and yield possible interferences.

Such a Schrödinger (or non-linear Schrödinger) form of motion equations can be obtained in at least two ways. One way is through the fractality of the space, and more generally of a medium [56]: this is the case of the application of the scale relativity approach o a turbulent fluid suggested by de Montera [10] and briefly considered in the present paper. Another way is through the emergence of macroscopic quantum-type potentials: this is the case of the application proposed here to superconductivity.

In this framework, one therefore expects a fundamentally wave-like and often quantized character of numerous processes implemented in self-organized systems. In the present contribution, we have concentrated ourselves mainly on the purely theoretical aspect of scale relativity. But many explicit applications to real systems in many different sciences have also been obtained with success (see [38, 44], [46, Chap. 14] and references therein).

Several properties that are considered to be specific of biological systems, such as self-organization, morphogenesis, ability to duplicate, reproduce and branch, confinement and multi-scale structuration and integration are naturally obtained in such an approach [4, 42]. Reversely, the implementation of this type of process in new technological devices involving intelligent feedback loops and quantum-type potentials could also lead to the emergence new forms of self-organizations [47].

Acknowledgment

The author gratefully thanks Dr. Philip Turner for his careful reading of the manuscript and for his useful comments, and Dr. Nicolas Mordant for having kindly provided us with his experimental turbulence data.

References

[1] Arimitsu T., Arimtsu,N., 2004, *Physica* D 193, 218.

[2] Beck C., 2001, *Phys. Rev. Lett.* 87, 180601.

[3] Beck C., 2003, *Europhys. Lett.* 64, 151.

[4] Auffray Ch. & Nottale L., 2008, *Progr. Biophys. Mol. Bio.* 97, 79.

[5] Barnsley M., 1988, *Fractals Everywhere*, Academic Press, Inc.

[6] Beck C., 2005, *"Superstatistical turbulence models"*, arXiv: physics/0506123.

[7] Berry M.V., 1996, *J. Phys. A: Math. Gen.* 29, 6617.

[8] Bohm D., 1954, *Quantum Theory*, Constable and Company Ltd, London.

[9] de Gennes G., 1989, *Superconductivity of metals and alloys*, Addison-Wesley.

[10] de Montera L., 2013, arXiv: 1303.3266.

[11] Falkovich G., 2012, *Phys. Fluids* 24, 055102.

[12] Forriez M., Martin P. & Nottale L., 2010, *L'Espace Gographique* 2, 97.

[13] Fratini M. et al., 2010, *Nature* 466, 841.

[14] Frisch U., 1995, *Turbulence: The Legacy of A.N. Kolmogorov*, Cambridge University Press.

[15] Grabert H., Hänggi P. & Talkner P., 1979, *Phys. Rev.* A19, 2440.

[16] Hermann R., 1997, *J. Phys. A: Math. Gen.* 30, 3967.

[17] Kolmogorov, A.N., 1941, *Proc. Acad. Sci. URSS., Geochem. Section* 30, 299-303.

[18] Landau L. and Lifchitz E, 1980, *Statistical Physics part 1* (Oxford: Pergamon Press)

[19] Lifchitz E. & Pitayevski L., 1980, *Statistical Physics part 2* (Oxford: Pergamon Press)

[20] La Porta A. et al., 2001, *Nature* 409, 1017.

[21] McElroy K. et al., 2005, *Science* 309, 1048.

[22] Mandelbrot B., 1975, *Les Objets Fractals*, Flammarion, Paris.

[23] Mandelbrot B., 1982, *The Fractal Geometry of Nature*, Freeman, San Francisco.

[24] Mordant N., Metz P., Michel O. and Pinton J.F., 2001, *Phys. Rev. Lett.* 87, 214501.

[25] Mordant N., 2001, *Ph. D. Thesis*, Ecole Normale Supérieure de Lyon.

[26] Mordant N., Crawford A.M., Bodenschatz E., 2004, *Physica* D 193, 245.

[27] Nelson E., 1966, *Phys. Rev.* 150, 1079.

[28] Noble D., 2002, *Science* 295, 1678.

[29] Noble D., 2008, *Exp Physiol* 93.1 pp 1626.

[30] Nore C Brachet ME Cerda E & Tirapegui E 1994 *Phys. Rev. Lett.* 72 2593

[31] Nottale L. & Schneider J., 1984, *J. Math. Phys.* **25**, 1296.

[32] Nottale L., 1989, *Int. J. Mod. Phys. A* 4, 5047.

[33] Nottale L., 1992, *Int. J. Mod. Phys. A* 7, 4899.

[34] Nottale L., 1993, *Fractal Space-Time and Microphysics: Towards a Theory of Scale Relativity*, World Scientific, Singapore.

[35] Nottale L., 1996, *Astron. Astrophys. Lett.* 315, L9.

[36] Nottale L., 1997, *Astron. Astrophys.* 327, 867.

[37] Nottale L., Schumacher G. & Gay J., 1997, *Astron. Astrophys.* 322, 1018.

[38] Nottale L., Chaline J. & Grou P., 2000, *Les arbres de l'évolution: Univers, Vie, Sociétés*, Hachette, Paris, 379 pp.

[39] Nottale L., Schumacher G. & Lefèvre E.T., 2000, *Astron. Astrophys.* 361, 379.

[40] Nottale, L., 2001, *Revue de Synthèse*, T. 122, 4e S., N.1 p. 93-116.

[41] Nottale L. & Célérier M.N., 2007, *J. Phys. A: Math. Theor.*, 40, 14471.

[42] Nottale L. & Auffray C., 2008, *Progr. Biophys. Mol. Bio.* 97, 115.

[43] Nottale L., 2009, *J. Phys. A: Math. Theor.* 42, 275306.

[44] Nottale L., Chaline J., Grou P., 2009, *Des fleurs pour Schrödinger: la relativité d'échelle et ses applications*, 421 p., Ellipses, Paris.

[45] Nottale L., 2009, "Quantum-like gravity waves and vortices in a classical fluid", arXiv: 0901.1270.

[46] Nottale L., 2011, *Scale Relativity and Fractal Space-Time: a New Approach to Unifying Relativity and Quantum Mechanics*, Imperial College Press, 762 p.

[47] Nottale L. & Lehner Th., 2012, *Int. J. Mod. Phys.* C 23 (5), 1250035 (1-27).

[48] Nottale L. & Célérier M.N., 2012, "Emergence of complex and spinor wave functions in scale relativity. I. Nature of scale variables", arXiv: 1211.0490.

[49] Nottale L., Martin P. & Forriez M. , 2012, *Revue Internationale de Géomatique*, 22, 103-134.

[50] Nottale L., Célérier M.-N., 2013, *J. Math. Phys.*, 54, 112102.

[51] Pan S.H. et al., 2001, *Nature* 413, 282.

[52] Pope, S.B., 2002, *Phys. Fluids* 14, 2360.

[53] Reynolds A.M. , 2003, *Phys. Fluids* 15, L1.

[54] Sawford, B.L., 1991, *Phys. Fluids* A3, 1577.

[55] Tanner D.B. et al., 1998, *Physica* (Amsterdam) 244B, 1.

[56] Turner P., Kowalczyk M. and Reynolds A., 2011, COST Action E54 Book, *"New insights into the micro-fibril architecture of the wood cell wall"*.

[57] Voth G.A., La Porta A., Crawford A.M., Alexander J., Bodenschatz E., 2002, *J. Fluid Mech.* 469, 121.

[58] Wang M.S. & Liang W.K., 1993, *Phys. Rev.* D 48, 1875.

[59] Weisskopf V., 1989, *La révolution des quanta*, Hachette.

In: Space-Time Geometry and Quantum Events
Editor: Ignazio Licata, pp. 197-213
ISBN: 978-1-63117-455-1

Chapter 6

THREE POSSIBLE IMPLICATIONS OF SPACE-TIME DISCRETENESS

Shan Gao*
Institute for the History of Natural Sciences,
Chinese Academy of Sciences, Beijing, China

Abstract

We analyze the possible implications of the discreteness of space-time, which is defined here as the existence of a minimum *observable* interval of space-time. First, it is argued that the discreteness of space-time may result in the existence of a finite invariant speed when combining with the principle of relativity. Next, it is argued that when combining with the uncertainty principle, the discreteness of space seems to require that space-time is curved by matter, and the dynamical relationship between matter and space-time holds true not only for macroscopic objects but also for microscopic particles. Moreover, the Einstein gravitational constant can also be determined in terms of the minimum size of discrete space-time. Thirdly, it is argued that the discreteness of time may result in the dynamical collapse of the wave function, and the minimum size of discrete space-time also yields a plausible collapse criterion consistent with experiments. These heuristic arguments might provide a deeper understanding of the special and general relativity and quantum theory, and also have implications for the solutions to the measurement problem and the problem of quantum gravity.

Keywords: Space-time discreteness, Planck scale, speed of light, gravity, wavefunction collapse,quantum gravity

1. Introduction

It has been widely argued that the existence of a minimum *observable* interval of space-time (MOIST) is a model-independent result of the proper combination of quantum

* E-mail address: gaoshan@ihns.ac.cn; Address: Institute for the History of Natural Sciences, Chinese Academy of Sciences, Beijing 100190, China.

mechanics (QM) and general relativity (GR) (see [1] for a review)[1]. This strongly suggests that the existence of a MOIST is a more fundamental postulate which has a firmer basis beyond the existing theories, and it reflects a more fundamental characteristic of nature, which may be called discreteness of space-time (in the observational sense). On the other hand, the existing theories are still based on some *unexplained* postulates. For example, special relativity, the common basis of quantum field theory and general relativity, postulates the invariance of the speed of light in all inertial frames[2], but the theory does not explain why. In the long run, these postulates need to be explained and replaced by some more fundamental ones. Therefore, it may be necessary to examine the relationship between the MOIST postulate and the existing theories from the opposite direction.

In this paper,we will investigate the implications of the MOIST postulate for the understandings of the special and general relativity and quantum theory.Concretely speaking, we will argue that the postulate may help explain why the speed of light is invariant in all inertial frames and why matter curves space-time and why the wave function collapses. The plan of this paper is as follows. In Section 2, we will first formulate the MOIST postulate, explain its physical meaning, and define the discreteness of space-time in terms of this postulate. In Section 3, we will argue that when combining with the principle of relativity the discreteness of space-time may result in the existence of a finite invariant speed. In Section4, we will argue that when combining with the uncertainty principle the discreteness of space may imply that space-time is curved by matter. In particular, the dynamical relationship between matter and space-time holds true not only for macroscopic objects but also for microscopic particles. Moreover, the Einstein gravitational constant in GR can also be determined in terms of the minimum size of the discrete space-time. In Section5, we will argue that the discreteness of time may result in the dynamical collapse of the wave function in quantum mechanics, and the minimum size of the discrete space-time also yields a plausible collapse criterion consistent with existing experiments and our macroscopic experience. This may provide a plausible solution to the quantum measurement problem, and might also have implications for a complete theory of quantum gravity. Conclusions are given in the last section.

2. The MOIST Postulate

Although QM and GR are both based on the concept of continuous space-time, it has been argued that their proper combination leads to the existence of the Planck scale, a lower bound to the uncertainty of distance and time measurements [1]. For example, when we measure a space interval near the Planck length the measurement will inevitably introduce an uncertainty comparable to the Planck length, and as a result, we cannot accurately measure a space interval shorter than the Planck length. This is clearly expressed by the generalized uncertainty principle [3]:

---

[1] The minimum observable space and time intervals are often loosely called minimum length and minimum time in the literature.

[2] Although Einstein originally based special relativity on two postulates: the principle of relativity and the constancy of the speed of light, he later thought that the universal principle of the theory is contained only in the postulate: The laws of physics are invariant with respect to Lorentz transformationsbetween inertial frames [2]. Note that the constancy of the speed of lightdenotes that the speed of light in vacuum is constant, independently of the motion of the source, in at least one inertial frame.

$$\Delta x = \Delta x_{QM} + \Delta x_{GR} \geq \frac{\hbar}{2\Delta p} + \frac{2l_P{}^2 \Delta p}{\hbar} \quad (1)$$

where Δx is the total position uncertainty of the measured system, Δx_{QM} is the position uncertainty resulting from QM, Δx_{GR} is the position uncertainty resulting from GR, $\hbar$ is the reduced Planck constant, Δp is the momentum uncertainty of the system, and l_P is the Planck length. Moreover, different approaches to quantum gravity also lead to the existence of a minimum length, a resolution limit in any experiment [1]. In this paper, we will promote this result to a fundamental postulate[3]:

> The MOIST Postulate: There are minimum observable space and time intervals, which are two times the Planck length and Planck time respectively[4].

The physical meaning of this postulate can be understood as follows. First of all, the existence of a minimum observable interval of time, denoted by $T_U \equiv 2t_P$, where t_P is the Planck time, implies that any physical change during a time interval shorter than it is unobservable, or in other words, a physically observable change only happens during a time interval not shorter than the minimum observable time. Otherwise we can measure a time interval shorter than the minimum observable time by observing the physical change, which contradicts the MOIST postulate. However, the postulate does not require that a nonphysical change (e.g. movement of a shadow) or an unobservable physical change (e.g. see below) cannot happen during a time interval shorter than the minimum observable time. An example of an observable physical change is the change of a light pulse from being absent to being present. The transmission of an observable physical change usually corresponds to the transmission of information or energy, which is also called the transmission of a physical signal[5].

Next, in a similar way, the existence of a minimum observable interval of space, denoted by $L_U \equiv 2l_P$, implies that a physically observable entity only exists in a region of space whose size is larger than it; otherwise we can measure a space interval smaller than the minimum observable length by observing the physical entity, which contradicts the MOIST postulate.

---

[3] It is worth pointing out that this postulate is not implied but only motivated by the existing arguments for the existence of minimum observable space and time intervals. One reason is that these arguments implicitly assume the (approximate) validity of both quantum theory and general relativity down to the Planck scale (e.g. [3]), but this assumption may be debatable (see also [4]).

[4] Note that the minimum observable space and time intervals might be other multiple of the Planck length and Planck time. However, the black hole entropy formula and the hypothetical holographic principle also support that the minimum observable space and time intervals are two times the Planck length and Planck time.

[5] It is worth pointing out that defining the transmission speed of a physical signal is not so simple. An actual physical signal with a finite extent, e.g. a pulse of light, travels at different speeds in a media. Roughly speaking, the largest part of the pulse travels at the group velocity, and its earliest part travels at the front velocity. Under conditions of normal dispersion, the group velocity can represent the signal speed, namely the actual propagation speed of information or energy. In particular, for a microscopic particle moving in vacuum as a physical signal, the signal velocity can be defined as the group velocity of its wave packet. But in an anomalously dispersive medium where the group velocity exceeds the speed of light in vacuum [5], the group velocity no longer represents the signal velocity. For these situations, the signal velocity is usually defined as the front velocity, namely the speed of the leading edge of the signal [6]. However, this definition is not operational in actual experiments. An operational definition of signal velocity may be based on the signal-to-noise ratio, which closely relates to quantum fluctuations [7].

Moreover, only the transmission of a physical signal over a distance not shorter than the minimum observable length is observable[6]. The transmission of a physical signal over a distance shorter than the minimum observable length is unobservable[7], but the happening of such transmissions is not prohibited by the postulate.

According to the above understandings, the MOIST postulate can be reformulated as follows:

(1) Observable physical entities exist in a region of space whose size is not smaller than the minimum observable interval of space, which is two times the Planck length;
(2) Observable physical processes happen during a time interval not shorter than the minimum observable interval of time, which is two times the Planck time.

The relationship between the MOIST postulate and discrete space-time can be understood as follows. On the one hand, the postulate requires that a space and time interval shorter than the Planck scale is unobservable, and thus it can be regarded as a minimum requirement of space-time discreteness in the observational sense. If an arbitrarily short interval of space and time is always observable, then space and time will be infinitely divisible and cannot be discrete. On the other hand, the MOIST postulate does not imply that space-time is discrete in the ontological sense[8]. It is also possible that space-time itself is still continuous but physical laws do not permit the resolution of space-time structures below the Planck scale. By comparison, there is a stronger requirement of space-time discreteness, namely that space-time itself is discrete. For instance, one may further impose a limitation stronger than the MOIST postulate, e.g., that an unobservable change does not happen (or no change happens) during a time interval shorter than the minimum observable time interval. However, there are at least two worries about such an extension. First, it can never be tested whether any change happens within the time interval as it is already shorter than the minimum observable time interval. Next, it seems that continuous space-time may be still useful as a description framework, even though all observable physical changes satisfy the requirements of the MOIST postulate[9].

Lastly, we note that the MOIST postulate implicitly assumes the validity of the principle of relativity. It means that the minimum observable length and the minimum observable time interval are the same in all inertial frames. If the minimum observable space and time intervals are different in different inertial frames, then there will exist a preferred Lorentz

[6] For a signal with a position uncertainty much larger than its transmission distance, which is not shorter than the minimum observable space interval, the transmission is still observable at the ensemble level.

[7] This kind of unobservability is not only at the individual level but also at the statistical level. We may understand this result by thinking that the signal (e.g. the wave function of a microscopic particle) has a spatial fuzziness not smaller than one L_U.

[8] There are already some models of discrete space-time in the ontological meaning. For example, in loop quantum gravity, the discreteness of space is represented by the discrete eigenvalues of area and volume operators [8,9], while in the causal set approach to quantum gravity, one has a direct discretization of the causal structure of continuum Lorentzian manifolds [10, 11].

[9] In our opinion, there are at least two reasons to support this view. First, it seems that there is no physical limitation on the difference of the happening times of two causally independent events, e.g., the difference is not necessarily an integral multiple of the Planck time. Next, it seems that continuous space-time is still needed to describe nonphysical superluminal motion, e.g., superluminal light pulse propagation in an anomalously dispersive media. During such superluminal propagations, moving one L_U will correspond to a time interval smaller than one T_U.

frame, while this contradicts the principle of relativity. One support for this assumption is that the generalized uncertainty principle is independent of the choice of inertial frames, and thus the MOIST postulate, which is directly motivated by the principle, should also be independent of the choice of inertial frames. In the following, the discreteness of space-time will be analyzed only in terms of the MOIST postulate.

3. There Is a Maximum Signal Speed

Now we will investigate the implications of the discreteness of space-time for the understandings of special relativity. By analyzing the continuous transmission of a physical signal, we will argue that when combining with the principle of relativity the discreteness of space-time may result in the existence of a finite invariant speed. This may help explain why the speed of light is invariant in all inertial frames.

Consider the continuous transmission of a physical signal in an inertial frame[10]. If the signal moves with a speed larger than $c = L_U / T_U$, then it will move more than one L_U during one T_U, and thus moving one L_U, which is physically observable in principle, will correspond to a time interval shorter than one T_U during the transmission. This contradicts the MOIST postulate, which requires that a physically observable change can only happen during a time interval not shorter than T_U.

By comparison, the continuous transmission of a physical signal with a speed smaller than c is permitted, as during the transmission the signal will move less than one L_U during a time interval shorter than one T_U, while the displacement smaller than one L_U is physically unobservable according to the MOIST postulate[11].

This argument shows that the MOIST postulate leads to the existence of a maximum signal speed for the continuous transmission of a physical signal, which is equal to the ratio of the minimum observable length to the minimum observable time interval, namely $v_{\max} = L_U / T_U = c$.

Since the minimum observable time interval and the minimum observable length are the same in all inertial frames, the maximum signal speed for the continuous transmission of a physical signal will be c in every inertial frame. Now we will argue that this maximum speed c is invariant in all inertial frames. Suppose a physical signal moves in the x direction with speed c in an inertial frame S. Then its speed will be either equal to c or larger than c in

---

[10] In the discussions of this section the speed of signal always denotes the two-way speed, and the speed of light is always the two-way speed of light. This two-way speed can be experimentally measured independent of any clock synchronization scheme. For an arbitrary convention of simultaneity, the Lorentz transformations in special relativity, which is based on the Einstein synchronization, will be replaced by the Edwards-Winnie transformations [12, 13].

[11] This suggests that space and time must both have a minimum observable interval; otherwise either the continuous transmission of a physical signal is impossible or the signal speed can be infinite, both of which contradict experience. For instance, consider the situation that time has a minimum observable interval but space has not. Then a physical signal can move an arbitrarily short distance that is physically observable. But for a signal moving with any finite speed v, moving an observable distance shorter than vT_U will correspond to a time interval shorter than one T_U, while this contradicts the existence of a minimum observable time interval.

another inertial frame S' with a velocity in the $-x$ direction relative to S. Since c is the maximum signal speed in every inertial frame, the speed of the signal in S' can only be equal to c. This result also means that when the signal moves in the x direction with speed c in the inertial frame S', its speed will be also c in the inertial frame S with a velocity in the x direction relative to S'. Since the inertial frames S and S' are arbitrary, we can reach the conclusion that if a signal moves with the speed c in an inertial frame, it will also move with the same speed c in all other inertial frames. This proves the invariance of speed c.

Here is another argument for the invariance of speed c. Suppose a signal moves in the x direction with speed c in an inertial frame S. Then its speed will be either c or smaller than c in another inertial frame S' with a velocity in the x direction relative to S. If its speed is smaller than c in S', say c-v, then there must exist a speed larger than c-v and a speed smaller than c-v in S' that correspond to the same speed in S due to the continuity of velocity transformation and the maximum of c. This means that when the signal moves with a certain speed in frame S its speed in frame S' will have two possible values, which is impossible. Thus the signal moving with speed c in S also moves with speed c in S', which has a velocity in the x direction relative to S. This result also means that when a signal moves in the x direction with speed c in S', its speed is also c in S with velocity in the $-x$ direction relative to S'. Since the inertial frames S and S' are arbitrary, this also proves that the maximum signal speed c is invariant in all inertial frames[12].

To sum up, we have argued that the MOIST postulate (i.e. assuming the existence of minimum observable intervals of space and time at the Planck scale) leads to the existence of a maximum signal speed, $v_{max} = L_U / T_U = c$, which is invariant in every inertial frame[13]. This may help understand the most puzzling aspect of special relativity, the invariance of the speed of light[14].

---

[12] Here one may object that we should first state clearly the space-time transformations before the analysis of speed transformation. However, on the one hand, we are trying to derive the fundamental postulate that determines the space-time transformations, and on the other hand, as we have argued above, the space-time transformations are not needed to derive the relation between the maximum speeds in two inertial frames, which can already be obtained from some basic requirements such as the continuity of speed transformation etc. Besides, it is worth noting that a similar argument for the invariance of a maximum speed was also given by Rindler [14].

[13] Note that the invariance of the speed of light has been confirmed by experiments with very high precision [15], and no violation of Lorentz invariance has been found either [16].

[14] In special relativity, the speed of light in vacuum, denoted by c, is invariant in all inertial frames. This postulate is not a result of logical analysis, but a direct representation of experience. The theory itself does not answer why the speed of light is invariant in all inertial frames (or equivalently why the space-time transformations are the Lorentz transformations). On the other hand, the suggested theory of relativity without light suggests that c is not (merely) the speed of light, but a universal constant of nature, an invariant speed (see, e.g. [17-19]). Furthermore, the theory also suggests that the existence of an invariant speed partly results from the properties of space and time, e.g. homogeneity of space and time and isotropy of space. However, this theory cannot determine whether the invariant speed is finite or infinite and thus cannot establish a real connection between its invariant speed with c [19]. Anyway, we need to explain exactly why there is a finite invariant speed. Since speed is essentially the ratio of space interval and time interval, it seems to be a natural conjecture that the existence of a finite invariant speed may result from some undiscovered property of space and time, as the existing theory of relativity without light has suggested. What we have argued above is just that the undiscovered property is certain discreteness of space-time.By comparison, if space and time are continuous, then no characteristic space and time sizes exist, and thus it seems unnatural that there exists a characteristic speed. If our argument is valid, then the existence of an invariant speed c may be regarded as a firm experimental confirmation of the MOIST postulate.

On this view, the speed constant c in special relativity (as well as in quantum field theory and general relativity) is not the actual speed of light in vacuum (though which may be also equal to c), but the ratio of the minimum observable length to the minimum observable time interval.

Two comments are in order before concluding this section. First of all, although the MOIST postulate leads to the existence of a maximum signal speed c for continuous transmissions, it does not preclude the superluminal continuous transmissions that do not correspond to actual information or energy transmissions. Two well-known examples are superluminal light pulse propagation and the hypothetical tachyons. Experiments have shown that the group velocity of a light pulse in an anomalously dispersive media (e.g. atomic caesium gas) can be much larger than c [5].But the superluminal light pulse propagation does not correspond to the superluminal transmission of a physical signal, and it can be shown that the signal speed is still equal to or smaller than c in this case [7].

Similarly, a consistent theory of tachyons also requires that the tachyons cannot be used to send signals with a speed larger than c from one place to another. Besides, the MOIST postulate does not preclude the existence of superluminal nonlocal signals either. If there is some mechanism to realize nonlocal signal transmission, then its signal speed can be larger than c, and the nonlocal process may also violate the Lorentz invariance [20]. But the signal speed in this case also has an upper limit depending on the distance due to the limitation of the MOIST postulate, which is equal to the ratio of transmission distance to the minimum observable time interval.

Next, we note that if the above argument is valid, then the theory of relativity will be based on two postulates: (1) the principle of relativity; and (2) the MOIST postulate, which states that the minimum observable space and time intervals are invariant in all inertial frames[15].

Moreover, the constancy of the speed of light is a consequence of these two postulates. Then there will exist three invariant scales in the theory: the Planck length, the Planck time, and the speed of light. It is an interesting issue how the space-time transformations will be in the theory. It is obvious that the transformations cannot be the strict Lorentz transformations, although which must be a good approximate on the scale much larger than the Planck scale. It seems that several existing variants of relativity in discrete space-time may provide useful clues for the MOIST space-time transformations.

For example, doubly special relativity assumes two invariant scales, the speed of light c and a minimum length λ [21-23][16], while triply special relativity assumes three invariant scales, the speed of light c, a mass κ and a length R [24]. In these theories, the classical Minkowski space-time is replaced by a quantum space-time, such as κ-Minkowski noncommutative space-time etc.[17]. Another possibility is that the MOIST postulate may not require such quantum space-time s. The reason is that space-time is no longer flat when considering gravity, and thus the Lorentz contraction, which applies to flat space-time, does not hold true (especially at a very small spatial scale such as the Planck scale). For example,

[15] Note that the minimum observable space and time intervals cannot depend on the choice of inertial frames, and in particular, it is impossible that only the ratio of the minimum observable space interval to the minimum observable time interval is constant in all inertial frames; otherwise the constancy of the speed of light, which has been tested by precise experiments, will be violated.

[16] There was a recent debate on whether the model of deformed special relativity is consistent [25,26].

[17] For a more philosophical discussion of these theories see Reference [27].

when considering the influence of gravity, Heisenberg's uncertainty principle will be replaced by the generalized uncertainty principle as shown by Equation (1), and thus the increase of energy will not decrease the position uncertainty at the Planck scale. In this way, gravity may help resolve the apparent contradiction between the MOIST postulate and Lorentz contraction. We will discuss this in more detail in the next section.

4. Matter Curves Space-Time

The origin of gravity is still a controversial issue. The solution of this problem may have important implications for a complete theory of quantum gravity. In this section, we will analyze the possible implications of the discreteness of space-time for the origin of gravity.

According to the Heisenberg uncertainty principle in quantum mechanics (QM) we have

$$\Delta x \geq \frac{\hbar}{2\Delta p} \tag{2}$$

The momentum uncertainty of a particle, Δp, will result in the uncertainty of its position, Δx. This poses a limitation on the localization of a particle in nonrelativistic domain. There is a more strict limitation on Δx in relativistic QM. A particle at rest can only be localized within a distance of the order of its reduced Compton wavelength, namely

$$\Delta x \geq \frac{\hbar}{2m_0 c} \tag{3}$$

where m_0 is the rest mass of the particle. The reason is that when the momentum uncertainty Δp is greater than $2m_0 c$ the energy uncertainty ΔE will exceed $2m_0 c^2$, but this will create a particle anti-particle pair from the vacuum and make the position of the original particle invalid. It then follows that the minimum position uncertainty of a particle at rest can only be the order of its reduced Compton wavelength as denoted by Equation (3). Using the Lorentz transformations, the minimum position uncertainty of a particle moving with (average) velocity v is

$$\Delta x \geq \frac{\hbar}{2mc} \text{ or } \Delta x \geq \frac{\hbar c}{2E} \tag{4}$$

where $m = m_0 / \sqrt{1 - v^2 / c^2}$ is the relativistic mass of the particle, and $E = mc^2$ is the total energy of the particle. This means that when the energy uncertainty of a particle is of the order of its (average) energy, it has the minimum position uncertainty. Note that Equation (4) also holds true for particles with zero rest mass such as photons.

According to Equation (4), when the energy and energy uncertainty of a particle becomes arbitrarily large, the uncertainty of its position Δx can be arbitrarily small, which means that the particle may exist in an arbitrarily small region of space. According to the MOST postulate, however, observable physical entities including the above particle can only exist in a region of space whose size is not smaller than the minimum observable length. Then the localization of any particle should have a minimum value L_U, namely Δx should satisfy the limiting relation: $\Delta x \geq L_U$. In order to satisfy this relation, the r.h.s of Equation (4) should at least contain another term proportional to the (average) energy of the particle[18], namely in the first order of E it should be

$$\Delta x \geq \frac{\hbar c}{2E} + \frac{L_U{}^2 E}{2\hbar c} \tag{5}$$

This new inequality, which can be regarded as one form of generalized uncertainty principle[19], can satisfy the limitation relation imposed by the MOIST postulate. It means that the localization length of a point like particle has a minimum value L_U.

How to understand the new term demanded by the discreteness of space-time? Obviously it indicates that the (average) energy of a particle increases the size of its localized state, and the increase is proportional to the energy. Since there is only one particle here, the increase of the size cannot result from any interaction between it and other particles such as electromagnetic interaction. Besides, since the increased part, which is proportional to the energy, is very distinct from the original quantum part, which is inversely proportional to the energy, it is a reasonable assumption that the increased size of the localized state of the particle does not come from its quantum motion. As a result, it seems that there is only one possibility, namely that the (average) energy of the particle influences the geometry of its background space-time and further results in the increase of the size of its localized state.

We can also give an estimate of the strength of this influence in terms of the new term $\frac{L_U{}^2 E}{2\hbar c}$. This term shows that theenergy E will lead to an inherentlength increase $\Delta L = \frac{L_U T_U E}{2\hbar}$.
In other words, the energy E contained in a region with size L will change the proper size of the region to

$$L' \approx L + \frac{L_U T_U E}{2\hbar} \tag{6}$$

---

[18] Note that if a constant term such as L_U is added to the r.h.s of the inequality, it may also satisfy the limitation relation imposed by the discreteness of space in terms of the MOIST postulate. However, it seems difficult to explain the origin of the constant term. The reason is that the Heisenberg uncertainty principle in QM may have a deeper basis in flat space-time, and if energy does not influence the background space-time, then no additional constant term will appear in the inequality.

[19] The argument here might be regarded as a reverse application of the generalized uncertainty principle. But it should be stressed that the existing arguments for the principle are based on the analysis of measurement process, and their conclusion is that it is impossible to *measure* positions to better precision than a fundamental limit. On the other hand, in the above argument, the uncertainty of position is objective, and the MOIST postulate requires that the objective localization length of a particle has a minimum value, which is independent of measurement.

When the energyis equal to zero or there are no particles, the background space-time will not be changed. Since what changes space-time here is the average energy, this change is irrelevant to the quantum fluctuations, and thus the relation between energy and proper size increase holds true in the classical domain.

Based on this result, there are some common steps to "derive" the Einstein field equations, the concrete relation between the geometry of space-time and the energy-momentum contained in that space-time, in terms of Riemann geometry and tensor analysis as well as the conservation of energy and momentum etc.

For example, it can be shown that there is only one symmetric second-rank tensor that will satisfy the following conditions: (1) Constructed solely from the space-time metric and its derivatives; (2) Linear in the second derivatives; (3) The four-divergence of which is vanishes identically (this condition guarantees the conservation of energy and momentum); (4) Is zero when space-time is flat (i.e. without cosmological constant). These conditions will yield a tensor capturing the dynamics of the curvature of space-time, which is proportional to the stress-energy density, and we can then obtain the Einstein field equations[20]

$$R_{\mu\nu} - \frac{1}{2} g_{\mu\nu} R = \kappa T_{\mu\nu} \tag{7}$$

where $R_{\mu\nu}$ the Ricci curvature tensor, R the scalar curvature, $g_{\mu\nu}$ the metric tensor, κ is the Einstein gravitational constant, and $T_{\mu\nu}$ the stress-energy tensor.

The left thing is to determine the value of the Einstein gravitational constant κ. It is usually derived by requiring that the weak and slow limit of the Einstein field equations must recover Newton's theory of gravitation. In this way, the gravitational constant is determined by experience as a matter of fact. If the above argument based on the MOIST postulate is valid, the Einstein gravitational constant can also be determined in theory in terms of the minimum observable space and time intervals.

Consider an energy eigenstate limited in a region with radius *R*. The space-time outside the region can be described by the Schwarzschild metric by solving the Einstein field equations:

$$ds^2 = (1 - \frac{r_S}{r})^{-1} dr^2 + r^2 d\theta^2 + r^2 \sin\theta^2 d\phi^2 - (1 - \frac{r_S}{r}) c^2 dt^2 \tag{8}$$

where $r_S = \frac{\kappa E}{4\pi}$ is the Schwarzschild radius. By assuming $r_S < R$ and the metric tensor inside the region *R* is the same order as that on the boundary, the proper size of the region is

$$L \approx 2\int_0^R (1 - \frac{r_S}{R})^{-1/2} dr \approx 2R + \frac{\kappa E}{4\pi} \tag{9}$$

[20] Another approach to deriving the Einstein field equations is through an action principle using a gravitational Lagrangian.

Therefore, the change of the proper size of the region due to the contained energy E is

$$\Delta L \approx \frac{\kappa E}{4\pi} \tag{10}$$

By comparing with Equation (6) we find $\kappa = 2\pi \frac{L_U T_U}{\hbar}$ in Einstein's field equations. It seems that this formula itself also suggests that gravity originates from the discreteness of space-time (together with the quantum principle that requires $\hbar \neq 0$). In continuous space-time where $T_U = 0$ and $L_U = 0$, we have $\kappa = 0$, and thus gravity does not exist.

In conclusion, we have argued that the discreteness of space-time (in the sense of the MOIST postulate) seems to imply that matter curves space-time, and the dynamical relationship between matter and space-time, which is governed by the Einstein field equations in the classical domain, holds true not only for macroscopic objects, but also for microscopic particles.

Moreover, the Einstein gravitational constant in the equations can also be determined by the minimum observable size of the discrete space-time. This provides an argument for the fundamental existence of gravity, which might then have further implications for a complete theory of quantum gravity. Let's give a little more detailed discussion.

It is well known that there exists a fundamental conflict between the superposition principle of QM and the general covariance principle of GR[21] [28,29]; QM requires a presupposed fixed space-time structure to define quantum state and its evolution, but the space-time structure is dynamical and determined by the state according to GR. The conflict indicates that at least one of these basic principles must be compromised in order to combine into a coherent theory of quantum gravity. But there has been a hot debate on which one should yield to the other. The problem is actually two-fold.

On the one hand, QM has been plagued by the measurement problem, and thus it is still unknown whether its superposition principle is universally valid, especially for macroscopic objects. On the other hand, it is not clear whether or not gravity as a geometric property of space-time described by GR is emergent either. The existing heuristic derivation of GR based on Newton's theory cannot determine whether gravity is fundamental.

If gravity is really emergent, for example, GR is treated as an effective field theory, then the dynamical relation between the geometry of space-time and the energy-momentum contained in that space-time, which is described by Einstein'sfield equations, will be not fundamental. As a consequence, different from the superposition principle of QM, the general covariance principle of GR will be not a basic principle, and thus no conflict will exist between quantum and gravity and we may directly extend the quantum field theory to include gravity (e.g. in string theory). In fact, the general covariance principle of GR has been compromised here because it is not fundamental. Note that besides the string theory, there are also some interesting suggestions that gravity may be emergent, such as Sakharov's induced

[21] This conflict between QM and GR can be regarded as a different form of the problem of time in quantum gravity. It is widely acknowledged that QM and GR contain drastically different concepts of time (and space-time), and thus they are incompatible in nature. In QM, time is an external (absolute) element (e.g. the role of absolute time is played by the external Minkowski space-time in quantum field theory). In contrast, space-time is a dynamical object in GR. This then leads to the notorious problem of time in quantum gravity [30,31].

gravity [32,33], Jacobson's gravitational thermodynamics [34], and Verlinde's idea of gravity as an entropic force [35] (see also [36]). On the other hand, if gravity is not emergent but fundamental as the above argument in terms of the MOIST postulate seems to imply, then quantum and gravity may be combined in a way different from the string theory. Now that the general covariance principle of GR is universally valid, the superposition principle of QM probably needs to be compromised when considering the fundamental conflict between them [28,37,38]. We will further analyze this possibility in terms of the discreteness of space-time in the next section.

5. The Wave Function Collapses

It is an important issue in the foundations of QM whether the wave function really collapses. In this section, we will argue that the discreteness of space-time may result in the dynamical collapse of the wave function, and the minimum observable size of the discrete space-time also yields a plausible collapse criterion consistent with experiments. This might provide a promising solution to the notorious measurement problem.

Consider a quantum superposition of two different energy eigenstates. Each eigenstate has a well-defined static mass distribution in the same spatial region with radius R. For example, they are rigid balls of radius R with different uniform mass density. The initial state is

$$\psi(x,0)=\frac{1}{\sqrt{2}}[\varphi_1(x)+\varphi_2(x)] \tag{11}$$

where $\varphi_1(x)$ and $\varphi_2(x)$ are two energy eigenstates with energy eigenvalues E_1 and E_2 respectively. According to the linear Schrödinger evolution, we have:

$$\psi(x,t)=\frac{1}{\sqrt{2}}[e^{-iE_1t/\hbar}\varphi_1(x)+e^{-iE_2t/\hbar}\varphi_2(x)] \tag{12}$$

and

$$\rho(x,t)=|\psi(x,t)|^2=\frac{1}{2}[\varphi_1^2(x)+\varphi_2^2(x)+2\varphi_1(x)\varphi_2(x)\cos(\Delta E/\hbar\cdot t)] \tag{13}$$

This result indicates that the density $\rho(x,t)$ will oscillate with a period $T=h/\Delta E$ in each position of space, where $\Delta E=E_2-E_1$ is the energy difference. This has no problem when the energy difference is small as in usual situations. But when the energy difference ΔE exceeds the Planck energy E_p [22], $\rho(x,t)$ will oscillate with a period shorter than the minimum

[22] Note that there is no limitation on the maximum value of the energy of each eigen state in the superposition in principle. For example, the energy of a macroscopic object in a stationary state can be larger than the Planck energy (cf. [28]). On the other hand, if the energy of a microscopic particle cannot be larger than the Planck energy and QM indeed fails at the energy scale larger than the Planck energy, then there will be no quantum superposition of different space-time geometries (as defined later) either, which is also consistent with the latter conclusion of this section.

observable time interval T_U that is of the order of T_p [23]. This is inconsistent with the requirement of the discreteness of space-time (in the sense of the MOIST postulate), according to which observable physical changes such as the above oscillation cannot happen during a time interval shorter than the minimum observable time interval[24]. In other words, the MOIST postulate requires that the superposition of two energy eigenstates with an energy difference larger than the Planck energy cannot exist or hold during a time interval longer than the minimum observable time interval[25], and must evolve to another state without the oscillation instantaneously or during a time interval shorter than the minimum observable time interval[26].

Then what state will the superposition evolve to? If the principle of conservation of energy (for an ensemble of identical systems) in quantum mechanics is still valid for the evolution[27], then the superposition can only evolve to one of the energy eigenstates in the superposition, which has no density oscillation, and the probability of evolving to each state satisfies the Born rule. This means that the superposition will collapse to one of the energy eigenstates in the superposition. By continuity, the superposition of energy eigenstates with an energy uncertainty smaller than the Planck energy will also undergo the collapse process[28].Moreover, the dynamical wavefunction collapse will satisfy the following criterion: when the energy uncertainty of a superposition of energy eigenstates is about the Planck energy, the collapse time is about the Planck time.

It can be argued that the MOIST postulate may impose more restrictions for the dynamical collapse of the wave function[29]. Since the effect of a dynamical collapse evolution

---

[23] Here we ignore the gravitational fields in the superposition, as their existence does not influence our conclusion. When the energy difference is very tiny such as for a microscopic particle, the corresponding gravitational fields in the superposition are almost the same and not orthogonal, and the interference effect or the oscillation can be detected in experiment, while when the energy difference become larger and larger such as approaching the Planck energy, the gravitational fields in the superposition are not orthogonal either, and thus the oscillation can also be detected in principle. Moreover, as we will argue later, the superposition state can still be defined when considering the existence of the gravitational fields (see also [39]).

[24] Note that the oscillation of $\rho(x,t)$ can be measured at least at the ensemble level, e.g. for a large number of bosons in the same initial superposition state. Moreover, protective measurements can also measure the density $\rho(x,t)$ of a single quantum system and its time evolution in principle [40-42].

[25] It is worth pointing out that the existence of a minimum observable interval of space does not demand that spatial oscillation cannot exist for the superposition of two momentum bases when their momentum difference exceeds the Planck energy divided by the speed of light. The reason is that the superposition state does not exist in a region of space whose size is smaller than the minimum observable interval of space.

[26] This means that the MOIST postulate entails that the superposition principle must be violated. Note also that Penrose's gravity-induced collapse argument strongly depends on the assumption that gravity is not emergent but fundamental and the general covariance principle of GR is universally valid [28], and thus even if the argument is valid (cf. [39]), it does not refute other theories without quantum collapse such as string theory which reject this assumption. By comparison, the argument given here only depends on the existence of a minimum observable space-time interval.

[27] Although no violation of the principle of conservation of energy has been found, it seems that there is no *a priori* reason why this principle must be universally true either [43].

[28] This conclusion has another support. Even for the superposition of two energy eigenstates with an energy difference smaller than the Planck energy, the density $\rho(x,t)$ also observably changes during a time interval smaller than the minimum observable time interval T_U, though the period of the oscillation is longer than T_U. Therefore, the MOIST postulate also requires that the whole superposition will collapse into one of the energy eigenstates in the superposition, though the collapse process is relatively slower.

[29] The existence of an invariant speed can be regarded as one implication of the MOIST postulate for the continuous, linear evolution of the wave function. Here is one of its implications for the discontinuous and nonlinear evolution of the wave function.

depends not only on time duration but also on the wave function itself (e.g. its energy distribution) in general, during an arbitrarily short time interval the effect can always be observable at the ensemble level for some wave functions. However, the MOIST postulate demands that all observable processes should happen during a time interval not smaller than the minimum T_U, and thus each tiny collapse must happen during one T_U or more.

Moreover, if there are infinitely many possible states toward which the collapse tends at any time, the duration of each tiny collapse will be exactly one T_U for most time; when the time interval becomes larger than one T_U the tiny collapse will happen in other states with a probability almost equal to one. This means that the dynamical collapse of the wave function will be basically a discrete process. It has been recently shown that such a discrete model of energy-conserved wavefunction collapse which satisfies the above criterion can be consistent with existing experiments and our macroscopic experience [43].

Since there is a connection between the difference of energy distribution and the difference of space-time geometries according to GR, the above result also suggests that the quantum superposition of two different space-time geometries cannot exist and must collapse into one of the definite space-time geometries in the superposition. In order to make this argument more precise, we need to define the difference between two space-time geometries here. As suggested by the generalized uncertainty principle denoted by Equation (1), the energy difference ΔE corresponds to the space-time geometry difference $\frac{L_U^{\,2}\Delta E}{2\hbar c}$. The physical meaning of this quantity can be further clarified as follows. Let the two energy eigenstates in the superposition be limited in the regions with the same radius R (they may locate in different positions in space). Then the space-time geometry outside the region can be described by the Schwarzschild metric denoted by Equation (8). By assuming that the metric tensor inside the region R is the same order as that on the boundary, the proper size of the region is

$$L \approx 2\int_0^R (1-\frac{r_S}{R})^{-1/2} dr \tag{14}$$

where $r_S = \frac{2GE}{c^4}$ is the Schwarzschild radius. Then the spatial difference of the two space-time geometries in the superposition inside the region R can be characterized by

$$\Delta L \approx \int_0^R \frac{\Delta r_S}{R} dr = \Delta r_S = \frac{2L_p^{\,2}\Delta E}{\hbar c} \tag{15}$$

This result is consistent with the generalized uncertainty principle. Therefore, as to the two energy eigenstate in a superposition, we can define the difference of their corresponding space-time geometries as the difference of the proper spatial sizes of the regions occupied by these states. Such difference represents the fuzziness of the point-by-point identification of the spatial section of the two space-time geometries (cf. [28]).

The space-time geometry difference defined above can be rewritten in the following form:

$$\frac{\Delta L}{L_U} \approx \frac{\Delta E}{E_P} \tag{16}$$

This relation indicates one kind of equivalence between the difference of energy and the difference of space-time geometries for the superposition of two energy eigenstates. Therefore, we can also give a collapse criterion in terms of space-time geometry difference. If the difference of the space-time geometries in the superposition ΔL is close to L_U, the superposition state will collapse to one of the definite space-time geometries in one T_U. If ΔL is smaller than L_U, the superposition state will collapse after a finite time interval longer than T_U.

As a result, the superposition of space-time geometries can only possess space-time uncertainty smaller than the minimum observable size of discrete space-time. If the uncertainty limit is exceeded, the superposition will collapse to one of the definite space-time geometries instantaneously. This will ensure that the wave function and its evolution can still be consistently defined during the process of wavefunction collapse, as the space-time geometries with a difference smaller than the minimum observable size can be regarded as physically identical according to the MOIST postulate (cf. [28])[30].

To sum up, the existence of a minimum observable interval of space-time may result in the dynamical collapse of the wave function and prohibit the existence of quantum superposition of different space-time geometries. This seems to provide a plausible solution to the measurement problem. Moreover, quantum and gravity might be unified with the help of the resulting wavefunction collapse [44]. In this way, there will be no quantized gravity in its usual meaning.

In contrast to the semiclassical theory of quantum gravity, however, the theory may naturally include the back-reactions of quantum fluctuations to gravity (e.g. the influence of wavefunction collapse to the geometry of space-time), as well as the reactions of gravity to quantum evolution. Therefore, it might be able to provide a consistent framework for a fundamental theory of quantum gravity. Certainly, the details of the theory such as the law of dynamical wavefunctionneed to be further studied. Our analysis suggests that space-time is not a pure quantum dynamical entity, but it is not wholly classical either.

Conclusion

We have argued that the existence of a minimum observable interval of space-time mayhelp explain why the speed of light is invariant in all inertial frames and why matter curves space-time and why the wave function collapses. These heuristic arguments might provide a deeper understanding of the special and general relativity and quantum theory, and may also have implications for the solutions to the measurement problem and the problem of quantum gravity.

---

[30] Certainly, the states of matter corresponding to these space-time geometries can still be distinguished in general.

References

[1] Garay, L. J. (1995).Quantum gravity and minimum length. *Int. J. Mod.Phys.* A10, 145.

[2] Einstein, A. (1949).Autobiographical Notes. In: *Albert Einstein: Philosopher-Scientist*. P. A. Schilpp, ed. *The Library of Living Philosophers*, vol. 7. Evanston, Illinois: Northwestern University Press.

[3] Adler, R. J. and Santiago, D. I. (1999). On gravity and the uncertainty principle. *Mod. Phys. Lett. A*14, 1371.

[4] Calmet, X., Hossenfelder, S. and Percacci, R. (2010). Deformed special relativity from asymptotically safe gravity, *Phys. Rev. D* 82, 124024.

[5] Wang, L. J., Kuzmich, A. and Dogariu, A. (2000).Gain-assisted superluminal light propagation, *Nature* 406, 277.

[6] Brillouin, L. (1960). *Wave Propagation and Group Velocity*. New York: Academic Press.

[7] Kuzmich, A., et al. (2001). Signal velocity, causality, and quantum noise in superluminal light pulse propagation. *Phys. Rev. Lett.*86, 3925.

[8] Rovelli, C. and Smolin, L. (1995). Discreteness of area and volume in quantum gravity, *Nucl.Phys.*, B 442, 593.

[9] Rovelli, C. (2004). *Quantum Gravity*. Cambridge: Cambridge University Press.

[10] Bombelli, L., Lee, J., Meyer, D., and Sorkin, R. D. (1987). Space-time as a causal set, *Phys. Rev. Lett.* 59, 521.

[11] Dowker, F. (2006). Causal sets as discrete space-time, *Contemporary Physics*, 47(1), 1.

[12] Edwards, W. (1963). Special *relativity* in anisotropic space. *Am. J. Phys.* 31, 482.

[13] Winnie, J. (1970). Special *relativity* without one-way velocity assumptions, Part I; Part II. *Philosophy of Science*, 37, 81, 223.

[14] Rindler, W. (1982) *Introduction to Special Relativity*. Oxford: Oxford University Press, p.22.

[15] Zhang, Y. Z. (1997). *Special Relativity and its Experimental Foundations*. Singapore: World Scientific.

[16] Mattingly, D. (2005). Modern tests of Lorentz invariance, *Living Rev. Rel.* 8, 5.

[17] Torretti, R. (1983). *Relativity and Geometry*. Oxford: Pergamon Press.

[18] Pal, P. B. (2003). Nothing but relativity. *Eur. J. Phys.* 24, 315-319.

[19] Brown, H. (2005). *Physical Relativity: Space-time structure from a dynamical perspective*. Oxford: Clarendon Press.

[20] Gao, S. (2004). Quantum collapse, consciousness and superluminal communication, *Found. Phys. Lett.*, 17, 167-182.

[21] Amelino-Camelia, G. (2000). Relativity in space-times with short-distance structure governed by an observer-independent (Planckian) length scale. *Int. J. Mod. Phys. D* 11, 35.

[22] Kowalski-Glikman, J. (2005). Introduction to doubly special relativity. *Lect. Notes Phys.*, 669, 131.

[23] Amelino-Camelia, G. (2010). Doubly-special relativity: facts, myths and some key opens issues, *Symmetry* 2, 230-271.

[24] Kowalski-Glikman, J. and Smolin, L. (2004). Triply special relativity. *Phys. Rev. D* 70, 065020.

[25] Hossenfelder, S. (2010). Bounds on an energy-dependent and observer-independent speed of light from violations of locality. *Phys. Rev. Lett.*104, 140402. And arXiv:1005.0535, arXiv:1006.4587, arXiv:1008.1312.

[26] Smolin, L. (2010). Classical paradoxes of locality and their possible quantum resolutions in deformed special relativity. arXiv:1004.0664, arXiv:1007.0718. Amelino-Camelia, G., et al. (2010).T*aming nonlocality in theories with deformed Poincare symmetry*. arXiv:1006.2126.

[27] Hagar, A. (2009). Minimal length in quantum gravity and the fate of Lorentz invariance. *Studies in the History and Philosophy of Modern Physics* 40, 259.

[28] Penrose, R. (1996). On gravity's role in quantum state reduction. *Gen. Rel. Grav.*28, 581.

[29] Rovelli, C. (2004). *Quantum Gravity*. Cambridge: Cambridge University Press.

[30] Isham, C. J. and Butterfield, J. (1999). On the emergence of time in quantum gravity. In: *The Arguments of Time*, ed. J. Butterfield. Oxford: Oxford University Press.

[31] Kiefer, C. (2004). *Quantum Gravity*. Oxford: Oxford University Press.

[32] Sakharov, A. D. (1968/2000). Vacuum quantum fluctuations in curved space and the theory of gravitation. *Sov. Phys. Dokl.* 12, 1040 [*Dokl. Akad. Nauk Ser. Fiz.* 177, 70]. Reprinted in *Gen. Rel. Grav.*32, 365-367 (2000).

[33] Visser, M. (2002). Sakharov's induced gravity: A modern perspective. *Mod. Phys. Lett. A* 17, 977.

[34] Jacobson, T. (1995). Thermodynamics of space-time: the Einstein equation of state. *Phys. Rev. Lett.* 75, 1260.

[35] Verlinde, E. P. (2011). On the origin of gravity and the laws of Newton, *JHEP* 04, 29.

[36] Gao, S. (2011). Is gravity an entropic force? *Entropy* special issue "*Black Hole Thermodynamics*", Jacob D. Bekenstein (eds). 13, 936-948.

[37] Christian, J. (2001). Why the quantum must yield to gravity. In: *Physics Meets Philosophy at the Planck Scale*, ed. C. Callender and N. Huggett. Cambridge: Cambridge University Press, p.305.

[38] Gao, S. (2006). A model of wavefunction collapse in discrete space-time, *Int. J. Theor. Phys.* 45 (10), 1965-1979.

[39] Gao, S. (2013). Does gravity induce wavefunction collapse? An examination of Penrose's conjecture. *Studies in History and Philosophy of Modern Physics*, 44, 148-151.

[40] Aharonov, Y. and Vaidman, L. (1993). Measurement of the Schrödinger wave of a single particle, *Phys. Lett.* A 178, 38.

[41] Aharonov, Y., Anandan, J. and Vaidman, L. (1993). Meaning of the wave function, *Phys. Rev.* A 47, 4616.

[42] Gao, S. (2014). Protective measurement and the meaning of the wave function. Book chapter in *Protective Measurement and Quantum Reality: Toward a New Understanding of Quantum Mechanics.* Gao, S. (eds.), Cambridge University Press, forthcoming.

[43] Gao, S. (2013). A discrete model of energy-conserved wavefunction collapse, *Proceedings of the Royal Society* A 469, 20120526.

[44] Gao, S. (2006). *Quantum Motion: Unveiling the Mysterious Quantum World.* Bury St Edmunds, Suffolk UK: Arima Publishing.

In: Space-Time Geometry and Quantum Events
Editor: Ignazio Licata, pp. 215-227
ISBN: 978-1-63117-455-1

Chapter 7

Quantum Computing Space-Time

P. A. Zizzi*
Department of Brain and Behavioural Sciences,
University of Pavia, Piazza Botta, Pavia, Italy

Abstract

A causal set C can describe a discrete space-time, but this discrete space-time is not quantum, because C is endowed with a Boolean logic, as it does not allow cycles. In a quasi-ordered set Q cycles are allowed. In this paper we consider a subset QC of a quasi-ordered set Q, whose elements are all the cycles. In QC, which is endowed with a quantum logic, each cycle of maximal outdegree N in a node is associated with N entangled qubits. Then QC describes a quantum computing space-time. This structure, which is non-local and non-causal, can be understood as a proto-space-time. Micro-causality and locality can be restored in the subset U of Q whose elements are unentangled qubits which we interpret as the states of quantum space-time. The mapping of quantum space-time into proto-space-time is given by the action of the XOR gate. Moreover, a mapping is possible from the Boolean causal set into U by the action of the Hadamard gate. In particular, the causal order defined on the elements of U induces the causal evolution of spin networks.

1. Introduction

Discreteness of space-time at the Planck scale seems to be one of the most compelling requirements of quantum gravity [1], the theory which should reconcile and unify Quantum Mechanics and General Relativity. In fact, both loop quantum gravity [2] and superstrings/M-theory [3], the two major candidates for quantum gravity, strongly suggest that space-time at the Planck scale must have a discrete structure. In particular, in loop quantum gravity non-perturbative techniques have led to a picture of quantum geometry, which is rather of a polymer type, and geometrical quantities such area and volume have discrete spectra. In quantum geometry spin networks play a very important role. They were invented by Penrose [4] and lead to a drastic change in the concept of space-time, going from that of a smooth

*E-mail address: paola.zizzi@unipv.it.

manifold to that of a discrete, purely combinatorial structure. Then, spin networks were rediscovered by Rovelli and Smolin [5] in the context of loop quantum gravity, where they are eigenstates of the area and volume operators [6].However, this theory of quantum geometry does not reproduce classical General Relativity in the continuum limit. Recent models of quantum gravity called "spin foam models" [7] seem to have continuum limits. Anyway, as spin foam models are Euclidean, they are not suitable to recover causality at the Planck scale. For this purpose the theory should be intrinsically Lorentzian. However, the very concept of causality becomes uncertain at the Planck scale, when the metric undergoes quantum fluctuations as Penrose [8] argued. So, one should consider a discrete alternative to the Lorentzian metric, which is the causal set (a partially ordered set-or poset- whose elements are events of a discrete space-time). Such theories of quantum gravity based on the causal set were formulated by Sorkin et al. [9]. Rather recently a further effort in trying to recover causality at the Planck scale has been undertaken by Markopoulou and Smolin [10]. They considered the evolution of spin networks in discrete time steps, and they claimed that the evolution is causal because the history of evolving spin networks is a causal set. Of course, also the causal set approach [9] to quantum gravity relies on a discrete structure of space-time at the fundamental level. Finally, the quantum computational approach to quantum gravity [11] suggests as well that space-time is discrete at the Planck scale. This fourth approach has been applied in particular to quantum cosmology, resulting in a model of quantum inflation describing the very early universe as a growing quantum network [12].

In this paper, we will investigate the structure of quantum space-time by applying the tools of quantum information and quantum computation [13] to an extended version of the causal set theory. Actually, we will not consider the partially ordered set (poset) on which the causal set C is based, but the quasi-ordered set Q, which allows closed loops. The poset was chosen to describe a discrete space-time with micro-causality, just because a poset does not allow cycles. However, we show that this restriction is in disagreement with the very nature of quantum space-time which should be endowed with a quantum logic. On the contrary, a poset is endowed with a Boolean logic. We interpret the one cycle graph in Q as the one-qubit state (or quantum bit, the unit of quantum information). There are two important subsets of Q, one whose elements are entangled qubits (which we call QC, i.e. quantum computing space-time), and one whose elements are unentangled qubits (which we call U). In QC micro-causality is missing, as well aslocality. Instead, we show that in U it is possible to define a causal order. This is due to the fact that unentangled qubits are product states, and then it is possible to define an increase of information entropy which induces an arrow of discrete time. We interpret QC as a proto-space-time, while U plays the role of quantum space-time itself. The elements of U are qubits, i.e. superposed states, and then U is endowed with a quantum logic. The elements of U are the "quantum events". Instead, in the Boolean causal set C considered by Sorkin and coworkers [9] the events of discrete space-time are points and not cycles. However, we show that the causal set C (or, better, its subset B-where B stands for Boolean- whose elements are classical bits) can be mapped into the set U by a quantum logic gate (the Hadamard gate). Also, the mapping from U to the proto-space-time QC is made by the XOR gate.

Finally, we show that there is a one-to-one relation between the "quantum events" (elements of U) and the punctures of spin networks' edges. In fact, the elements of U can be interpreted as pairs of virtual events which are the birth and death of a Planckian black hole, which has a horizon area of one pixel (one unit of Planck area). By the quantum version [11] of the holographic principle [14] each pixel of area encodes one qubit. Moreover, we know from loop quantum gravity that, if a 2-surface is punctured by a spin network edge in one point, it acquires an area of one pixel. So, each (extended) quantum event corresponds to one point of discrete space-time, i.e. to one puncture of a spin networks' edge. Then, the causal relation defined on the elements of U induces the causal evolution of spin networks.

2. A Brief Review of Ordered Sets

2.1. Partially Ordered Set P

A partially ordered set (or poset) **P** is a set S plus a relation $\leq$ on the set, with the following properties:

1. Reflexivity: $\alpha \leq \alpha$ for all aϵS
2. Antisymmetry: $\alpha \leq b$ and $b \leq \alpha$ implies $\alpha = b$
3. Transitivity: $\alpha \leq b$ and $b \leq c$ implies $\alpha \leq c$

2.2. Totally Ordered Set T

A totally ordered set **T** is a set S plus a relation *R* on the set called total order, that satisfies the conditions for a partial order plus the comparability condition (or trichotomy law):

1. Reflexivity: $a \leq a$ for all aϵS
2. Antisymmetry: $a \leq b$ and $\boldsymbol{b} \leq \boldsymbol{a}$ implies $\boldsymbol{a} = b$
3. Transitivity: $a \leq b$ and $b \leq c$ implies a $\leq c$
4. Comparability: either $a \leq b$ or $b \leq a$ for any $a,b \epsilon$S

2.3. Quasi-ordered Set Q

A quasi-ordered set **Q** is a set S plus a relation $\leq$ on the set, which satisfies the properties of reflexivity and transitivity:

1. Reflexivity: $a \leq a$ for all $a \epsilon S$
2. Transitivity: $a \leq b$ and $b \leq c$ implies $a \leq c$

A quasi-order does not satisfy antisymmetry, so cycles are allowed in **Q.**

In **Q** we can have pairs (a,b) of three different types:

i) incomparable: neither $a\leq b$ nor $b\leq a$
ii) comparable: either $a\leq b$ _ or $b \leq a$ ($a\leq b$ and $b\leq a$ implies $a = b$)
iii) comparable but not equivalent: $a\leq b$ and $b\leq a$ ($a\leq b$ and $b\leq a$ with $a \neq b$)

Of course, any poset is also quasi-ordered.

3. The Causal Set C

A causal set C is a locally finite, partially ordered set, whose elements are events of a discrete space-time.

For a causal set C, the following properties hold:

1. Reflexivity: $p \leq p$ for all $p \epsilon C$
2. Antisymmetry: $p \leq q$ and $q \leq p$ implies $p = q$
3. Transitivity: $p\leq q$ and $q \leq r$ implies $p \leq r$
4. Local finiteness: $|A(p,q)| < \infty$

where $|A(p,q)|$ is the cardinality of the "Alexandrov set" $A(p, q)$ of two events p and q, which is the set of all events x such that $p\leq x \leq q$.

In particular, antisymmetry (or acyclicity) is needed to avoid closed time like loops.

It is generally believed that the causal set C can describe a quantum space-time endowed with micro-causality.

However, the underlying logic of the causal set C is classical, i.e, Boolean.

In fact, once we define: $p\leq q$ as "yes" and $q\leq p$ as "no", antisymmetry implies: either "yes" or "no". This means that the information stored in a causal set is given in terms of classical bits "0" and "1", as for example, in a classical computer. However, if the aim is to describe a quantum space-time, one should deal with a discrete structure whose underlying logic is quantum.

4. The Quantum Computing Set QC

A digraph (or directed graph) is a graph in which each edge is replaced by a directed edge.

A digraph G is transitive if any three vertices a,b,c such that edges (a,b), $(b,c) \in G$ imply $(a,c) \in G$.

An oriented graph is a digraph having no symmetric pair of directed edges.

Moreover, a simple graph is a graph in which each pair of vertices are connected by at most one edge, while in a non-simple graph multiple edges are also allowed. The indegree (outdegree) is the number of incoming (outgoing) directed edges in a node and the local degree is the total number of directed edges visiting a node.

A cycle: $p \leq q \leq p$, with $p \neq q$ in **Q,** implies "yes" and "no" at the same time, which is anon-Boolean proposition. The superposition of bits "0" and "1" is a quantum bit of information (or qubit).

The single qubit can be written as: $a|0\rangle + b|1\rangle$ where a and b are the complex amplitudes of the two states, with the condition: $|a|^2+|b|^2 = 1$

Then, some of the information stored in **Q** is given in terms of qubits as in a quantum computer. Given a quasi-ordered set **Q**, let us consider first only those pairs of elements (p,q) which are related but not equivalent (cycles): $p \leq q$ and $q \leq p$ with $p \neq q$.

We will denote this subset of **Q** as **QC**, where **QC** stands for "Quantum Computing".

Now, let us consider only those pairs of elements in **Q** which are related: either $p \leq q$ or $q \leq p$.

We will indicate this subset of **Q** as **B**, where **B** stands for "Boolean". **B** is also a subset of a causal set **C**. All the "events" of B are the classical bits "0" and "1". The sets **B** and **QC** are disjoint:

$$B \cap QC = \emptyset \text{ where } B \subset C \subset Q \text{ and } QC \subset Q.$$

In **Q** there are also pairs of elements which are cycles, but are not entangled with other cycles. For example, let us consider the cycle $p \leq q$ and $q \leq p$ with $p \neq q$, where one element of the cycle, let us say q, is related to a third element r:$q \leq r$. This subset of **Q** will be called **U** where **U** stands for "unentangled".

In **U,** as it will be showed in the following, the concepts of time flow, micro-causality, and locality are still valid. The same concepts are instead completely lost in **QC**, which is to be considered just as a proto-structure of quantum space-time.

The passage from the Boolean logic of **B** to the quantum logic of **QC** can be interpreted in terms of the action of two quantum logic gates, the Hadamard gate *H,* and the XOR gate (see figure 1). In fact, first the Hadamard gate transforms the classical bits into superposed states (qubits), then the XOR gate transforms the superposed states into entangled states, as it will be showed in what follows.

The Hadamard gate is:

$$H = \frac{1}{\sqrt{2}}\begin{pmatrix} 1 & 1 \\ -1 & 1 \end{pmatrix}.$$

Its action on bits $|0\rangle$ and $|1\rangle$ gives asymmetric and an antisymmetric 1-qubit state respectively:

$$H|0\rangle = \frac{1}{\sqrt{2}}\left(|1\rangle + |0\rangle\right) \text{ and } H|1\rangle = \frac{1}{\sqrt{2}}\left(|1\rangle - |0\rangle\right)$$

where we have represented the base states $|1\rangle$ and $|0\rangle$ as the vectors $\begin{pmatrix}1\\0\end{pmatrix}$ and $\begin{pmatrix}0\\1\end{pmatrix}$ respectively.

The quantum logic gate which transforms unentangled qubits (elements of **U**) into entangled qubits (elements of **QC**) is the XOR gate. The XOR gate (or controlled-NOT gate) is the standard 2-qubits gate, and illustrates the interactions between two quantum systems. Any quantum computation can be performed by using the XOR gate, and the set of one-qubit gates. The XOR gate flips the "target" input if its "control" input is $|1\rangle$ and does nothing if it is $|0\rangle$:

$|1\rangle$ -------------------- -------------------- $|1\rangle$

XOR i.e $|10\rangle \rightarrow |11\rangle$

$|0\rangle$ -------------------- -------------------- $|1\rangle$

$|0\rangle$ -------------------- -------------------- $|0\rangle$

XOR i.e $|00\rangle$ unchanged

$|0\rangle$ -------------------- -------------------- $|0\rangle$

Hence, a XOR gate can clone Boolean inputs. But, if one tries to clone a superposed state, one gets an entangled state:

$\frac{1}{\sqrt{2}}(|1\rangle + |0\rangle)$ -------------

XOR --------------- $\frac{1}{\sqrt{2}}(|00\rangle + |11\rangle)$

(entangled state)

$|0\rangle$ ----------------

Then, the XOR gate cannot be used to copy superposed states (impossibility of cloning an unknown quantum state).

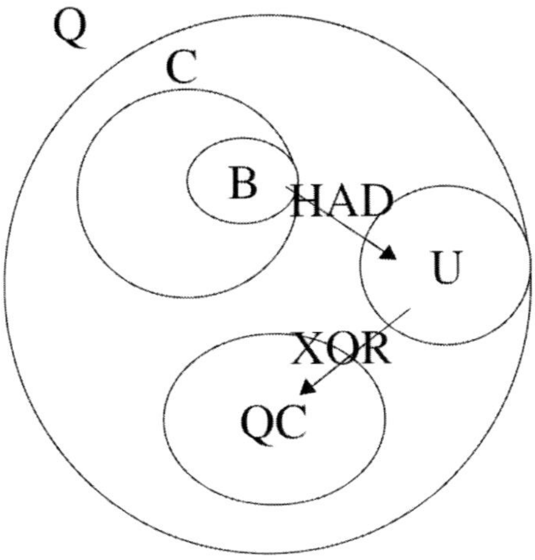

Figure 1.

5. Cycles as Qubits

Let us consider two "events" p and q in **QC**: $p \leq q \leq p$, with $p \neq q$.

This is a cycle graph, in particular it is the Z_2 graph {0,1} which is associated with the symmetric 1-qubit

$|Q_1\rangle^S = \frac{1}{\sqrt{2}}(|0\rangle + |1\rangle)$ if the orientation is clockwise and with the antisymmetric 1-qubit

$|Q_1\rangle^A = \frac{1}{\sqrt{2}}(|0\rangle - |1\rangle)$ if the orientation is anti-clockwise, as shown in figure 2.

Let us consider a third event r such that $p \leq q \leq p$ and $q \leq r \leq q$. The resulting cycle $p \leq q \leq r \leq q \leq p$ corresponds to four G' graphs, each one being the union of twoZ_2graphs joining in two nodes. Then the two nodes of the G' graphs are both 2-valued.

G'_1 = {0,1} ∪ {1,0} joining in the nodes (00) and (11)
G'_2 = {0,1} ∪ {0,1} joining in the nodes (01) and (10)
G'_3 = {1,0} ∪ {0,1} joining in the nodes (11) and (00)
G'_4 = {1,0} ∪ {1,0} joining in the nodes (10) and (01).

The above four G' graphs are associated with the four Bell states which form an entangled basis for 2-qubits:

$$|\Phi_\pm\rangle = \frac{1}{\sqrt{2}}(|11\rangle \pm |00\rangle); \qquad |\Psi_\pm\rangle = \frac{1}{\sqrt{2}}(|10\rangle \pm |01\rangle).$$

One example of such graphs is given in figure 3

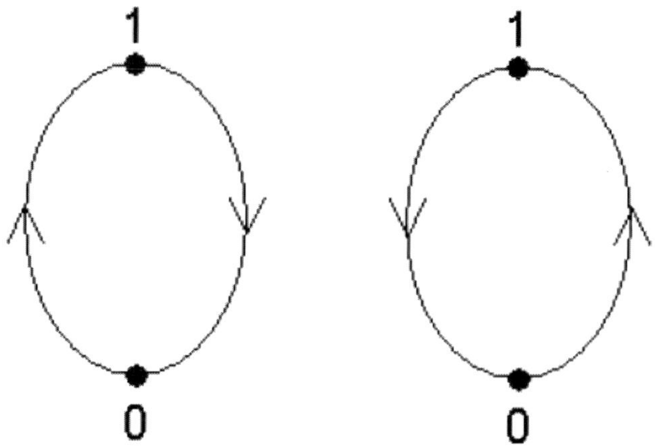

Figure 2.

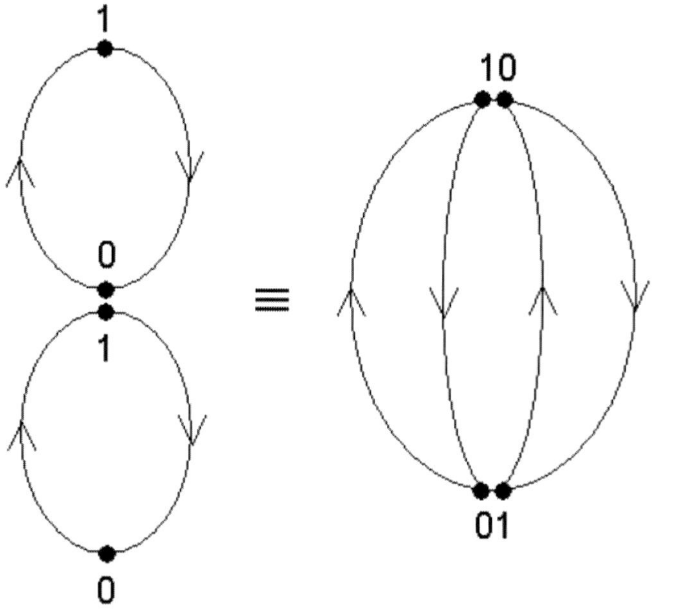

Figure 3.

As all the elements of **QC** are related to each other, but not equivalent, all they are entangled qubits, unless the dimension of the discrete space is two, in which case we have only 1-qubit state, as in figure 2.In general, a cycle in **QC**, with a node of maximal indegree (or outdegree) N, will be associated with N entangled qubits. The discrete space-time described by **QC** is then non-causal (because of cyclicity) and non-local (because of entanglement).**QC** then represents a proto-space-time endowed with quantum logic, whose events are entangled qubits.

6. Unentangled Qubits: Events of Quantum Space-Time

Let us now consider the three events p, q and r in the quasi-ordered set **Q**, such that $p \leq q \leq p$ with $p \neq q$ (which is associated with the 1-qubit) and either $q \leq r$ or $r \leq q$ which is associated with the classical bits 0 and 1). This describes four graphs G". Each G" has a one 2-valued node and one single- valued node, although both nodes have indegree (or outdegree) 2.

The first G" is a Z_2 graph {0,1} with one extra outgoing edge from the node 1, and incoming in the node 0. So, nodes 1 and 0 are identified (1,0). The resulting 2-valued node (1,0) is associated with the state $|10\rangle$.

The second G" is a Z_2 graph {0,1} with one extra outgoing edge from the node 0, and incoming in the node 1. The resulting 2-valued node (0,1) will be associated with the state $|01\rangle$. .

The third G" is a Z_2 graph {0,1} with a loop joining the node 1 to itself. The resulting 2-valued node (1,1) will be associated with the state $|11\rangle$.

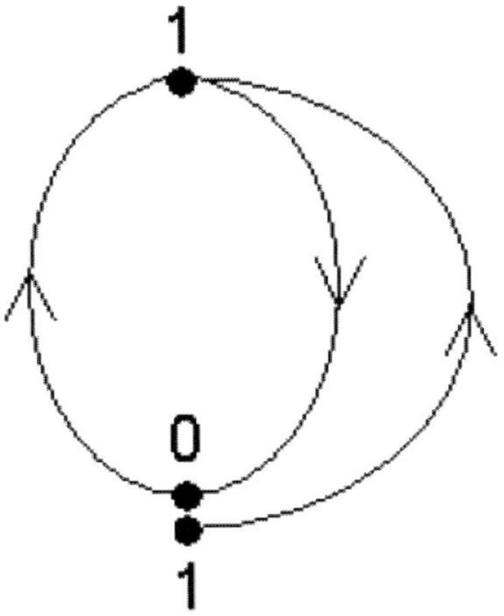

Figure 4.

The fourth G" is a Z_2 graph {0,1} with a loop joining the node 0 to itself. The resulting 2-valued node (0,0) is associated with the state $|00\rangle$.

Then the four G" graphs provide the unentangled basis $|11\rangle$, $|00\rangle$, $|01\rangle$, $|10\rangle$ for 2-qubits. One example of such G" graphs is given in figure 4.

7. Information Entropy, the Arrow of Discrete Time and Micro-Causality

The second law of thermodynamics states that an increase of entropy induces an arrow of time. An increase of information entropy will also induce an arrow of (discrete) time. The information entropy of N qubits is $S=N \ln 2$, and if $\Delta S>0$, we can define an arrow of discrete time t_N, where t_N should be directly proportional to some expression of N. However, one should be aware that there are two different situations in QC and in U. In QC, the N qubits are all entangled to each other once for all, and they influence each other simultaneously. The usual causal relation is then meaningless in this case. In fact, there is no increase of information entropy, and an arrow of time cannot be defined. Instead, in U, there are N unentangled qubits, which are product states. For each factor state we can define the information entropies S_1, S_2... S_n...S_N and define an increase of entropy $\Delta S_{n,n'} = (n'-n) \ln 2 > 0$ if n' > n. In this case an arrow of discrete time can be defined.

Let us consider the two elements p and q of Q with $p \leq q \leq p$ and $p \neq q$ (a cycle graph Z_2 which is associated with the 1-qubit state). This state is the ground state Ψ_0 of quantum space-time, with minimal information entropy $S = \ln 2 (N=1)$.

We interpret p and q as virtual events in the time interval $\Delta t = t_p$ where t_p is the Planck time.

The energy associated with this virtual process is:

$$\Delta E = \frac{\hbar}{\Delta t} = \frac{\hbar}{t_P} = E_P$$

where $Ep \approx 10^{19} GeV$ is the Planck energy.

This process describes a virtual Planckian blackhole whose birth and death are the virtual events p and q. The two virtual events are associated with one Z_2graph. In fact the graph Z_2 itself can be considered as the building block of quantum space-time. In this context, the "event" of quantum space-time, is the ensemble of two virtual events, and is not a point, but an extended object: a Planckian black hole.

The horizon area of the Planckian black hole is one pixel, i.e., one unit of Planck area L_P^2, (where $L_P \approx 10^{-35} m$ is the Planck length), and, in accordance with the quantum version [11] of the holographic principle [14], it encodes one qubit. In conclusion, an event of quantum space-time is an extended object which is endowed with the Planck energy, and encodes one unit of quantum information.

Let us now consider the four graphs in section 6, one of which is represented in figure 4.

This is in fact the ensemble of three virtual events which is associated with four Z_2 graphs, i.e., with four "events" of quantum space-time, encoding four (unentangled) qubits.

In general, a number n_V of virtual events (with n_V=2,3,4....) is associated with N= $(n_V - 1)^2$cycle graphs or "events" of quantum space-time (with N=1,4,9...) encoding N unentangled qubits.

$$|N\rangle = \frac{1}{\sqrt{2}^N} |Q_1\rangle^{\otimes N}$$

where in fact $|Q_1\rangle$ is the ground state $|\Psi_0\rangle$ of quantum space-time.

As we have seen, the uncertainty in the energy associated with one pair of such virtual events (nv= 2) is the Planck energy. The uncertainty in the energy in a process involving a total number nv of virtual events, is the Planck energy divided by the total number of pairs nv- 1:

$$\Delta E \equiv E_N = \frac{E_P}{n_V - 1} = \frac{E_P}{\sqrt{N}}$$

Moreover, the time-energy uncertainty relation should be saturated for every process involving n_V virtual events: $\Delta E \Delta t \equiv E_N t_N \approx \hbar$ from which it follows: $t_N = \sqrt{N}\, t_P$, which is in fact proportional to (the square root of) the information entropy. In summary, the information entropy of N events of quantum spacetime induces an arrow of discrete time which is quantized in Planck time units.

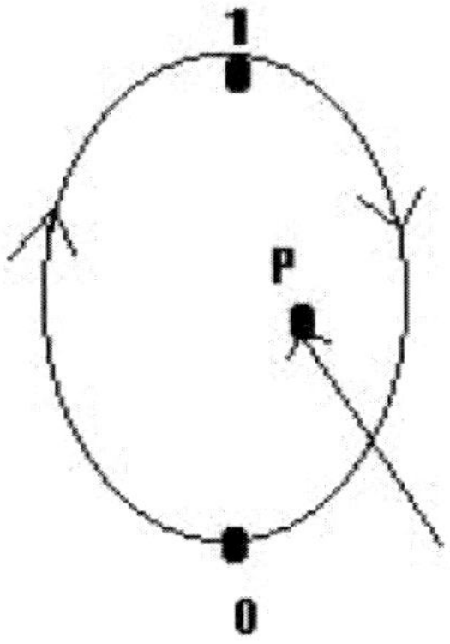

Figure 5.

The (unentangled) N-qubit states $|N\rangle$ form a causal set C_N:

$$|N\rangle \leq |M\rangle \text{ for } t_N \leq t_M$$

The $|N\rangle$ states satisfy reflexivity, antisymmetry, and transitivity.
Thus, in C_N, micro-causality is recovered.

However, C_N is just a subset of **Q**. In particular, in C_N all entangled states are missing. A similar attempt has been done in terms of evolving spin networks [10]. As the causal set of spin networks and the causal set C_N are strictly related to each other by the holographic principle (see [11] for more details on this), the above arguments hold for the spin networks' case as well. Let us now look in more detail for the relation between the causal order defined on N unentangled qubits and the causal evolution of spin networks. Basically, spin networks are graphs embedded in 3-space, with edges labeled by spins j=0, 1/2, 1, 3/2...and vertices labeled by intertwining operators. In loop quantum gravity, spin networks are eigenstates of the area and volume-operators [6]. If a single edge punctures a 2-surface transversely, it contributes an area proportional to:

$$L_P^2 \sqrt{j(j+1)}$$

where L_P is the Planck length.

The points where the edges end on the surface are called “punctures”. If the surface is punctured in n points, the area is proportional to:

$$L_P^2 \sum_n \sqrt{j_n(j_n+1)}\,.$$

The cycle graph Z_2 in figure 2 represents a virtual, Planckian black hole, whose surface horizon has an area of one pixel (one unit of Planck area, L_P^2), which, by the holographic principle, encodes one qubit. By the above arguments, it follows that there is only one puncture P_1giving rise to one pixel of area, associated with the 1-qubit state. This is the relation between the extended object (the quantum event) and the point P_1 (the classical event of a discrete space-time). In this way, a causal order on N (unentangled) qubits induces causal evolution of spin networks. It should be noticed that the causal evolution of spin networks was originally put "by hand", instead the arrow of discrete time naturally arises in our context as a consequence of the increasing of information entropy. The situation is schematised in figure 5, where the surface enclosed by the cycle graph Z_2 has an area of one pixel, due to one puncture P of a spin networks' edge.

8. Concluding Remarks

In summary, we found that space-time at the fundamental level shows a quite rich spectrum of different structures which can be mapped into each other by quantum logic gates. The starting point was the quasi-ordered set Q, which allows closed loops. Then we focused on an important subset of Q, the set QC, which is a quantum-computing space-time, or proto space-time. The set QC is a very peculiar type of space-time, or better is not really a true space-time, neither quantum nor classical, as it misses locality and micro-causality. The latter can be recovered by mapping QC into U-another subset of Q-which is instead a proper quantum space-time, endowed with quantum events. This hierarchy of quantum space-time s might be useful for a deeper understanding of the inner structure of quantum gravity.

Acknowledgments

I wish to thank M. Pregnolato and E. Pessa for careful revision and technical support.

References

[1] C. Rovelli, "*Notes for a brief history of quantum gravity*", gr-qc/0006061.

[2] C. Rovelli, "Loop Quantum Gravity", gr-qc/9710008; C. Rovelli and L. Smolin, "Loop representation of quantum general relativity", *Nucl. Phys.* B133 (1990) 80-152.

[3] J. H. Schwarz, "Introduction to Superstring Theory", hep-th/0008017; M. J. Duff, "*M-Theory (The Theory Formerly known as Strings)*", hep-th/9608117.

[4] R. Penrose, "Theory of quantised directions" in: *Quantum theory and beyond*, ed. T. Bastin, (Cambridge University Press, 1971).

[5] C. Rovelli and L. Smolin, "Spin networks and quantum gravity", gr-qc/9505006, *Phys. Rev.* D52 (1995) 5743-5759.

[6] C. Rovelli and L. Smolin, "Discreteness of area and volume in quantum gravity", *Nucl. Phys.* B442 (1995) 593-622.

[7] J. C. Baez, "Spin Foam Models", gr-qc/9709052, *Class. Quant. Gravity* 15 (1998) 1827-1858.

[8] R. Penrose, in: *Quantum Gravity, an Oxford Symposium*, ed. C. J. Isham, R. Penrose and D. W. Sciama (Clarendon Press, Oxford 1975).

[9] L. Bombelli, J. Lee, D. Meyer and R. Sorkin, "Space-time as a causal set", *Phys. Rev. Lett.* 59 (1987) 521-524.

[10] F. Markopoulou and L. Smolin, "Causal evolution of spin networks", gr-qc/9702025, *Nucl. Phys.* B508 (1997).

[11] P. A. Zizzi, "Holography, Quantum Geometry and Quantum Information Theory", gr-qc/9907063, Entropy 2 (2000) 39-69; "*Quantum Computation toward Quantum Gravity*", gr-qc/0008049, *Gen. Rel. Grav*.33 (2001)1305-1318.

[12] P. A. Zizzi, "The Early Universe as a Quantum Growing Network", gr-qc/0103002; "*Ultimate Internets*", gr-qc/0110122.

[13] M. A. Nielsen and I. L. Chuang, "Quantum computation and quantum information", Cambridge University Press 2000; A. Ekert, P. Hayden, and H. Inamori, "*Basic Concepts in Quantum Computation*", quant-ph/0011013.

[14] G. 'Htooft, "Dimensional Reduction in Quantum Gravity", gr-qc/9310026; "The Holographic Principle", hep-th/0003004; L. Susskind, "*The World as a Hologram*", hep-th/9409089.

In: Space-Time Geometry and Quantum Events
Editor: Ignazio Licata, pp. 229-247
ISBN: 978-1-63117-455-1

Chapter 8

Metric Gauge Fields in Deformed Special Relativity

Roberto Mignani[1,2,3] ***Fabio Cardone***[2,4] ***and Andrea Petrucci***[5]
[1]Dipartimento di Matematica e Fisica, Sezione di Fisica
Università degli Studi "Roma Tre", Roma, Italy
[2]GNFM, Istituto Nazionale di Alta Matematica "F.Severi"
Città Universitaria, Roma, Italy
[3]I.N.F.N. - Sezione di Roma III, Roma, Italy
[4]Istituto per lo Studio dei Materiali Nanostrutturati (ISMN – CNR), Roma, Italy
[5]ENEA, Italian National Agency for new Technologies, Energy and sustainable economic development, Roma, Italy

Abstract

We show that, in the framework of Deformed Special Relativity (DSR), namely a (four-dimensional) generalization of the (local) space-time structure based on an energy-dependent "deformation" of the usual Minkowski geometry, two kinds of gauge symmetries arise, whose spaces either coincide with the deformed Minkowski space or are just internal spaces to it. This is why we named it *"metric gauge theories"*. In the case of the internal gauge fields, they are a consequence of the deformed Minkowski space (DMS) possessing the structure of a generalized Lagrange space. Such a geometrical structure allows one to define curvature and torsion in the DMS.

1. Introduction

It is well known that gauge theories play presently a basic role in describing all the known interactions. In all cases, gauge symmetries are related to physical fields directly arising from the symmetries ruling some given interaction; on one side, this leads to the rising of a new, dynamical gauge field; on the other hand, if the gauge symmetry is broken, such a circumstance provides one with new — often unforeseen — informations about the structural properties of the interaction considered.

Often, as known as well, in spite of the fact that the physical world is the usual Minkowski space-time, the gauge manifold is *not* the usual, Minkowski one. For instance, in the case of the usual Minkoski space, the gauge symmetry of electrodynamics does actually work in an auxiliary space (Weyl charge space). It is therefore worth to investigate when and where gauge symmetries can be introduced in a Minkowski space, and to lead to significant physical results.

It is just the purpose of the present paper to show that this circumstance occurs in the framework of *Deformed Special Relativity (DSR),* namely a (four-dimensional) generalization of the (local) space-time structure based on an energy-dependent "deformation" of the usual Minkowski geometry [1, 2]. As we shall see, in DSR two kinds of gauge symmetries arise, whose spaces either coincide with the deformed Minkowski space (DMS) $\widetilde{M}$ or are just internal spaces to it. This is why we named it *"metric gauge theories".*

The paper is organized as follows. In Sect.2, we review the basic features of DSR that are relevant to our purposes. Sect.3 discuss DSR as a metric gauge theory. Metric gauge fields can be external (Subsect.3.1) or internal (Subsect.3.2). The last topic is related to the structure of DMS as generalized Lagrange space, whose main properties are summarized. Subsect. 3.2.2 deals with the structure of $\widetilde{M}$ as generalized Lagrange space. The internal gauge fields of $\widetilde{M}$ are discussed in Subsect.3.2.4. In 3.3 we present a possible experimental evidence for such metric gauge fields. Conclusions and perspectives are given in Sect.4.

2. Elements of Deformed Special Relativity

2.1. Energy and Geometry

The geometrical structure of the physical world — both at a large and a small scale — has been debated since a long. After Einstein, the generally accepted view considers the arena of physical phenomena as a four-dimensional spacetime, endowed with a *global,* curved, Riemannian structure and a *local,* flat, Minkowskian geometry.

However, an analysis of some experimental data concerning physical phenomena ruled by different fundamental interactions have provided evidence for a local departure from Minkowski metric [1, 2]: among them, the lifetime of the (weakly decaying) K_s^0 meson, the Bose-Einstein correlation in (strong) pion production and the superluminal propagation of electromagnetic waves in waveguides. These phenomena seemingly show a (local) breakdown of Lorentz invariance, together with a plausible inadequacy of the Minkowski metric; on the other hand, they can be interpreted in terms of a deformed Minkowski spacetime, with metric coefficients depending on the energy of the process considered [1, 2].

All the above facts suggested to introduce a (four-dimensional) generalization of the (local) space-time structure based on an energy-dependent "deformation" of the usual Minkowski geometry of M, whereby the corresponding deformed metrics ensuing from the fit to the experimental data seem to provide an *effective dynamical description of the relevant interactions* (*at the energy scale and in the energy range considered*).

An analogous energy-dependent metric seems to hold for the gravitational field (at least locally, i.e. in a neighborhood of Earth) when analyzing some classical experimental data concerning the slowing down of clocks.

Let us shortly review the main ideas and results concerning the (four-dimensional) deformed Minkowski spacetime $\widetilde{M}$.

The four-dimensional "deformed" metric scheme is based on the assumption that space-time, in a preferred frame which is *fixed* by the scale of energy E, is endowed with a metric of the form

$$\begin{gathered} ds^2 = b_0^2(E)c^2dt^2 - b_1^2(E)dx^2 - b_2^2(E)dy^2 - b_3^2(E)dz^2 = g_{DSR\mu\nu}(E)dx^\mu dx^\nu; \\ g_{DSR\mu\nu}(E) = \left(b_0^2(E), -b_1^2(E), -b_2^2(E), -b_3^2(E)\right), \end{gathered} \tag{1}$$

with $x^\mu = (x^0, x^1, x^2, x^3) = (ct, x, y, z)$, c being the usual speed of light in vacuum. We named "Deformed Special Relativity" (DSR) the relativity theory built up on metric (1).

Metric (1) is supposed to hold locally, i.e. in the spacetime region where the process occurs. It is supposed moreover to play a *dynamical* role, and to provide a geometric description of the interaction considered. In this sense, DSR realizes the so called *"Finzi Principle of Solidarity"* between space-time and phenomena occurring in it [1](see [3]). Futhermore, we stress that, from the physical point of view, *E is the measured energy of the system*, and thus a merely phenomenological (non-metric) variable[2].

We notice explicitly that the spacetime $\widetilde{M}$ described by (1) is flat (it has zero four-dimensional curvature), so that the geometrical description of the fundamental interactions based on it differs from the general relativistic one (whence the name "deformation" used to characterize such a situation). Although for each interaction the corresponding metric reduces to the Minkowskian one for a suitable value of the energy E_0 (which is characteristic of the interaction considered), the energy of the process is fixed and cannot be changed at will. Thus, in spite of the fact that *formally* it would be possible to recover the usual Minkowski space M by a suitable change of coordinates (e.g. by a rescaling), this would amount, in such a framework, to be a mere mathematical operation devoid of any physical meaning.

As far as phenomenology is concerned, it is important to recall that a local breakdown of Lorentz invariance may be envisaged for all the four fundamental interactions (electromagnetic, weak, strong and gravitational) whereby *one gets evidence for a departure of the spacetime metric from the Minkowskian one* (in the energy range examined). The explicit functional form of the metric (1) for all the four interactions can be found in [1, 2]. Here, we confine ourselves to recall the following basic features of these energy-dependent phenomenological metrics:

[1]Let us recall that in 1955 the Italian mathematician Bruno Finzi stated his *"Principle of Solidarity"*(PS), that sounds "*It's (indeed) necessary to consider space-time TO BE SOLIDLY CONNECTED with the physical phenomena occurring in it, so that its features and its very nature do change with the features and the nature of those. In this way not only (as in classical and special-relativistic physics) space-time properties affect phenomena, but reciprocally phenomena do affect space-time properties. One thus recognizes in such an appealing "Principle of Solidarity" between phenomena and space-time that characteristic of mutual dependence between entities, which is peculiar to modern science.*" Moreover, referring to a generic N-dimensional space: " *It can,* a priori, *be pseudoeuclidean, Riemannian, non-Riemannian. But* — he wonders — *how is indeed the space-time where physical phenomena take place? Pseudoeuclidean, Riemannian, non-Riemannian, according to their nature, as requested by the principle of solidarity between space-time and phenomena occurring in it.*"

Of course, Finzi's main purpose was to apply such a principle to Einstein's Theory of General Relativity, namely to the class of gravitational phenomena. However, its formulation is as general as possible, so to apply in principle to all the known physical interactions. Therefore, Finzi's PS is at the very ground of any attempt at geometrizing physics, i.e. describing physical forces in terms of the geometrical structure of space-time.

[2]As is well known, all the present physically realizable detectors work *via* their electromagnetic interaction in the usual space-time M. So, E is the energy of the system measured in *fully Minkowskian conditions.*

1) Both the electromagnetic and the weak metric show the same functional behavior, namely

$$g_{DSR\mu\nu}(E) = diag\left(1, -b^2(E), -b^2(E), -b^2(E)\right); \tag{2}$$

$$b^2(E) = \begin{cases} (E/E_0)^{1/3}, & 0 \le E \le E_0 \\ 1, & E_0 \le E \end{cases} \tag{3}$$

with the only difference between them being the threshold energy E_0 , i.e. the energy value at which the metric parameters are constant, i.e. the metric becomes Minkowskian; the fits to the experimental data yield

$$E_{0,e.m.} = 5.0 \pm 0.2\mu eV \; ; \; E_{0w} = 80.4 \pm 0.2GeV; \tag{4}$$

2) for strong and gravitational interactions, the metrics read:

$$g_{DSR}(E) = diag\left(b_0^2(E), -b_1^2(E), -b_2^2(E), -b_3^2(E)\right); \tag{5}$$

$$\begin{aligned} b_{0,strong}^2(E) &= b_{3,strong}^2(E) = \begin{cases} 1, & 0 \le E < E_{0strong} \\ (E/E_{0strong})^2, & E_{0strong} < E \end{cases} ; \\ b_{1,strong}^2(E) &= \left(\sqrt{2}/5\right)^2; b_{2,strong}^2 = (2/5)^2; \end{aligned} \tag{6}$$

$$b_{0,grav}^2(E) = \begin{cases} 1, & 0 \le E < E_{0grav} \\ \frac{1}{4}(1 + E/E_{0grav})^2, & E_{0grav} < E \end{cases} \tag{6'}$$

with

$$E_{0s} = 367.5 \pm 0.4GeV; E_{0grav} = 20.2 \pm 0.1\mu eV. \tag{7}$$

Let us stress that, in this case, contrarily to the electromagnetic and the weak ones, *a deformation of the time coordinate occurs;* moreover, *the three-space is anisotropic*[3], with two spatial parameters constant (but different in value) and the third one variable with energy in an *"over-Minkowskian"* way (namely it reaches the limit of Minkowskian metric for decreasing values of E , with $E > E_0$) [1, 2].

As a final remark, we stress that actually *the four-dimensional energy-dependent space-time $\widetilde{M}$ is just a manifestation of a larger, five-dimensional space in which energy plays the role of a fifth dimension.* Indeed, it can be shown that the physics of the interaction lies in the curvature of such a five-dimensional spacetime, in which the four-dimensional, deformed Minkowski space is embedded. Moreover, *all* the phenomenological metrics (2), (3) and (5), (6) can be obtained as solutions of the vacuum Einstein equations in this generalized Kaluza-Klein scheme [1, 2].

2.2. Field Deformation

We want now to show that the deformation of space-time, expressed by the metric g_{DSR} (Eq.(1)), does affect also the external fields applied to the physical system considered.

[3]At least for strong interaction; nothing can be said for the gravitational one.

Let us consider for instance the case of a physical process ruled by the electromagnetic interaction. Therefore, the Minkowski space M is endowed with the electromagnetic tensor $F_{\mu\nu}(x)$ (external e.m. field) acting on the system. Of course $F^{\mu}_{\nu}(x) = g^{\mu\rho}_{SR}F_{\rho\nu}(x)$.

In the deformed Minkowski space $\widetilde{M}$, the covariant components of the electromagnetic tensor read

$$\widetilde{F}_{\mu\nu} = g_{DSR\mu\rho}F^{\rho}_{\nu} = g_{DSR\mu\rho}g^{\mu\sigma}_{SR}F_{\sigma\nu}, \tag{8}$$

where

$$(g_{DSR\mu\rho}g^{\mu\sigma}_{SR}) = diag(b_0^2, b_1^2, b_2^2, b_3^2) = (b_\sigma^2 \delta^\sigma_\rho). \tag{9}$$

We have therefore

$$\widetilde{F}_{0\nu} = b_0^2 F_{0\nu}; \widetilde{F}_{1\nu} = b_1^2 F_{1\nu}; \widetilde{F}_{2\nu} = b_2^2 F_{2\nu}; \widetilde{F}_{3\nu} = b_3^2 F_{3\nu}, \tag{10}$$

or

$$\widetilde{F}_{\mu\nu} = b_\mu^2 F_{\mu\nu}, \qquad \mu, \nu = 0, 1, 2, 3 \tag{11}$$

(no sum on repeated indices!).

It follows that the tensor $\widetilde{F}_{\mu\nu}$ is not antisymmetric:

$$\widetilde{F}_{\mu\nu} \neq -\widetilde{F}_{\nu\mu}. \tag{12}$$

The result shown here for the electromagnetic interaction can be generalized to other fundamental interactions described by tensor fields.

On account of the well-known identification

$$\widetilde{F}_{0i} = \widetilde{E_i}, \widetilde{F}_{12} = -\widetilde{B_3}, \widetilde{F}_{23} = -\widetilde{B_1}, \widetilde{F}_{31} = -\widetilde{B_2} \tag{13}$$

(and analogously for $F_{\mu\nu}$), we can write, for the energy density $\widetilde{\mathcal{E}}$ of the deformed electromagnetic field:

$$\widetilde{\mathcal{E}} = \frac{\widetilde{\mathbf{E}}^2 + \widetilde{\mathbf{B}}^2}{8\pi} = \frac{b_0^4\mathbf{E}^2 + b_1^4 B_3^2 + b_2^4 B_1^2 + b_3^4 B_2^2}{8\pi}, \tag{14}$$

to be compared with the standard expression for the e.m. field $\mathbf{E}$, $\mathbf{B}$:

$$\mathcal{E} = \frac{\mathbf{E}^2 + \mathbf{B}^2}{8\pi}. \tag{15}$$

There is therefore a difference in the energy associated to the electromagnetic field in the deformed space-time region. We have, for the energy density

$$\Delta\mathcal{E} = \mathcal{E} - \widetilde{\mathcal{E}}. \tag{16}$$

We can state that the difference $\Delta\mathcal{E}$ represents *the energy spent by the interaction in order to deform the space-time geometry.*

We can therefore conclude that *the deformation of space-time does affect the field itself that deforms the geometry of the space.* There is therefore a feedback between space and interaction which fully implements the Solidarity Principle.

3. DSR as Metric Gauge Theory

3.1. External Metric Gauge Fields

It is clear from the discussion of the phenomenological metrics describing the four fundamental interactions in DSR that the Minkowski space M is the space-time manifold of background of any experimental measurement and detection (namely, of any process of acquisition of information on physical reality). In particular, we can consider this Minkowski space as that associated to the electromagnetic interaction above the threshold energy $E_{0,e.m.}$. Therefore, in modeling the physical phenomena, one has to take into account this fact. The geometrical nature of interactions, *i.e.* assuming the validity of the Finzi principle, means that one has to suitably *gauge* (with reference to M) the space-time metrics with respect to the interaction — and/or the phenomenon — under study. In other words, one needs to "adjust" suitably the local metric of space-time according to the interaction acting in the region considered. We can name such a procedure *"Metric Gaugement Process"* (M.G.P.). Like in usual gauge theories a different phase is chosen in different space-time *points*, in DSR different metrics are associated to different space-time *manifolds* according to the interaction acting therein. We have thus a gauge structure on the space of manifolds

$$\widetilde{\mathcal{M}} \equiv \cup_{g_{DSR} \in \mathcal{P}(E)} \widetilde{M}\left(g_{DSR}\right), \tag{17}$$

where $\mathcal{P}(E)$ is the set of the energy-dependent pseudoeuclidean metrics of the type (1). This is why it is possible to regard Deformed Special Relativity as a *Metric Gauge Theory*. In this case, we can consider the related fields as *external metric gauge fields*.

However, let us notice that DSR can be considered as a metric gauge theory from another point of view, on account of the dependence of the metric coefficients on the energy. Actually, once the MGP has been applied, by selecting the suitable gauge (namely, the suitable *functional form* of the metric) according to the interaction considered (thus implementing the Finzi principle), the metric dependence on the energy implies another different gauge process. Namely, the metric is gauged according to the process under study, thus selecting the *given* metric, with the *given values* of the coefficients, suitable for the given phenomenon.

We have therefore a *double* metric gaugement, according, on one side, to the interaction ruling the physical phenomenon examined, and on the other side to its energy, in which the metric coefficients are the analogous of the gauge functions[4] .

3.2. Internal Metric Gauge Fields

We want now to show that the deformed Minkowski space $\widetilde{M}$ of Deformed Special Relativity does possess another well-defined geometrical structure, besides the deformed

[4] The analogy of this second kind of metric gauge with the standard, non-abelian gauge theories is more evident in the framework of the five-dimensional space-time $\Re_5$ (with energy as extra dimension) embedding $\widetilde{M}$, on which Deformed Relativity in Five Dimensions (DR5) is based (see [1, 2]). In $\Re_5$, in fact, energy is no longer a parametric variable, like in DSR, but plays the role of fifth (metric) coordinate. The invariance under such a metric gauge, not manifest in four dimensions, is instead recovered in the form of the isometries of the five-dimensional space-time-energy manifold $\Re_5$.

metrical one. Precisely, we will show that $\widetilde{M}$ is a *generalized Lagrange space* [6]. As we shall see, this implies that DSR admits a different, *intrinsic* gauge structure.

3.2.1. Deformed Minkowski Space as Generalized Lagrange Space

Generalized Lagrange Spaces Let us give the definition of generalized Lagrange space [4], since usually one is not acquainted with it.

Consider a N-dimensional, differentiable manifold $\mathcal{M}$ and its (N-dimensional) tangent space in a point, $T\mathcal{M}_{\mathbf{x}}$ ($\mathbf{x} \in \mathcal{M}$). As is well known, the union

$$\bigcup_{\mathbf{x}\in\mathcal{M}} T\mathcal{M}_{\mathbf{x}} \equiv T\mathcal{M} \tag{18}$$

has a fibre bundle structure. Let us denote by $\mathbf{y}$ the generic element of $T\mathcal{M}_{\mathbf{x}}$, namely a vector tangent to $\mathcal{M}$ in $\mathbf{x}$. Then, an element $u \in T\mathcal{M}$ is a vector tangent to the manifold in some point $\mathbf{x} \in \mathcal{M}$. Local coordinates for $T\mathcal{M}$ are introduced by considering a local coordinate system $(x^1, x^2, ..., x^N)$ on $\mathcal{M}$ and the components of y in such a coordinate system $(y^1, y^2, ..., y^N)$. The $2N$ numbers $(x^1, x^2, ..., x^N, y^1, y^2, ..., y^N)$ constitute a local coordinate system on $T\mathcal{M}$. We can write synthetically $u = (\mathbf{x}, \mathbf{y})$. $T\mathcal{M}$ is a $2N$-dimensional, differentiable manifold.

Let π be the mapping (*natural projection*) $\pi : u = (\mathbf{x}, \mathbf{y}) \longrightarrow \mathbf{x}$. ($\mathbf{x} \in \mathcal{M}$, $\mathbf{y} \in T\mathcal{M}_{\mathbf{x}}$). Then, the tern $(T\mathcal{M}, \pi, \mathcal{M})$ is the *tangent bundle* to the base manifold $\mathcal{M}$. The image of the inverse mapping $\pi^{-1}(\mathbf{x})$ is of course the tangent space $T\mathcal{M}_{\mathbf{x}}$, which is called the *fiber corresponding to the point* $\mathbf{x}$ *in the fiber bundle* One considers also sometimes the manifold $\widetilde{T\mathcal{M}} = T\mathcal{M}/\{0\}$, where 0 is the zero section of the projection π. We do not dwell further on the theory of the fiber bundles, and refer the reader to the wide and excellent literature on the subject [5].

The natural basis of the tangent space $T_u(T\mathcal{M})$ at a point $u = (\mathbf{x}, \mathbf{y}) \in T\mathcal{M}$ is $\left\{\frac{\partial}{\partial x^i}, \frac{\partial}{\partial y^j}\right\}$, $i, j = 1, 2, ..., N$.

A local coordinate transformation in the differentiable manifold $T\mathcal{M}$ reads

$$\begin{cases} x'^i = x'^i(\mathbf{x}), \quad \det\left(\dfrac{\partial x'^i}{\partial x^j}\right) \neq 0, \\ y'^i = \dfrac{\partial x'^i}{\partial x^j} y^j. \end{cases} \tag{19}$$

Here, y^i is *the Liouville vector field* on $T\mathcal{M}$, i.e. $y^i \frac{\partial}{\partial y^i}$.

On account of Eq.(19), the natural basis of $T\mathcal{M}_{\mathbf{x}}$ can be written as

$$\begin{cases} \dfrac{\partial}{\partial x^i} = \dfrac{\partial x'^k}{\partial x^i}\dfrac{\partial}{\partial x'^k} + \dfrac{\partial y'^k}{\partial x^i}\dfrac{\partial}{\partial y'^k}, \\ \dfrac{\partial}{\partial y^j} = \dfrac{\partial y'^k}{\partial y^j}\dfrac{\partial}{\partial y'^k}. \end{cases} \tag{20}$$

Second Eq.(20) shows therefore that the vector basis $\left(\frac{\partial}{\partial y^j}\right)$, $j = 1, 2, ..., N$, generates a distribution $\mathcal{V}$ defined everywhere on $T\mathcal{M}$ and integrable, too (*vertical distribution on* $T\mathcal{M}$).

If $\mathcal{H}$ is a distribution on $T\mathcal{M}$ supplementary to $\mathcal{V}$, namely

$$T_u(T\mathcal{M}) = \mathcal{H}_u \oplus \mathcal{V}_u \quad , \ \forall u \in T\mathcal{M}, \tag{21}$$

then $\mathcal{H}$ is called a *horizontal distribution*, or a *nonlinear connection* on $T\mathcal{M}$. A basis for the distributions $\mathcal{H}$ and $\mathcal{V}$ are given respectively by $\left\{\frac{\delta}{\delta x^i}\right\}$ and $\left\{\frac{\partial}{\partial y^j}\right\}$, where the basis in $\mathcal{H}$ explicitly reads

$$\frac{\delta}{\delta x^i} = \frac{\partial}{\partial x^i} - H^j_i(\mathbf{x}, \mathbf{y})\frac{\partial}{\partial y^j}. \tag{22}$$

Here, $H^j_i(\mathbf{x}, \mathbf{y})$ are the *coefficients* of the nonlinear connection $\mathcal{H}$. The basis $\left\{\frac{\delta}{\delta x^i}, \frac{\partial}{\partial y^j}\right\} = \left\{\delta_i, \dot{\partial}_j\right\}$ is called the *adapted basis*.

The dual basis to the adapted basis is $\left\{dx^i, \delta y^j\right\}$, with

$$\delta y^j = dy^j + H^j_i(\mathbf{x}, \mathbf{y})dx^i. \tag{23}$$

A *distinguished tensor* (or *d-tensor*) *field of (r,s)-type* is a quantity whose components transform like a tensor under the first coordinate transformation (19) on $T\mathcal{M}$ (namely they change as tensor in $\mathcal{M}$). For instance, for a d-tensor of type (1,2):

$$R'^{i}{}_{jk} = \frac{\partial x'^i}{\partial x^s}\frac{\partial x^r}{\partial x'^j}\frac{\partial x^p}{\partial x'^k}R^s{}_{rp}. \tag{24}$$

In particular, both $\left\{\frac{\delta}{\delta x^i}\right\}$ and $\left\{\frac{\partial}{\partial y^j}\right\}$ are d-(covariant) vectors, whereas $\left\{dx^i\right\}$, $\left\{\delta y^j\right\}$ are d-(contravariant) vectors.

A *generalized Lagrange space* is a pair $\mathcal{GL}^N$=($\mathcal{M}$, $g_{ij}(\mathbf{x}, \mathbf{y})$), with $g_{ij}(\mathbf{x}, \mathbf{y})$ being a d-tensor of type (0,2) (covariant) on the manifold $T\mathcal{M}$, which is symmetric, non-degenerate[5] and of constant signature.

A function

$$L : (\mathbf{x}, \mathbf{y}) \in T\mathcal{M} \to L(\mathbf{x}, \mathbf{y}) \in \mathcal{R} \tag{25}$$

differentiable on $\widehat{T\mathcal{M}}$ and continuous on the null section of π is named a *regular Lagrangian* if the Hessian of L with respect to the variables y^i is non-singular.

A generalized Lagrange space $\mathcal{GL}^N$=($\mathcal{M}$, $g_{ij}(\mathbf{x}, \mathbf{y})$) is reducible to a *Lagrange space* $\mathcal{L}^N$ if there is a regular Lagrangian L satisfying

$$g_{ij} = \frac{1}{2}\frac{\partial^2 L}{\partial y^i \partial y^j} \tag{26}$$

[5]Namely it must be $rank \, \|g_{ij}(\mathbf{x}, \mathbf{y})\| = N$.

on $\widehat{T\mathcal{M}}$. In order that $\mathcal{GL}^N$ is reducible to a Lagrange space, a necessary condition is the total symmetry of the d-tensor $\frac{\partial g_{ij}}{\partial y^k}$. If such a condition is satisfied, and g_{ij} are 0-homogeneous in the variables y^i, then the function $L = g_{ij}(\mathbf{x}, \mathbf{y})y^i y^j$ is a solution of the system (26). In this case, the pair $(\mathcal{M}, L)$ is a *Finsler space*[6] $(\mathcal{M}, \Phi)$, with $\Phi^2 = L$. One says that $\mathcal{GL}^N$ is reducible to a Finsler space.

Of course, $\mathcal{GL}^N$ reduces to a pseudo-Riemannian (or Riemannian) space $(\mathcal{M}, g_{ij}(\mathbf{x}))$ if the d-tensor $g_{ij}(\mathbf{x}, \mathbf{y})$ does not depend on $\mathbf{y}$. On the contrary, if $g_{ij}(\mathbf{x}, \mathbf{y})$ depends only on $\mathbf{y}$ (at least in preferred charts), it is a generalized Lagrange space which is locally Minkowskian.

Since, in general, a generalized Lagrange space is not reducible to a Lagrange one, it cannot be studied by means of the methods of symplectic geometry, on which — as is well known — analytical mechanics is based.

A linear $\mathcal{H}-$connection on $T\mathcal{M}$ (or on $\widehat{T\mathcal{M}}$) is defined by a couple of geometrical objects $\mathcal{C}\Gamma(\mathcal{H}) = (L^i_{\ jk}, C^i_{\ jk})$ on $T\mathcal{M}$ with different transformation properties under the coordinate transformation (19). Precisely, $L^i_{\ jk}(\mathbf{x}, \mathbf{y})$ transform like the coefficients of a linear connection on $\mathcal{M}$, whereas $C^i_{\ jk}(\mathbf{x}, \mathbf{y})$ transform like a d-tensor of type (1,2). $\mathcal{C}\Gamma(\mathcal{H})$ is called *the metrical canonical* $\mathcal{H}-$*connection* of the generalized Lagrange space $\mathcal{GL}^N$.

In terms of $L^i_{\ jk}$ and $C^i_{\ jk}$ one can define two kinds of covariant derivatives: a *covariant horizontal (h-) derivative*, denoted by "$_{\shortmid}$", and a *covariant vertical (v-) derivative*, denoted by "|". For instance, for the d-tensor $g_{ij}(\mathbf{x}, \mathbf{y})$ one has

$$\begin{cases} g_{ij\shortmid k} = \dfrac{\delta g_{ij}}{\delta x^k} - g_{sj}L^s_{\ ik} - g_{is}L^s_{\ jk}; \\ \\ g_{ij|k} = \dfrac{\partial g_{ij}}{\partial x^k} - g_{sj}C^s_{\ ik} - g_{is}C^s_{\ jk}. \end{cases} \tag{27}$$

The two derivatives $g_{ij\shortmid k}$ and $g_{ij|k}$ are both d-tensors of type (0,3).

The coefficients of $\mathcal{C}\Gamma(\mathcal{H})$ can be expressed in terms of the following *generalized Christoffel symbols*:

$$\begin{cases} L^i_{\ jk} = \frac{1}{2}g^{is}\left(\dfrac{\delta g_{sj}}{\delta x^k} + \dfrac{\delta g_{ks}}{\delta x^j} + \dfrac{\delta g_{jk}}{\delta x^s}\right); \\ \\ C^i_{\ jk} = \frac{1}{2}g^{is}\left(\dfrac{\partial g_{sj}}{\partial x^k} + \dfrac{\partial g_{ks}}{\partial x^j} + \dfrac{\partial g_{jk}}{\partial x^s}\right). \end{cases} \tag{28}$$

[6]Let us recall that a Finsler space is a couple $(\mathcal{M}, \Phi)$, where $\mathcal{M}$ is be an N-dimensional differential manifold and $\Phi : T\mathcal{M} \Rightarrow \mathcal{R}$ a function $\Phi(\mathbf{x}, \xi)$ defined for $\mathbf{x} \in \mathcal{M}$ and $\xi \in T_{\mathbf{x}}\mathcal{M}$ such that $\Phi(\mathbf{x}, \cdot)$ is a possibly non symmetric norm on $T_{\mathbf{x}}\mathcal{M}$.

Notice that every Riemann manifold $(\mathcal{M}, \mathbf{g})$ is also a Finsler space, the norm $\Phi(\mathbf{x}, \xi)$ being the norm induced by the scalar product $\mathbf{g}(\mathbf{x})$.

A finite-dimensional Banach space is another simple example of Finsler space, where $\Phi(\mathbf{x}, \xi) \equiv \|\xi\|$.

Curvature and Torsion in a Generalized Lagrange Space

By means of the connection $C\Gamma(\mathcal{H})$ it is possible to define a *d-curvature* in $T\mathcal{M}$ by means of the tensors $R_{j\,kh}^{\,i}$, $S_{j\,kh}^{\,i}$ and $P_{j\,kh}^{\,i}$ given by

$$\begin{aligned}
R_{j\,kh}^{\,i} &= \frac{\delta L^i_{\ jk}}{\delta x^h} - \frac{\delta L^i_{\ jh}}{\delta x^k} + L^r_{\ jk}L^i_{\ rh} - L^r_{\ jh}L^i_{\ rk} + C^i_{\ jr}R^r_{\ kh};\\
S_{j\,kh}^{\,i} &= \frac{\partial C^i_{\ jk}}{\partial y^h} - \frac{\partial C^i_{\ jh}}{\partial y^k} + C^r_{\ jk}C^i_{\ rh} - C^r_{\ jh}C^i_{\ rk};\\
P_{j\,kh}^{\,i} &= \frac{\partial L^i_{\ jk}}{\partial y^h} - C^i_{\ j|h} + C^i_{\ jr}P^r_{\ kh}.
\end{aligned} \tag{29}$$

Here, the d-tensor $R^i_{\ jk}$ is related to the bracket of the basis $\left\{\frac{\delta}{\delta x^i}\right\}$:

$$\left[\frac{\delta}{\delta x^i}, \frac{\delta}{\delta xj}\right] = R^s_{\ ij}\frac{\partial}{\partial y^s} \tag{30}$$

and is explicitly given by[7]

$$R^i_{\ jk} = \frac{\delta H^i_{\ j}}{\delta x^k} - \frac{\delta H^i_{\ k}}{\delta x^j}. \tag{31}$$

The tensor $P^i_{\ jk}$, together with $T^i_{\ jk}$, $S^i_{\ jk}$, defined by

$$\begin{aligned}
P^i_{\ jk} &= \frac{\partial H^i_{\ j}}{\partial y^k} - L^i_{\ jk};\\
T^i_{\ jk} &= L^i_{\ jk} - L^i_{\ kj};\\
S^i_{\ jk} &= C^i_{\ jk} - C^i_{\ kj}
\end{aligned} \tag{32}$$

are *the d-tensors of torsion of the metrical connection* $C\Gamma(\mathcal{H})$.

¿From the curvature tensors one can get the corresponding Ricci tensors of $C\Gamma(\mathcal{H})$:

$$\begin{cases}
R_{ij} = R_{i\,js}^{\,s}; \quad S_{ij} = S_{i\,js}^{\,s};\\[2ex]
\overset{1}{P}_{ij} = P_{i\,js}^{\,s} \quad \overset{2}{P}_{ij} = P_{i\,sj}^{\,s},
\end{cases} \tag{33}$$

and the scalar curvatures

$$R = g^{ij}R_{ij}; \quad S = g^{ij}S_{ij}. \tag{34}$$

Finally, *the deflection d-tensors associated to the connection* $C\Gamma(\mathcal{H})$ are

$$\begin{cases}
D^i_{\ j} = y^i_{\ |j} = -H^i_{\ j} + y^sL^i_{\ sj};\\[1ex]
d^i_{\ j} = y^i_{\ |j} = \delta^i_{\ j} + y^sC^i_{\ sj},
\end{cases} \tag{35}$$

[7] R^i_{jk} plays the role of a curvature tensor of the nonlinear connection $\mathcal{H}$. The corresponding tensor of torsion is instead

$$t^i_{jk} = \frac{\partial H^i_j}{\partial y^k} - \frac{\partial H^i_k}{\partial y^j}.$$

namely the h- and v-covariant derivatives of the Liouville vector fields.

In the generalized Lagrange space $\mathcal{GL}^N$ it is possible to write the Einstein equations with respect to the canonical connection $C\Gamma(\mathcal{H})$ as follows:

$$\begin{cases} R_{ij} - \frac{1}{2}Rg_{ij} = \kappa \overset{H}{T}_{ij}; & \overset{1}{P}_{ij} = \kappa \overset{1}{T}_{ij}; \\ \\ S_{ij} - \frac{1}{2}Sg_{ij} = \kappa \overset{V}{T}_{ij}; & \overset{2}{P}_{ij} = \kappa \overset{2}{T}_{ij}, \end{cases} \tag{36}$$

where κ is a constant and $\overset{H}{T}_{ij}$, $\overset{V}{T}_{ij}$, $\overset{1}{T}_{ij}$, $\overset{2}{T}_{ij}$ are the components of the energy-momentum tensor.

3.2.2. Generalized Lagrangian Structure of $\widetilde{M}$

On the basis of the previous considerations, let us analyze the geometrical structure of the deformed Minkowski space of DSR $\widetilde{M}$, endowed with the by now familiar metric $g_{\mu\nu,DSR}(E)$. As said in Sect.2, E is the energy of the process measured by the detectors in Minkowskian conditions. Therefore, E is a function of the velocity components, $u^\mu = dx^\mu/d\tau$, where τ is the (Minkowskian) proper time[8]:

$$E = E\left(\frac{dx^\mu}{d\tau}\right). \tag{37}$$

The derivatives $dx^\mu/d\tau$ define a contravariant vector tangent to M at x, namely they belong to $TM_{\mathbf{x}}$. We shall denote this vector (according to the notation of the previous Subsubsection) by $\mathbf{y} = (y^\mu)$. Then, (**x**,**y**) is a point of the tangent bundle to M. We can therefore consider the generalized Lagrange space $\mathcal{GL}^4 = (M, g_{\mu\nu}(\mathbf{x},\mathbf{y}))$, with

$$\begin{cases} g_{\mu\nu}(\mathbf{x},\mathbf{y}) = g_{\mu\nu DSR}(E(\mathbf{x},\mathbf{y})), \\ \\ E(\mathbf{x},\mathbf{y}) = E(\mathbf{y}). \end{cases} \tag{38}$$

Then, it is possible to prove the following theorem [6]:

The pair $\mathcal{GL}^4 = (M, g_{DSR,\mu\nu}(\mathbf{x},\mathbf{y})) \equiv \widetilde{M}$ is a generalized Lagrange space which is not reducible to a Riemann space, or to a Finsler space, or to a Lagrange space.

Notice that such a result is strictly related to the fact that the deformed metric tensor of DSR is diagonal.

If an external electromagnetic field $F_{\mu\nu}$ is present in the Minkowski space M, in $\widetilde{M}$ the deformed electromagnetic field is given by $\widetilde{F}^\mu_\nu(\mathbf{x},\mathbf{y}) = g^{\mu\rho}_{DSR}F_{\rho\nu}(\mathbf{x})$ (see Eq.(8)). Such a field is a d-tensor and is called *the electromagnetic tensor of the generalized Lagrange space*. Then, the nonlinear connection $\mathcal{H}$ is given by

$$H^\mu_\nu = \left\{ \begin{matrix} \mu \\ \nu\rho \end{matrix} \right\} y^\rho - \widetilde{F}^\mu_\nu(\mathbf{x},\mathbf{y}), \tag{39}$$

[8]Contrarily to ref.[6], we shall not consider the restrictive case of a classical (non-relativistic) expression of the energy, but assume a general dependence of E on the velocity (eq.(38)).

where $\left\{ \begin{matrix} \mu \\ \nu\rho \end{matrix} \right\}$, the Christoffel symbols of the Minkowski metric $g_{\mu\nu}$, are zero, so that

$$H^{\mu}_{\ \nu} = -\widetilde{F}^{\mu}_{\ \nu}(\mathbf{x}, \mathbf{y}), \tag{40}$$

namely, *the connection coincides with the deformed field.*

The adapted basis of the distribution $\mathcal{H}$ reads therefore

$$\frac{\delta}{\delta x^{\mu}} = \frac{\partial}{\partial x^{\mu}} + \widetilde{F}^{\nu}_{\ \mu}(\mathbf{x}, \mathbf{y})\frac{\partial}{\partial y^{\nu}}. \tag{41}$$

The local covector field of the dual basis (cfr. Eq.(23)) is given by

$$\delta y^{\mu} = dy^{\mu} - \widetilde{F}^{\mu}_{\ \nu}(\mathbf{x}, \mathbf{y})dx^{\nu}. \tag{42}$$

3.2.3. Canonical Metric Connection of $\widetilde{M}$

The derivation operators applied to the deformed metric tensor of the space $\mathcal{GL}^4 = \widetilde{M}$ yield

$$\frac{\delta g_{DSR\mu\nu}}{\delta x^{\rho}} = \frac{\partial g_{DSR\mu\nu}}{\partial x^{\rho}} + \widetilde{F}^{\sigma}_{\ \rho}\frac{\partial g_{DSR\mu\nu}}{\partial y^{\sigma}} = \widetilde{F}^{\sigma}_{\ \rho}\frac{\partial g_{DSR\mu\nu}}{\partial E}\frac{\partial E}{\partial y^{\sigma}}, \tag{43}$$

$$\frac{\partial g_{DSR\mu\nu}}{\partial y^{\sigma}} = \frac{\partial g_{DSR\mu\nu}}{\partial E}\frac{\partial E}{\partial y^{\sigma}}. \tag{44}$$

Then, the coefficients of the canonical metric connection $\mathcal{C}\Gamma(\mathcal{H})$ in $\widetilde{M}$ (see Eq.(28)) are given by

$$\begin{cases} L^{\mu}_{\ \nu\rho} = \frac{1}{2}g^{\mu\sigma}_{DSR}\frac{\partial E}{\partial y^{\alpha}}\left(\frac{\partial g_{DSR\sigma\nu}}{\partial E}\widetilde{F}^{\alpha}_{\ \rho} + \frac{\partial g_{DSR\sigma\rho}}{\partial E}\widetilde{F}^{\alpha}_{\ \nu} - \frac{\partial g_{DSR\nu\rho}}{\partial E}\widetilde{F}^{\alpha}_{\ \sigma}\right), \\ C^{\mu}_{\ \nu\rho} = \frac{1}{2}g^{\mu\sigma}_{DSR}\frac{\partial E}{\partial y^{\alpha}}\left(\frac{\partial g_{DSR\sigma\nu}}{\partial E}\delta^{\alpha}_{\ \rho} + \frac{\partial g_{DSR\sigma\rho}}{\partial E}\delta^{\alpha}_{\ \nu} - \frac{\partial g_{DSR\nu\rho}}{\partial E}\delta^{\alpha}_{\ \sigma}\right). \end{cases} \tag{45}$$

The vanishing of the electromagnetic field tensor, $F^{\alpha}_{\ \rho} = 0$, implies $L^{\mu}_{\ \nu\rho} = 0$.

One can define the deflection tensors associated to the metric connection $\mathcal{C}\Gamma(\mathcal{H})$ as follows (cfr. Eq.(36)):

$$\begin{aligned} D^{\mu}_{\ \nu} &= y^{\mu}_{\ \shortmid\nu} = \frac{\delta y^{\mu}}{\delta x^{\nu}} + y^{\alpha}L^{\mu}_{\ \alpha\nu} = \widetilde{F}^{\mu}_{\ \nu} + y^{\alpha}L^{\mu}_{\ \alpha\nu}; \\ d^{\mu}_{\ \nu} &= y^{\mu}_{\ |\nu} = \delta^{\mu}_{\ \nu} + y^{\alpha}C^{\mu}_{\ \alpha\nu}. \end{aligned} \tag{46}$$

The covariant components of these tensors read

$$\begin{aligned} D_{\mu\nu} &= g_{\mu\sigma,DSR}D^{\sigma}_{\ \nu} = g_{\mu\sigma,DSR}\left(\widetilde{F}^{\sigma}_{\ \nu} + y^{\alpha}L^{\sigma}_{\ \alpha\nu}\right) = \\ &= F_{\mu\nu}(\mathbf{x}) + \frac{1}{2}y^{\sigma}\frac{\partial E}{\partial y^{\alpha}}\left(\frac{\partial g_{DSR\mu\sigma}}{\partial E}\widetilde{F}^{\alpha}_{\ \nu} + \frac{\partial g_{DSR\mu\nu}}{\partial E}\widetilde{F}^{\alpha}_{\ \sigma} - \frac{\partial g_{DSR\sigma\nu}}{\partial E}\widetilde{F}^{\alpha}_{\ \mu}\right); \end{aligned}$$

$$\begin{aligned} d_{\mu\nu} &= g_{\mu\sigma,DSR}d^{\sigma}_{\ \nu} = \\ &= g_{DSR,\mu\nu} + \frac{1}{2}y^{\sigma}\frac{\partial E}{\partial y^{\alpha}}\left(\frac{\partial g_{DSR\mu\sigma}}{\partial E}\delta^{\alpha}_{\ \nu} + \frac{\partial g_{DSR\mu\nu}}{\partial E}\delta^{\alpha}_{\ \sigma} - \frac{\partial g_{DSR\sigma\nu}}{\partial E}\delta^{\alpha}_{\ \mu}\right). \end{aligned} \tag{47}$$

It is important to stress explicitly that, on the basis of the results of 3.2.1, *the deformed Minkowski space* $\widetilde{M}$ *does possess curvature and torsion,* namely it is endowed with a very rich geometrical structure. This permits to understand the variety of new physical phenomena that occur in it (as compared to the standard Minkowski space) [1, 2].

Following ref.[6], let us show how the formalism of the generalized Lagrange space allows one to recover some results on the phenomenological energy-dependent metrics discussed in Sect.2.

Consider the following metric ($c = 1$):

$$ds^2 = a(E)dt^2 + (dx^2 + dy^2 + dz^2) \tag{48}$$

where $a(E)$ is an arbitrary function of the energy and spatial isotropy ($b^2 = 1$) has been assumed. In absence of an external electromagnetic field ($F_{\mu\nu} = 0$), the non-vanishing components $C'^{\mu}_{\nu\rho}$ of the canonical metric connection $C\Gamma(\mathcal{H})$ (see Eq.(46)) are

$$\begin{cases} C^0_{00} = \dfrac{a'}{a}y^0, \quad C^0_{01} = -\dfrac{a'}{a}y^1, \quad C^0_{02} = -\dfrac{a'}{a}y^2, \quad C^0_{03} = \dfrac{a'}{a}y^3, \\ \\ C^1_{00} = -a'y^1, \quad C^2_{00} = -a'y^2, \quad C^0_{00} = -a'y^3, \end{cases} \tag{49}$$

where the prime denotes derivative with respect to E: $a' = \dfrac{da}{dE}$.

According to the formalism of generalized Lagrange spaces, we can write the Einstein equations in vacuum corresponding to the metrical connection of the deformed Minkowski space (see Eqs.(37)). It is easy to see that the independent equations are given by

$$a' = 0; \tag{50}$$

$$2aa'' - (a')^2 = 0. \tag{51}$$

The first equation has the solution $a = const.$, namely we get the Minkowski metric. Eq.(52) has the solution

$$a(E) = \frac{1}{4}\left(a_0 + \frac{E}{E_0}\right)^2, \tag{52}$$

where a_0 and E_0 are two integration constants.

This solution represents the time coefficient of an over-Minkowskian metric. For $a_0 = 0$ it coincides with (the time coefficient of) the phenomenological metric of the strong interaction, Eq.(6). On the other hand, by choosing $a_0 = 1$, one gets the time coefficient of the metric for gravitational interaction, Eq.(6').

In other words, *considering* $\widetilde{M}$ *as a generalized Lagrange space permits to recover (at least partially) the metrics of two interactions (strong and gravitational) derived on a phenomenological basis.*

It is also worth noticing that this result shows that *a spacetime deformation (of over-Minkowskian type) exists even in absence of an external electromagnetic field* (remember that Eqs.(51),(52) have been derived by assuming $F_{\mu\nu} = 0$).

3.2.4. Intrinsic Physical Structure of a Deformed Minkowski Space: Gauge Fields

As we have seen, the deformed Minkowski space $\widetilde{M}$, considered as a generalized Lagrange space, is endowed with a rich geometrical structure. But the important point, to our purposes, is the presence of a physical richness, intrinsic to $\widetilde{M}$. Indeed, let us introduce the following *internal electromagnetic field tensors* on $\mathcal{GL}^4 = \widetilde{M}$, defined in terms of the deflection tensors:

$$\mathcal{F}_{\mu\nu} \equiv \frac{1}{2}\left(D_{\mu\nu} - D_{\nu\mu}\right) =$$
$$= F_{\mu\nu}(\mathbf{x}) + \frac{1}{2}y^{\sigma}\frac{\partial E}{\partial y^{\alpha}}\left(\frac{\partial g_{DSR\mu\sigma}}{\partial E}\widetilde{F}^{\alpha}_{\ \nu} - \frac{\partial g_{DSR\nu\sigma}}{\partial E}\widetilde{F}^{\alpha}_{\ \mu}\right) \tag{53}$$

(*horizontal electromagnetic internal tensor*) and

$$f_{\mu\nu} \equiv \frac{1}{2}\left(d_{\mu\nu} - d_{\nu\mu}\right) =$$
$$= \frac{1}{2}y^{\sigma}\frac{\partial E}{\partial y^{\alpha}}\left(\frac{\partial g_{DSR\mu\sigma}}{\partial E}\delta^{\alpha}_{\nu} - \frac{\partial g_{DSR\nu\sigma}}{\partial E}\delta^{\alpha}_{\mu}\right) \tag{54}$$

(*vertical electromagnetic internal tensor*).

The internal electromagnetic h- and v-fields $\mathcal{F}_{\mu\nu}$ and $f_{\mu\nu}$ satisfy the following *generalized Maxwell equations*

$$\begin{aligned} 2\left(\mathcal{F}_{\mu\nu|\rho} + \mathcal{F}_{\nu\rho|\mu} + \mathcal{F}_{\rho\mu|\nu}\right) &= y^{\alpha}\left(R^{\beta}_{\ \mu\nu}C_{\beta\alpha\rho} + R^{\beta}_{\ \nu\rho}C_{\beta\alpha\mu} + R^{\beta}_{\ \rho\mu}C_{\beta\alpha\nu}\right), \\ R^{\beta}_{\ \mu\nu} &= g^{\beta\sigma}\frac{\partial F_{\mu\nu}}{\partial x^{\sigma}}; \end{aligned} \tag{55}$$

$$\mathcal{F}_{\mu\nu|\rho} + \mathcal{F}_{\nu\rho|\mu} + \mathcal{F}_{\rho\mu|\nu} = f_{\mu\nu|\rho} + f_{\nu\rho|\mu} + f_{\rho\mu|\nu}; \tag{56}$$

$$f_{\mu\nu|\rho} + f_{\nu\rho|\mu} + f_{\rho\mu|\nu} = 0. \tag{57}$$

Let us stress explicitly the different nature of the two internal electromagnetic fields. In fact, the horizontal field $\mathcal{F}_{\mu\nu}$ is strictly related to the presence of the external electromagnetic field $F_{\mu\nu}$, and vanishes if $F_{\mu\nu} = 0$. On the contrary, *the vertical field $f_{\mu\nu}$ has a geometrical origin, and depends only on the deformed metric tensor $g_{DSR\mu\nu}(E(\mathbf{y}))$ of $\mathcal{GL}^4 = \widetilde{M}$ and on $E(\mathbf{y})$. Therefore, it is present also in space-time regions where no external electromagnetic field occurs.* As we shall see, this fact has deep physical implications.

A few remarks are in order. First, the main results obtained for the (abelian) electromagnetic field can be probably generalized (with suitable changes) to non-abelian gauge fields. Second, the presence of the internal electromagnetic h- and v-fields $\mathcal{F}_{\mu\nu}$ and $f_{\mu\nu}$, intrinsic to the geometrical structure of $\widetilde{M}$ as a generalized Lagrange space, is the cornerstone to build up a *dynamics (of merely geometrical origin) internal to the deformed Minkowski space.*

The important point worth emphasizing is that *such an intrinsic dynamics springs from gauge fields.* Indeed, the two internal fields $\mathcal{F}_{\mu\nu}$ and $f_{\mu\nu}$ (in particular the latter one) do satisfy equations of the gauge type (cfr. Eqs.(57)-(58)). Then, we can conclude that *the (energy-dependent) deformation of the metric of $\widetilde{M}$, which induces its geometrical* *structure as generalized Lagrange space, leads in turn to the appearance of (internal) gauge fields*.

Such a fundamental result can be schematized as follows:

$$\widetilde{M} = (M, g_{DSR\mu\nu}(E)) \Longrightarrow \mathcal{GL}^4 = (M, g_{\mu\nu}(\mathbf{x}, \mathbf{y})) \Longrightarrow \left(\widetilde{M}, \mathcal{F}_{\mu\nu}, f_{\mu\nu}\right) \tag{58}$$

(with self-explanatory meaning of the notation).

We want also to stress explicitly that this result follows by the fact that, in deforming the metric of the space-time, *we assumed the energy as the physical (non-metric) observable on which letting the metric coefficients depend* . This is crucial in stating the generalized Lagrangian structure of $\widetilde{M}$, as shown above.

3.3. Possible Evidence for DSR Internal Gauge Fields: Shadow of Light

We want now to discuss some results on anomalous interference effects, which admit a quite straightforward interpretation in terms of the intrinsic gauge fields of DSR.

In double-slit-like experiments in the infrared range, we collected evidences of an anomalous behaviour of photon systems under particular (energy and space) constraints [7, 8, 9, 10]. The experimental set-up is reported in Fig.1.

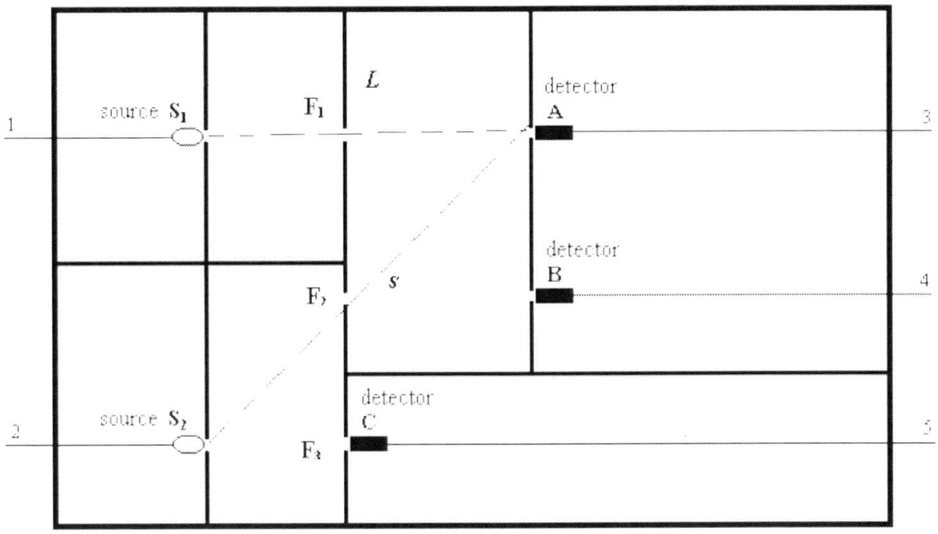

Figure 1. Schematic layout of the box used to detect the anomalous interference effect.

This layout shows the horizontal view of the interior of a closed box divided into different rooms by panels. The box was 20 cm long, 12 cm large and 7 cm high. It contained two infrared LEDs S_1 and S_2, three detectors A, B and C (either photodiodes or phototransistors) and three apertures F_1, F_2 and F_3. The source S_1 was aligned with the detector A through the aperture F_1, the source S_2 was aligned with the detector C which was right on the aperture F_3. The detector B was in front of the aperture F_2 and did not receive any photon directly. The position of the detectors, the sources and the apertures was designed so that the detector A was not influenced by the lighting state of the source S_2 according to the laws of physics governing photons propagation. In other words, A did not have to distinguish whether S_2 was on or off. Besides, in order to prevent reflections of photons, the internal surfaces of the box had been coated by an absorbing material. While the detectors B and C were controlling detectors, A was devoted to perform the actual experiment. In particular, we compared the signal, measured on A when S_1 was on and S_2 was off, with the signal on A when both sources S_1 and S_2 were on. As to what it has been said about the incapability of A to distinguish between S_2 off or on, these two compared conditions were expected to produce compatible results. However, it turned out that the sampling of the signal on A with S_1 on and S_2 on and the sampling of the signal on A when only S_1 was on do not belong to the same population and are represented by two different gaussian distributions whose mean values are significantly different. Besides, the difference between the two mean values was less than 4.5 μeV , as predicted by the theory of Deformed Space-time [1, 2]. Since it was experimentally verified that no photons passed through the aperture F_2, this result shows an anomalous behaviour of the photon system. The same experiment was carried out by different sources, detectors, by two different boxes and different measuring systems. Nevertheless, every time we obtained the same anomalous result [8, 9, 10]. Moreover, the same kind of geometrical structure and the same spatial distances were used in other kind of experiments carried out in the microwave region of the spectrum and by a laser system [11, 12, 1, 2]. Although these experiments had completely different experimental set-ups from our initial one, they succeeded in finding out the same kind of anomalous behaviour that we had found out by the box experiments.

The anomalous effect in photon systems, at least in those experimental set-ups that were used, disagrees both with standard quantum mechanics (Copenhagen interpretation) and with classical and quantum electrodynamics. Some possible interpretations can be given in terms of either the existence of de Broglie–Bohm pilot waves associated to photons, and/or the breakdown of local Lorentz invariance (LLI) [7, 8, 9, 10]. Besides, it turns out that it is also possible to move a step forward and hypothesise the existence of an intriguing connection between the pilot wave interpretation and that involving LLI breakdown. One might assume that the pilot wave is, in the framework of LLI breakdown, a local deformation of the flat Minkowskian spacetime.

The interpretation in terms of DSR is quite straightforward. Under the energy threshold $E_{0,em}$=4.5 μeV, the metric of the electromagnetic interaction is no longer Minkowskian. The corresponding space-time is deformed. Such a space-time deformation shows up as the hollow wave accompanying the photon, and is able to affect the motion of other photons. This is the origin of the anomalous interference observed (*shadow of light*). The difference of signal measured by the detector A in all the double-slit experiments can be regarded as the energy spent to deforme space-time. In space regions where the external electromagnetic field is present (regions of "standard" photon behavior), we can associate such energy to the difference $\Delta\mathcal{E}$, Eq.(16), between the energy density corresponding to the external e.m. field $F_{\mu\nu}$ and that of the deformed one $\widetilde{F}_{\mu\nu}$ given by Eq.(8).

But it is known from the experimental results that the anomalous interference effects observed can be explained in terms of the shadow of light, namely in terms of the hollow waves present in space regions where no external e.m. field occurs. How to account for this anomalous photon behavior within DSR? The answer is provided by the internal structure of the deformed Minkowski space discussed above. In fact, we have seen that the structure of the deformed Minkowski space $\widetilde{M}$ as Generalized Lagrange Space implies the presence of two internal e.m. fields, the horizontal field $\mathcal{F}_{\mu\nu}$ and the vertical one, $f_{\mu\nu}$. Whereas $\mathcal{F}_{\mu\nu}$ is strictly related to the presence of the external electromagnetic field $F_{\mu\nu}$, vanishing if $F_{\mu\nu} = 0$, the vertical field $f_{\mu\nu}$ is geometrical in nature, depending only on the deformed metric tensor $g_{DSR,\mu\nu}(E)$ of $GL^4 = \widetilde{M}$ and on E. Therefore, it is present also in space-time regions where no external electromagnetic field occurs. In our opinion, the arising of the internal electromagnetic fields associated to the deformed metric of $\widetilde{M}$ as Generalized Lagrange space is at the very physical, *dynamic* interpretation of the experimental results on the anomalous photon behavior. Namely, *the dynamic effects of the hollow wave of the photon, associated to the deformation of space-time* — which manifest themselves in the photon behavior contradicting both classical and quantum electrodynamics —*, arise from the presence of the internal v-electromagnetic field* $f_{\mu\nu}$ (in turn strictly connected to the geometrical structure of $\widetilde{M}$).

Moreover, as is well known, in relativistic theories, the vacuum is nothing but Minkowski geometry. A LLI breaking connected to a deformation of the Minkowski space is therefore associated to a lack of Lorentz invariance of the vacuum. Then, the view by Kostelecky [13] that the breakdown of LLI is related to the lack of Lorentz symmetry of the vacuum accords with our results in the framework of DSR, provided that the quantum vacuum is replaced by the geometric vacuum.

4. Conclusion and Perspectives

As is well known, successfully embodying gauge fields in a space-time structure is one of the basic goals of the research in theoretical physics starting from the beginning of the XX century. The almost unique tool to achieve such objective is increasing the number of space-time dimensions. In such a kind of theories (whose prototype is the celebrated Kaluza-Klein formalism), one preserves the usual (special-relativistic or general-relativistic) structure of the four-dimensional space-time, and gets rid of the non-observable extra dimensions by compactifying them (for example to circles). Then the motions of the extra metric components over the standard Minkowski space satisfy identical equations to gauge fields. The gauge invariance of these fields is simply a consequence of the Lorentz invariance in the enlarged space. In this framework, gauge fields are *external* to the space-time, because they are *added* to it by the hypothesis of extra dimensions.

In the case of the DSR theory, gauge fields arise from the very geometrical, basic structure of $\widetilde{M}$, namely they are a consequence of the metric deformation. The arising gauge fields are *intrinsic and internal to the deformed space-time, and do not need to be added from the outside.* As a matter of fact, *DSR is the first theory based on a four-dimensional space-time able to embody gauge fields in a natural way.*

Such a conventional, intrinsic gauge structure is related to a *given* deformed Minkowski space $\overline{\widetilde{M}}$, in which the deformed metric is fixed:

$$\overline{\widetilde{M}} = (M, \bar{g}_{DSR\mu\nu}(E)) \,. \tag{59}$$

On the contrary, with varying g_{DSR}, we have another gauge-like structure — as already stressed in Sect.3 — namely what we called an external metric gauge. In the latter case, the gauge freedom amounts to choosing the metric according to the interaction considered.

The circumstance that the deformed Minkowski space $\widetilde{M}$ is endowed with the geometry of a generalized Lagrange space testifies the richness of non-trivial mathematical properties present in the seemingly so simple structure of the deformation of the Minkowski metric. In this connection, let us recall that $\widetilde{M}$ (contrarily to the usual Minkowski space) is not flat, but does possess curvature and torsion (see 3.2.1).

Let us stress that — as already mentioned — the deformed Minkowski space $\widetilde{M}$ can be naturally embedded in a five-dimensional Riemannian space $\Re_5$ (see [1, 2]). We denoted by DR5 the generalized theory based on this five-dimensional space.

In embedding the deformed Minkowski space $\widetilde{M}$ in $\Re_5$, *energy does lose its character of dynamic parameter* (the role it plays in DSR), *by taking instead that of a true metrical coordinate,* $E = x^5$, on the same footing of the space-time ones. This has a number of basic implications. In such a change of role of energy, with the consequent passage from $\widetilde{M}$ to $\Re_5$, some of the geometrical and dynamic features of DSR are lost, whereas others are still present and new properties appear. The first one is of geometrical nature, and is just the passage from a (flat) pseudoeuclidean metric to a genuine (curved) Riemannian one. The other consequences pertain to both symmetries and dynamics. Among the former, we recall the basic one — valid at the slicing level $x^5 = const.$ $(dx^5 = 0)$ —, related to the Generalized Lagrange Space structure of $\widetilde{M}$, which implies *the natural arising of gauge fields*, intimately related to the inner geometry of the deformed Minkowski space. Let us also stress that, in the framework of $\Re_5$, the dependence of the metric coefficients on a true metric coordinate make them fully analogous to the gauge functions of non-abelian gauge theories, thus implementing DR5 as a metric gauge theory (in the sense specified in Subsect.3.1). Let us recall that the metric homomorphisms of $\Re_5$ are strictly connected to the invariance under what we called the Metric Gaugement Process of DSR (see Subsect.3.1).

Concerning the influence of the extra dimension on the physics in the four-dimensional deformed space-time, points worth investigating are the possible connection between Lorentz invariance in DR5 and the usual gauge invariance, and the occurrence of parity violation as consequence of space anisotropy when viewed from the standpoint of the space-time-energy manifold $\Re_5$.

A further basic topic deserving study in DSR is the extension to the non-abelian case of the results obtained for the abelian gauge fields (like the e.m. one), based on the structure of the deformed Minkowski space $\widetilde{M}$ as Generalized Lagrange Space (see Subsubsect.3.2.1). In other words, it would be worth verifying if also non-abelian internal gauge fields can exist in absence of external fields, due to the intrinsic geometry of $\widetilde{M}$.

References

[1] F. Cardone and R. Mignani: *Energy and Geometry – An Introduction to Deformed Special Relativity* (World Scientific, Singapore, 2004).

[2] F. Cardone and R. Mignani: *Deformed Spacetime – Geometrizing Interactions in Four and Five Dimensions* (Springer, Heidelberg, Dordrecht, 2007); and references therein.

[3] F. Cardone, R. Mignani and A. Petrucci: *"The principle of solidarity: Geometrizing interactions"*, in *Einstein and Hilbert: Dark matter*, V.V. Dvoeglazov ed. (Nova Science, Commack, N.Y., 2011), p.19.

[4] See *e.g.* R. Miron and M. Anastasiei: *The Geometry of Lagrange Spaces: Theory and Applications* (Kluwer, 1994); R. Miron, D. Hrimiuc, H. Shimada and S.V. Sabau: *The Geometry of Hamilton and Lagrange Spaces* (Kluwer, 2002); and references therein.

[5] See *e.g.* N. Steenrod: *The Topology of Fibre Bundles* (Princeton Univ. Press, 1951).

[6] R. Miron, A. Jannussis and G. Zet: in *Proc. Conf. Applied Differential Geometry - Gen. Rel. and The Workshop on Global Analysis, Differential Geometry and Lie Algebra*, 2001., Gr. Tsagas ed. (Geometry Balkan Press, 2004), p.101, and refs. therein.

[7] F. Cardone, R. Mignani, W. Perconti and R. Scrimaglio, *Phys. Lett. A*, 326, 1 (2004).

[8] F. Cardone, R. Mignani, W. Perconti, A. Petrucci and R. Scrimaglio, *Int. J. Mod. Phys. B*, 20, 85 (2006).

[9] F. Cardone, R. Mignani, W. Perconti, A. Petrucci and R. Scrimaglio, *Int. J. Mod. Phys. B*, 20, 1107 (2006).

[10] F. Cardone, R. Mignani, W. Perconti, A. Petrucci and R. Scrimaglio, *Annales Fond. L. de Broglie*, 33, 319, (2008).

[11] A. Ranfagni, D. Mugnai and R. Ruggeri, *Phys. Rev. E*, 69, 027601 (2004).

[12] A. Ranfagni and D. Mugnai, *Phys. Lett. A* 322, 146 (2004).

[13] See *e.g. CPT and Lorentz Symmetry I, II, III*, V.A. Kostelecky ed. (World Scientific, Singapore, 1999, 2002, 2004).

In: Space-Time Geometry and Quantum Events
Editor: Ignazio Licata, pp. 249-276
ISBN: 978-1-63117-455-1

Chapter 9

NON-LOCAL GRANULAR SPACE-TIME FOAM AS AN ULTIMATE ARENA AT THE PLANCK SCALE

Davide Fiscaletti *
Space Life Institute, San Lorenzo in Campo (PU), Italy

Abstract

In this paper the suggestive perspective of a granular, non-local, holographic space-time foam of Planck's scale as an ultimate background of processes is analysed. After reviewing the fundamental theoretical results about the discreteness of the spatial geometry, the relevant role of holography in the mapping of this geometry and making some considerations of quantum gravity in the quantum potential approach, an entropic bohmian model of a fundamental space-time foam is introduced where the fundamental entities are non-local weaves of Planck's scale whose behaviour is determined by the quantum entropy.

1. Introduction

Before last century, space-time was regarded as nothing more than a passive and static arena of physics. From Newton's physics to Einstein's special relativity, a fixed space-time serves as the arena for all of physics: particles and fields are actors that cannot influence this arena.

The most profound changes as regards the background of physical processes happened in the first quarter of the XIX century with the advent of Einstein's general relativity and of quantum mechanics. On one hand, general relativity promoted space-time to an active and dynamical entity. In general relativity, the space-time metric is no longer fixed, immune to change: gravity has to be encoded into the very geometry of space-time, thereby making this geometry dynamical. The central discovery of general relativity is that Newton's space-time and the gravitational field are the same thing and that the dynamics of the gravitational field and any other dynamical object is fully relational, in the Aristotelian-Cartesian sense. Matter curves space-time, geometry is a physical entity that interacts with matter. On the other hand,

* E-mail address: spacelife.institute@gmail.com

quantum mechanics showed that any dynamical entity has not a continuous spectrum but is made up of quanta and can be in probabilistic superposition states. As regards the geometry of the physical world, quantum theory introduces much more spread perspectives than those offered by every previous physical theory. In particular, the most fundamental and intriguing element lies in the fact that subatomic particles are non-locally connected and can be in entangled states. The non-local correlations between subatomic particles which characterize EPR-type experiments turn out to be inexplicable and understandable inside a classical picture. In the Copenhagen interpretation non-locality emerges indeed as an unexpected host which lies behind the purely probabilistic interpretation of the wave function and the mechanism of "casuality" associated with it. However, on the basis of the experimental results of many authors (such as those of A. Aspect and A. Zeilinger, just to name a few),one has to acknowledge that non-locality is the ultimate visiting card of the geometry of quantum physics and it should be introduced from the very beginning within the theory structure as a fundamental principle. The experimental results suggest that quantum non-locality must be considered as the essential property which is at the basis of the behaviour of subatomic particles and of the geometry of the quantum world [1]. Since the non-local correlations do not transport energy, they do not violate relativity but nonetheless remain outside the bounds of the classical picture of the world. Hence one has to face the problem of reconciling the classical image of Einstein's space-time and quantum non-locality, namely of constructing a dynamic model of space-time in which the quantum processes also find a place, in other words of developing a quantum geometrodynamics.

Moreover, in a fundamental quantum geometrodynamics describing the small scales, one expects that there should be quanta of space and quanta of time and quantum superposition of spaces. But, if we live in a space-time background with quantum properties (and that exhibits non-local features), how can we describe it from the mathematical point of view?

The challenge to understand and describe mathematically the quantum nature of the space-time arena of processes was not taken up seriously until theoretical physicists attempted to develop models describing the so-called "quantum gravity domain", able to provide an unitary picture incorporating the principles of both quantum mechanics and general relativity, and reducing to them in appropriate limits. Now, in the light of the attempts to find a consistent unification of general relativity and quantum mechanics, many physicists believe that space-time, like all matter and energy, undergoes quantum fluctuations and thus that the fundamental arena of physics is an arrangement of bubbles. These quantum fluctuations imply that space-time is foamy on small space-time scales, in particular at the Planck scale.

In this paper we propose a model of a quantum space-time foam, which is based on a granular structure of the space background at the Planck scale and is composed by weaves having non-local features. The paper is structured in the following manner. In chapter 2 we will review the fundamental results of loop quantum gravity about the discreteness of spatial geometry. In chapter 3 we will analyse the connection between the discrete and quantized geometry of the fundamental foam predicted by loop quantum gravity and the smooth physical geometry perceived in the everyday life and the relevant role of holography in the mapping of the geometry of the space-time foam in the context of some current research. In chapter 4 we will review some relevant results about the space-time background in the quantum gravity domain – obtained recently by the author – in a re-reading of a bohmian model, based on a physical quantity describing the geometry called quantum entropy. In chapter 5 we will introduce the considerations made in chapter 4 about the entropic bohmian model of quantum gravity in the picture of the quantized geometry analysed in chapter 3, introducing in this way an entropic bohmian version of a fundamental space-time foam based on non-local weaves whose behaviour is determined by the quantum entropy.

2. The Granularity of the Geometry of Space in Loop Quantum Gravity

Twenty-six years ago Abhay Ashtekar, on the basis of earlier work by Sen, laid the foundations of loop quantum gravity by reformulating general relativity in terms of canonical connection and triad variables. The completion of the kinematics – the quantum theory of spatial geometry – led to the prediction of a granular structure of space, described by specific discrete spectra of geometric operators: area [2], volume [2-4], length [5-7] and angle [8] operators.

The discreteness of quantum geometry predicted by loop quantum gravity can be considered as a genuine property of space, independent of the strength of the actual gravitational field at any given location. Thus the interesting perspective is opened to observe quantum gravity effects even without strong gravitational field, in the flat space limit. In 1998 Amelino-Camelia and other authors proposed that the granularity of space influences the propagation of particles, when their energy is comparable with the quantum gravity energy scale [9]. Further, the assumed invariance of this energy scale, or the length scale, respectively, are in apparent contradiction with special relativity. So it is expected that the energy-momentum dispersion relation could be modified to include dependence on the ratio of the particle's energy and the quantum gravity energy.

In the years following this work, the nascent field of quantum gravity phenomenology developed from ad hoc effective theories, like isolated isles lying between the developing quantum gravity theories and reality, linked to the former ones loosely by plausibility arguments [10]. Today the main efforts of quantum gravity phenomenology go in two directions: to establish a bridge between the intermediate effective theories and the fundamental quantum gravity theory and the refinement of observational methods, through new effective theories and experiments that could shed new light on quantum gravity effects. These are exceptionally healthy developments for the field. The development of physical theory relies on the link between theory and experiment. Now these links between current observation and quantum gravity theory are possible and under active development.

Loop quantum gravity hews close to the classical theory of general relativity, taking the notion of background independence and apparent four dimensionality of space-time seriously. Loop quantum gravity takes the novel view of the world provided by general relativity and quantum theory, by incorporating the notions of space and time from general relativity directly into quantum field theory. The quantization is performed in stages: kinematics, quantization of spatial geometry, dynamics and finally the full description of space-time.

The kinematics of the theory is given by a Hilbert space carrying an algebra of operators that have a physical interpretation in terms of observable quantities of the system considered. In loop quantum gravity the operators representing area, angle, length, and volume have discrete spectra, so discreteness is naturally incorporated into loop quantum gravity.

The Hilbert space $\widetilde{H}$ on which the theory is defined is given by relation

$$\widetilde{H} = \underset{\Gamma}{\oplus} H_\Gamma \tag{1}$$

where the Γ are abstract graphs defined by sets of links $l$and sets of nodes n. Spin network states represent a convenient basis in $\widetilde{H}$. Spin network states are eigenstates of the area and volume operators. A spin network state can receive a simple geometrical interpretation. It represents a “granular” space where each node n represents a “grain” or “chunk” of space. Two grains n and n' are adjacent if there is a link l connecting the two, and in this case the area of the elementary surface separating the two grains is $8\pi\gamma\hbar G\sqrt{j_l(j_l+1)}$. In the recent Rovelli’s paper [11], the states in $\widetilde{H}$ are interpreted as “boundary states”, in the sense that they describe the quantum space surrounding a given finite four-dimensional region of space-time.

The area operator depends on a surface Σ cutting the links l_1, ..., l_S. The area operator is defined as

$$A_\Sigma = \sum_{l \in \Sigma} \sqrt{L_l^i L_l^i} \tag{2}$$

$$\vec{L}_l = \{L_l^i\}, \text{ i=1, 2, 3} \tag{3}$$

being the gravitational field operator corresponding to the flux of Ashtekar’s electric field, or to the flux of the inverse triad, across “an elementary surface cut by the link l”. The action of the area operator (2) on a spin network function ψ_Γ is

$$A_\Sigma |\psi_\Gamma\rangle = \frac{8\pi\gamma}{\kappa^2} \sum_{p \in \Gamma \cap \Sigma} \sqrt{j_p(j_p+1)} |\psi_\Gamma\rangle \tag{4}$$

where j_p is the spin, or color, of the link that intersects Σ at p and γ is the Barbero-Immirzi parameter.

The area operator acts only on the intersection points of the surface with the spin network graph, $\gamma \cap \Sigma$ and so provides a finite number of contributions. Spin network functions are eigenfunctions of the area operator.

The eigenvalues are obviously discrete. The expression (4) gives the “spectrum of the area” of loop quantum gravity. The quanta of area live on the edges of the graph and are the simplest elements of quantum geometry. There is a minimal eigenvalue, the so-called area gap, which is the area when a single edge with j= ½ intersects Σ ,

$$\Delta A = 4\sqrt{3}\pi\hbar G c^{-3} \approx 10^{-70} m^2 \tag{5}.$$

This is the minimal quantum of area, which can be carried by a link.

The eigenvalues (4) form only the main sequence of the spectrum of the area operator. When nodes of the spin network lie on Σ and some links are tangent to it the relation is modified (about this topic the reader can find details in the paper [12]). As regards the area operator, the important fact is that discreteness of area with the spin network links, carrying its quanta, emerges in a natural way.

The interpretation of discrete geometric eigenvalues as observable quantities goes back to early work in [13]. This discreteness allowed the calculation of black hole entropy possible by counting the number of microstates of the gravitational field that lead to a given area of the horizon within some small interval. Moreover, area operators acting on surfaces that intersect in a line fail to commute, when spin network nodes line in that intersection [14]. In the recent paper [15] additional insight into this non-commutativity comes from the formulation of discrete classical phase space of loop quantum gravity, in which the flux operators also depend on the connection.

As regards the volume there are two definitions of the operator, one due to Rovelli and Smolin [2] and the other due to Ashtekar and Lewandowski [16]. In Rovelli's paper [11], the operator volume of a region R (which is a collection of nodes) is given by

$$V_R = \sum_{n\in R}' V_n \tag{6}$$

where

$$V_n^2 = \frac{2}{9}\left|\varepsilon_{ijk} L_{la}^i L_{lb}^j L_{lc}^k\right| \tag{7}$$

where l_a, l_b, l_c are any three distinct of the four links of n. In Ashtekar's and Lewandowski's definition, for a given spin network function based on a graph Γ, the operator $\hat{V}_{R,\Gamma}$ is

$$\hat{V}_{R,\Gamma} = \left(\frac{l_p}{2}\right)^3 \sum_{v\in R} \sqrt{\left|\frac{i}{3!\cdot 8}\sum_{I,J,K} s(e_I, e_J, e_K)\varepsilon_{ijk}\hat{X}^i_{v,e_I}\hat{X}^j_{v,e_J}\hat{X}^k_{v,e_K}\right|} \tag{8}$$

where l_p is Planck length, the three derivative operators $\hat{X}^i_{v,e_I}$ act at every node or vertex $v$on each triple of adjacent edges e_I and the dependence on the tangent space structure of the embedding is manifest in $s(e_I, e_J, e_K)$. This is +1 or -1 whether e_I, e_J and e_K are positive or negative oriented, and is zero when the edges are coplanar. The action of the operators $\hat{X}^i_{v,e_I}$ on a spin network function $\psi_\Gamma = \psi(h_{e_1}(A),\ldots,h_{e_n}(A))$ based on the graph Γ is

$$\hat{X}^i_{v,e_I}\,\psi_\Gamma = i\,tr\left(h_{e_I}(A)\tau^i \frac{\partial\psi}{\partial h_{e_I}(A)}\right) \tag{9}$$

when e_I is outgoing at v. Equation (9) expresses the action of the left-invariant vector field on SU(2) in the direction of τ^i; for ingoing edges it would be the right-invariant vector field. In virtue of the "triple-product" action of the operator (8), vertices carry discrete quanta of volume. The volume operator of a small region containing a node does not change the graph, nor the colors of the adjacent edges, it acts in the form of a linear transformation in the space of intertwiners at the vertex for given colors of the adjacent edges. This space of intertwiners forms the "atoms of quantum geometry".

The spectrum of the volume operator (8) has been investigated in the papers [17-22]. In the thorough analysis of [17, 18], Brunnemann and Rideout showed that the volume gap, namely the lowest boundary for the smallest non-zero eigenvalue, depends on the geometry of the graph and does not in general exist. In the simplest nontrivial case, for a four-valent vertex, the existence of a volume gap is demonstrated analytically.

The two volume operators (6) and (8) are in equivalent, leading to different spectra. While the details of the spectra of the Rovelli-Smolin and the Ashtekar-Lewandowski definitions of the volume operator differ, they do share the property that the volume operator vanishes on all gauge invariant trivalent vertices [4]. According to an analysis performed in the papers [23, 24] the Ashtekar-Lewandowski operator is compatible with the flux operators, on which it is based, and the Rovelli-Smolin operator is not. On the other hand, thanks to its topological structure the Rovelli-Smolin volume does not depend on tangent space structure; the operator is "topological" in that is invariant under spatial homeomorphisms. It is also covariant also under "extended diffeomorphisms", which are everywhere continuous mappings that are invertible everywhere except at a finite number of isolated points; the Ashtekar-Lewandowski operator is invariant under diffeomorphisms.

Physically, the distinction between the two operators is the role of the tangent space structure at spin network nodes. There is some tension in the community about the role of this structure. Recent developments in twisted discrete geometries [25] and the polyhedral point of view [26] may help resolve these issues. In [27] Bianchi and Haggard show that the volume spectrum of the four-valent node may be obtained by direct Bohr-Sommerfeld quantization of geometry. The description of the geometry goes all the way back to Minkowski, who showed that the shapes of convex polyhedra are determined from the areas and unit normals of the faces. Kapovich and Millson showed that this space of shapes is a phase space, and it is this phase space – the same as the phase space of intertwiners – that Bianchi and Haggard used for the Bohr-Sommerfeld quantization. The agreement between the spectra of the Bohr-Sommerfeld and loop quantum gravity volume turns out to be quite good [27].

A fundamental result of the geometric operators described above (as well as of length and angle) is their discrete spectra. In other words, the geometry associated with spin network states is characterized by a discrete quantized three-dimensional (3D) metric. It is natural to ask whether this discreteness of loop quantum gravity operators and thus the granularity of spatial geometry is physical. Can it be used as a basis for the phenomenology of quantum geometry? Using examples, Dittrich and Thiemann [28] argue that the discreteness of the geometric operators, being gauge non-invariant, may not survive implementation in the full dynamics of loop quantum gravity. On the contrary, Rovelli [29] argues in favour of the plausibility and full coherence of physical geometric discreteness, showing in one case that the preservation of discreteness in the generally covariant context is immediate. In phenomenology this discreteness has been a source of inspiration for models. Nonetheless as the discussion of these operators makes clear, there are subtleties that wait to be resolved, either through further completion of the theory or, perhaps, through observational constraints on phenomenological models. As regards the phenomenology of loop quantum gravity and its physical consequences and applications, in the recent paper *Loop quantum gravity phenomenology: linking loops to observational physics*, Girelli, Hinterleitner and Major [30] described ways in which loop quantum gravity, mainly by means of discreteness of the spatial geometry, may lead to experimentally viable predictions and explored the possibility of a large variety of modifications of special relativity, particle physics and field theory in the weak field limit on the basis of the discreteness of the spatial geometry.

3. About the Holographic and Non-Local Features of the Ultimate Texture of Space-Time Foam

What is the connection between the discrete and quantized geometry of the fundamental foam predicted by loop quantum gravity and the smooth physical geometry that we perceive in our everyday life? In this chapter we review some relevant results as regards the texture of spin network weaves of loop quantum gravity and of other similar models of spacetime foams characterized by an holographic mapping.

By following Rovelli in his book *Quantum Gravity* [31], let us start by considering a classical macroscopic 3D gravitational field which determines a 3D metric $g_{\mu\nu}$ and in this metric let us fix a region R of area S with a size much larger than $l >> l_P$ and slowly varying at this scale. A spin network state $|S\rangle$, if is an eigenstate of the volume operator $V(R)$ (and of the area operator $A(S)$), with eigenvalues equal to the volume of R (and of the area of S) determined by the metric $g_{\mu\nu}$, up to small corrections in l/l_P, namely if satisfies the following equations

$$V(R)|S\rangle = \left(V[g_{\mu\nu}, R] + O\left(\frac{l}{l_P}\right)\right)|S\rangle \tag{10}$$

$$A(S)|S\rangle = \left(A[g_{\mu\nu}, S] + O\left(\frac{l}{l_P}\right)\right)|S\rangle \tag{11}$$

is called a weave state of the metric g. The texture of reality which emerges from loop quantum gravity is a weave of spin network states satisfying equations (10) and (11).

At large scale, the state $|S\rangle$ determines precisely the same volumes and areas as the metric g. Equations (10) and (11) are valid also to the diffeomorphism invariant level: the *s*-knot state $|s\rangle = P_{diff}|S\rangle$ is called the weave state of the 3D geometry $[g]$, namely the equivalence class of 3-metric to which the metric g belongs.

Several weave states were constructed and analysed in the early days of loop quantum gravity, for various 3D metrics, including those for flat space, Schwarzschild and gravitational waves. They satisfied equations (10) and (11) or equations similar to these. Weave states have played an important role as regards the historical development of loop quantum gravity. In particular, they provided an explanation of the emergence of the Planck-scale discreteness. The intuition was that a macroscopic geometry could be built by taking the limit of an infinitely dense lattice of loops, as the lattice size goes to 0. With increasing density of loops, the eigenvalue of the operator turned out to increase.

In order to define a weave on a 3D manifold with coordinates $\vec{x}$ that approximates the flat 3D metric $g^{(0)}{}_{\mu\nu}(\vec{x}) = \eta_{\mu\nu}$, one can build a spatially uniform weave state $|S_{\mu_0}\rangle$ constituted by a set of loops of coordinate density $\rho = \mu_0^{-2}$. The loops are then at an average distance μ_0 from each other. By decreasing the "lattice spacing" μ_0, namely by increasing the coordinate density of the loops, one obtains

$$A(S)|S_\mu\rangle \approx \frac{\mu_0^2}{\mu^2}\left(A[g^{(0)}, S] + O\left(\frac{l}{l_P}\right)\right)|S\rangle \tag{12}$$

which indicates an increasing of the area. Since

$$\frac{\mu_0^2}{\mu^2} A[g^{(0)}, S] = \left(A\left[\frac{\mu_0}{\mu} g^{(0)}, S\right] + O\left(\frac{l}{l_P}\right)\right) = A[g^{(\mu)}, S] \tag{13}$$

the weave with increased loop density approximates the metric

$$g^{(\mu)}{}_{\mu\nu}(\vec{x}) = \frac{\mu_0^2}{\mu^2}\eta_{\mu\nu} \tag{14}$$

At the same time, however, the physical density ρ_μ, defined as the ratio between the total length of the loops and the total volume, determined by the metric $g^{(\mu)}$, remains μ_0, irrespective of the density of the loops μ chosen:

$$\rho_\mu = \frac{L_\mu}{V_\mu} = \frac{(\mu_0/\mu)L}{(\mu_0/\mu)^3 V} = \frac{\mu^2}{\mu_0^2}\rho = \frac{\mu^2}{\mu_0^2}\mu^{-2} = \mu_0^{-2} \tag{15}$$

Equation (15) implies that, if μ_0 is not determined by the density of the loops, it must be given by the only dimensional constant of the theory, namely the Planck length:

$$\mu_0 \approx l_P \tag{16}$$

By substituting equation (16) into equation (14), the metric approximated by the increased loop density becomes

$$g^{(\mu)}{}_{\mu\nu}(\vec{x}) = \frac{l_P^2}{\mu^2}\eta_{\mu\nu} \tag{17}$$

The physical meaning of the approach based on equations (12)-(17) is that, in loop quantum gravity, a smooth geometry cannot be approximated at a physical scale lower than the Planck length. Each loop carries a quantum geometry of the Planck scale: more loops give more size, not a better approximation to a given geometry.

Another significant feature of the texture of the weaves of the fundamental quantum geometry of the Planck scale is its holographic nature. In this regard, it is important to underline that in the paper [32] Gambini and Pullin showed that from the framework of loop quantum gravity in spherical symmetry an holography emerges in the form of an uncertainty in the determination of volumes that grows radially. Gambini and Pullin found that, in the kinematical structure of spherical loop quantum gravity, the minimal increment in the volume, which derives from the Hamiltonian constraint due to the fact that one cannot take the continuum limit in the loop representation, is given by relation

$$\Delta V = 8\pi\gamma\rho l_P(x+2M) \tag{18}$$

where $x+2M$ is the radial coordinate in Schwarzschild coordinate, ρ is the coordinate density of the loops (whose minimum value possible leads to interpret (18) as the elementary volume). As a consequence, the number ΔN of elementary volumes in a shell with width Δx is

$$\Delta N = \frac{x\Delta x}{2\gamma\rho l_P^2} \tag{19}$$

and its entropy is

$$\Delta S = \nu_V \frac{x\Delta x}{2\gamma\rho l_P^2} \tag{20}$$

where ν_V is the mean entropy per unit volume. According to equations (18)-(20) of Gambini's and Pullin's approach, the kinematical structure of loop quantum gravity in spherical symmetry implies holography. This is a very general result which stems from the fact that the elementary volume that any dynamical operator may involve behaves as xl_P^2.

An interesting model of fundamental spacetime foam of holographic nature has been recently proposed by Jack Ng in the papers [33-36], in which the concept of entropy plays a crucial role. By following Ng's treatment in the papers [33-36], let us review the fundamental features and results of this model. In Ng's model, spacetime undergoes quantum fluctuations which appear when we measure a distance *l*, in the form of uncertainties in the measurement. The quantum fluctuations of spacetime manifest themselves in the form of uncertainties in the geometry of spacetime and thus the structure of spacetime foam can be inferred from the accuracy with which we can measure its geometry. Let us consider mapping out the geometry of spacetime for a spherical volume of radius *l* over the amount of time $T = 2l/c$ it takes light to cross the volume. The total number of operations, including the ticks of the clocks and the measurements of signals, is determined by the Margolus-Levitin theorem in quantum computation, which establishes that the rate of operations for any computer cannot exceed the amount of energy *E* that is available for computation divided by $\pi\hbar/2$ [37]. A total mass *M* of clocks then yields, via the Margolus-Levitin theorem, the bound on the total number of operations given by $\left(2Mc^2/\pi\hbar\right)\cdot 2l/c$. But to prevent black hole formation, the mass *M* must be less than $lc^2/2G$. Together, these two limits imply that the total number of operations that can occur in a spatial volume of radius *l* for a time period $2l/c$ is no greater than $2\left(l/l_P\right)^2/\pi$. To maximize spatial resolution, each clock must tick only once during the entire time period. The operations can be regarded as partitioning the spacetime volume into "cells", yielding an average separation between neighbouring cells no less than $\left(2\pi^2/3\right)^{1/3} l^{1/3} l_P^{2/3}$. This spatial separation can be interpreted as the average minimum uncertainty [38], and thus the accuracy in the measurement of a distance *l*, namely

$$\delta l \geq \left(2\pi^2/3\right)^{1/3} l^{1/3} l_P^{2/3} \tag{21}$$

One can easily understand in what sense this quantum foam model turns out to be a holographic model. Dropping the multiplicative factor of order 1, since on the average each cell occupies a spatial volume of ll_P^2, a spatial region of size *l* can contain no more than $l^3/\left(ll_P^2\right) = \left(l/l_P\right)^2$ cells. Thus this model corresponds to the case of maximum number of bits of information $\left(l/l_P\right)^2$ in a spatial region of size *l*, that is allowed by the holographic principle [39-44] which implies that, although the world around us appears to have three spatial dimensions, its contents can actually be encoded on a two-dimensional surface, like a hologram. In other words, the maximum entropy, i.e., the maximum number of degrees of freedom, of a region of space is given by its surface area in Planck units. In order to see explicitly in what sense the holographic principle has its origin in the quantum fluctuations of spacetime, namely how spacetime foam manifests itself holographically, let us consider a cubic region of space with linear dimension *l*. In this case, from conventional wisdom the number of degrees of freedom of the region should be bounded by $(l/l_P)^3$, namely the volume of the region in Planck units. But conventional wisdom is wrong, because, on the basis of equation (21), the smallest cubes into which one can partition the region cannot have a linear dimension smaller than $(ll_P^2)^{1/3}$. Therefore, the number of degrees of freedom of the region is bounded by $\left[l/\left(ll_P^2\right)^{1/3}\right]^3$, i.e., the area of the region in Planck units, in agreement with the holographic principle [39-44].

Assuming that there is unity of physics connecting the Planck scale to the cosmic scale, Ng also applied his holographic spacetime foam model to cosmology [33, 45]. In this regard, the consistency of the holographic spacetime foam model is assured if the average minimum uncertainty (21) corresponds to a maximum energy density

$$\rho = \frac{3}{8\pi}\left(ll_P\right)^{-2} \tag{22}$$

for a sphere of radius l that does not collapse into a black hole. Hence, according to the holographic spacetime foam cosmology proposed by Ng, one obtains that the cosmic energy density is given by

$$\rho = \frac{3}{8\pi}\left(R_H l_P\right)^{-2} \tag{23}$$

where R_H is the Hubble radius. The energy density (23) is the critical cosmic energy density as observed.

Moreover, if one divides a large distance l into l/λ equal parts each of which has length λ (that can be taken as small as l_P), the cumulative factor C characterizing the cumulative effects of the spacetime fluctuations over this large distance defined by $C = \frac{\delta l}{\delta\lambda}$ turns out to be $C = (l/\lambda)^{1/3}$. Because of its holographic features and quantum-gravity effects, the individual fluctuations cannot be completely random: successive fluctuations appear to be entangled and somewhat anti-correlated in such a way that one obtains a cube root dependence in the number l/l_P of fluctuations for the total fluctuation of l.

In Ng's model, as a consequence of the holographic principle, the physical degrees of freedom of the spacetime foam, at the Planck scale, must be considered as infinitely correlated, with the result that the spacetime location of an event may lose its invariant significance. In other words, in virtue of its holographic nature, the spacetime foam gives rise to non-locality. This argument is also supported by the observation that the long-wavelength (hence "non-local") "particles" constituting dark energy in the holographic spacetime foam cosmology obey an exotic statistics which has attributes of non-locality [33].

In this regard, let us consider a perfect gas of N particles obeying Boltzmann statistics at temperature T in a volume V. At the lowest-order approximation, since one can neglect the contributions from matter and radiation to the cosmic energy density for the recent and present eras, the Friedmann equations (for the relativistic case) yield the partition function

$$Z_N = (N!)^{-1}(V/\lambda^3)^N \tag{24}$$

where $\lambda = (\pi)^{2/3}/T$, and the entropy

$$S = N\left[\ln\frac{V}{N\lambda^3} + \frac{5}{2}\right] \tag{25}$$

Here, since $V \approx \lambda^3$, the entropy S becomes nonsensically negative unless $N \approx 1$ which is equally nonsensical because $N \approx (R_H/l_P)^2 >> 1$. The solution comes with the observation that the N inside the log term for S somehow must be absent. Then $S \approx N \approx (R_H/l_P)^2 >> 1$ without N being small (of order 1) and S is non-negative as physically required. In this case, the Gibbs $1/N!$ factor is absent from the partition function (24) and thus the entropy (25) becomes

$$S = N\left[\ln\frac{V}{\lambda^3} + \frac{3}{2}\right] \tag{26}$$

Taking account of equation (26), the only consistent statistics in greater than two space dimensions without the Gibbs factor is the infinite statistics, called also "quantum Boltzmann statistics", which is characterized by a q deformation of the commutation relations of the oscillators:

$$a_k a_l^+ - q a_l^+ a_k = \delta_{kl} \tag{27}$$

with q between -1 and 1. Infinite statistics can be thought of as corresponding to the statistics of identical particles with an infinite number of internal degrees of freedom, which is equivalent to the statistics of non-identical particles since they are distinguishable by their internal states. As shown in [46-49], a theory of particles obeying infinite statistics turns out to be explicitly non-local. In Ng's model, since the holographic principle is believed to be an important ingredient in the formulation of quantum gravity, it is just the non-local features of the spacetime foam that make it easier to incorporate gravitational interactions in the theory. In this approach, quantum gravity and infinite statistics appear to fit together nicely, and non-locality seems to be a common feature of both of them [33].

On the other hand, other recent research seem to support the idea of non-locality as a fundamental feature of the quantum gravity domain. Using the Matrix theory approach, Jejjala, Kavic and Minic showed that dark energy quanta obey infinite statistics and also concluded that the non-locality present in systems obeying infinite statistics and the non-locality present in holographic theories may be related [50-52]. Giddings remarked that the non-perturbative dynamics of gravity is non-local [53]. His argument is based on several reasons: lack of a precise definition in quantum gravity, connected with the apparent absence of local observables; indications from high-energy gravitational scattering; hints from string theory, particularly the AdS/CFT correspondence; conundrums of quantum cosmology; and the black hole information paradox. Horowitz [54] remarked that quantum gravity may need some violation of locality, in particular if one reconstructs the string theory from the gauge theory (in the AdS/CFT correspondence), physics may not be local on all length scales.

Violation of locality in the form of non-local links between the fundamental weaves of the spacetime foam appear also in quantum graphity models, which are spin system toy models for emergent geometry and gravity. They are based on quantum, dynamical graphs whose adjacency is dynamical: their edges can be on (connected), off (disconnected), or in a superposition of on and off. In these approaches, one can interpret the graph as a pregeometry (and the connectivity of the graph tells us who is neighbouring whom). A particular graphity model is given by such graph states evolving under a local Ising-type Hamiltonian. For example, in the reference [55] a graphity toy model was proposed for interacting matter and geometry, characterized by a Bose Hubbard model where the interactions are quantum variables.

In [56], Caravelli, Hamma, Markopoulou and Riera solved the model of [55] in the limit of no back-reaction of the matter on the lattice, and for states with certain symmetries called rotationally invariant graphs. In this case, the problem reduces to an one-dimensional Hubbard model on a lattice with variable vertex degree and multiple edges between the same two vertices. The probability density for the matter obeys a (discrete) differential equation closed in the classical regime. This is a wave equation in which the vertex degree is related to the local speed of propagation of probability. This approach allows thus an interpretation of the probability density of particles similar to what is usually considered in similar gravity systems in the sense that here matter sees a curved spacetime.

In the recent paper *Disordered locality and Lorentz dispersion relations: an explicit model of quantum foam* [57], Caravelli and Markopoulou, using the framework of Quantum Graphity, suggested an explicit model of a quantum foam, a quantum spacetime with spatial non-local links. The quantum states describing this spacetime foam depend on two parameters: the minimal size of the link and their density with respect to this length. In particular, Caravelli and Markopoulou considered the case in which the quantum state of the spacetime background is a superposition of many graph states. In this picture, the minimum scale of the spacetime foam is provided by the intrinsic discreteness of the graph picture. Caravelli and Markopoulou also studied the effect of the discreteness of the graph picture on the dispersion relations and showed that these states with non-local links violate macrolocality and give corrections to Lorentz invariance.

Finally, as regards the violation of locality and of Lorentz invariance characterizing the fundamental foam of physical processes, it is also interesting to remark that Mignani, Cardone and Petrucci recently suggested that the local violation of Lorentz invariance can be interpreted in terms of a energy-dependent deformation of the Minkowski geometry, whose corresponding metrics provide an effective dynamical description of the fundamental interactions (at the energy scale and in the energy range considered). In Mignani's, Cardone's and Petrucci's approach, the deformation of space-time affects the external fields that deform the geometry of space and, differently from multi-dimensional theories (whose prototype is the well known

Kaluza-Klein formalism), gauge fields do not need to be added from the outside but emerge in a natural way as a direct consequence of the metric deformation [58, 59].

In the light of the current research, in particular of Ng, of Caravelli and Markopoulou, and of Mignani, Cardone and Petrucci the suggestive perspective is opened that the fundamental space-time background of processes is a non-local, holographic quantum foam characterized by a Lorentz invariance and a deformation of the geometry. In the following chapters we will explore how to express the deformation characterizing this fundamental holographic non-local foam at the Planck scale in the context of a bohmian framework.

4. The Non-Local Features of Quantum Gravity in an Entropic Bohmian Approach

In a generalized geometric picture of Bohm's interpretation one can unify the quantum effects and gravity in the context of a non-local model of quantum gravity. In particular, F. Shojai and A. Shojai of the Teheran school recently demonstrated that the non-local quantum potential regarding the behaviour of spinless particles in a curved space-time expresses a conformal degree of freedom of the space-time metric and developed a toy model of quantum gravity in which the form of the quantum potential and its relation to the conformal degree of freedom of the space-time metric can be derived using the equations of motion. To review the most relevant results about these topics the reader can find details in the papers [60-73].

By the analysis of the quantum effects of matter in the framework of bohmian mechanics, A. Shojai's and F. Shojai's toy model shows that the motion of a spinless particle with quantum effects is equivalent to its motion in a curved space-time. The quantum effects of matter as well as the gravitational effects of matter have geometrical nature and are highly related.

The presence of the quantum force is just like having a curved space–time which is conformally flat and the conformal factor is expressed in terms of the quantum potential. All this is expressed by an equation of motion of the form

$$\tilde{g}^{\mu\nu}\tilde{\nabla}_{\mu}S\tilde{\nabla}_{\nu}S = m^2c^2 \tag{28}$$

where S is the phase of the wave function ψ, $\tilde{\nabla}_{\mu}$ is the covariant differentiation with respect to the metric

$$\tilde{g}_{\mu\nu} = \frac{M^2}{m^2} g_{\mu\nu} \tag{29}$$

which is a conformal metric) where the quantum mass is

$$M^2 = m^2 \exp Q \tag{30}$$

where

$$Q = \frac{\hbar^2}{m^2 c^2} \frac{\left(\nabla^2 - \frac{1}{c^2} \frac{\partial^2}{\partial t^2} \right)_g |\psi|}{|\psi|} \tag{31}$$

Is the quantum potential (in (31), of course, c is the light speed and $\hbar$ is Planck's reduced constant).

It is also interesting to remark that, if according to F. Shojai's and A. Shojai's bohmian model in a relativistic curved space-time the presence of the quantum potential is equivalent to a conformal mapping of the metric of space-time, in conformally related frames one measures different quantum masses and different curvatures. Considering the quantum force, the conformally related frames are not distinguishable, just like the coordinate systems when we consider gravity. Since the conformal transformation changes the length scale locally, one measures different quantum forces in different conformal frames. This is analogous to what happens in general relativity in which general coordinate transformation changes the gravitational force at any arbitrary point. Then, the following basic question becomes natural. Does applying the above correspondence, between quantum and gravitational forces, and between the conformal and general coordinate transformations, means that the geometrization of quantum effects implies conformal invariance just as gravitational effects imply general coordinate invariance?

In a similar way to what happens in general relativity, according to A. Shojai's and F. Shojai's bohmian approach to quantum gravity at any point (or even globally) the quantum effects of matter can be removed by a suitable conformal transformation. Thus in that point(s) matter behaves classically. In this way F. Shojai's and A. Shojai's approach introduces a new quantum equivalence principle, similar to the standard equivalence principle of general relativity, that can be called the conformal equivalence principle. According to this quantum equivalence principle gravitational effects can be removed by going to a freely falling frame while quantum effects can be eliminated by choosing an appropriate scale. The latter interconnects gravity and general covariance while the former has the same role about quantum and conformal covariance. Both these principles state that there is no preferred frame, either coordinate or conformal. And these aspects of the geometry of physical space regarding frames characterized by quantum and conformal covariance derive just from the quantum potential.

In the reference [74], the author demonstrated that, inside F. Shojai's and A. Shojai's model, the effects of gravity on geometry and the quantum effects on the geometry of space-time are highly coupled, and the quantum potential emerges as the conformal degree of freedom of the space-time metric, as a consequence of the fact that the ensemble of particles associated with the wave function of the physical system under consideration determines a modification of the geometry expressed by the quantum entropy

$$S_Q = -\frac{1}{2} \ln \rho \tag{32}$$

where $\rho = |\psi(\vec{x},t)|^2$ is the probability density associated with the wave function $\psi(\vec{x},t)$ of the physical system. In the light of Sbitnev's results [75], the quantum entropy (32) can be interpreted as the physical entity that, in the quantum domain, describes the degree of order and chaos of the vacuum supporting the density ρ describing the space-temporal distribution of the ensemble of particles associated with the wave function of the physical system under consideration.

In the entropic approach, the equations of motion for a spinless particle in a curved background assume the following form:

$$\tilde{g}^{\mu\nu}\frac{1}{c}\frac{\partial S_Q}{\partial t} = \tilde{g}^{\mu\nu}\left[-\left(p_\mu \tilde{\nabla}^\mu S_Q\right) + \frac{1}{2}\nabla_\mu p^\mu\right] \tag{33}$$

and

$$\tilde{g}^{\mu\nu}\tilde{\nabla}_\mu S \tilde{\nabla}_\nu S = m^2 c^2 \tag{28}$$

where

$$\tilde{g}_{\mu\nu} = \frac{M^2}{m^2} g_{\mu\nu} \tag{29}$$

is a conformal transformation that indeed is determined by the quantum entropy (32),

$$M^2 = m^2 \exp\left[-\frac{\hbar^2}{m^2c^2}\left(\nabla_\mu S_Q\right)^2 + \frac{\hbar^2}{m^2c^2}\left(\left(\nabla^2 - \frac{1}{c^2}\frac{\partial^2}{\partial t^2}\right)_g S_Q\right)\right], \tag{34}$$

and the quantum potential is

$$Q = -\frac{\hbar^2}{m^2c^2}\left(\nabla_\mu S_Q\right)^2 + \frac{\hbar^2}{m^2c^2}\left(\left(\nabla^2 - \frac{1}{c^2}\frac{\partial^2}{\partial t^2}\right)_g S_Q\right). \tag{35}$$

In the entropic approach, the presence of the non-local quantum potential is equivalent to a curved space-time with its metric being given by (29) where the mass square is determined by the quantum entropy on the basis of equation (34). In this way, the entropic approach suggests a geometrization of the quantum aspects of matter in a picture based on the idea that the density of particles associated with a given wave function determines a non-local modification of the geometry.

In the entropic approach, the real key of reading of the link between gravity and quantum behaviour of matter lies just in the quantum entropy: the effects of gravity on geometry and the quantum effects on the geometry of space-time are highly coupled because they are both determined by the non-local geometry of the background described by the quantum entropy.

The quantum entropy appears indeed as a real non-local intermediary between gravitational and quantum effects of matter.

Moreover, in the context of this entropic approach to F. Shojai's and A. Shojai's model of relativistic spinless particles in a curved space-time, the geometrical properties can be characterized by introducing a quantum length associated with the conformal metric (26) determined by the quantum entropy:

$$L_{quantum} = \frac{1}{\sqrt{(\nabla_\mu S_Q)^2 - \left(\nabla^2 - \frac{1}{c^2}\frac{\partial^2}{\partial t^2}\right)_g S_Q}} \quad (36).$$

The quantum length (36) can be used to evaluate the strength of quantum effects and, therefore, the modification of the geometry – introduced in a relativistic curved space-time regime – with respect to the Euclidean geometry characteristic of classical physics. Once the quantum length (36) becomes non-negligible the spinless particle into consideration goes into a quantum regime where the quantum and gravitational effects are highly related. In this picture, Heisenberg's uncertainty principle derives from the fact that we are unable to perform a classical measurement to distances smaller than this quantum length. A quantum regime where there is a coupling between gravitational and quantum effects is entered when the quantum length (36) must be taken under consideration.

Moreover, F. Shojai's and A. Shojai's model demonstrates that it is just the quantum gravity equations of motion which make the quantum potential the entity expressing the geometrical properties which influence the behaviour of the particles and which are related to the space-time metric. In this way, it suggests an unification of the gravitational and quantum aspects of matter in the quantum gravity domain in a non-local picture. By following F. Shojai's and A. Shojai's treatment in the papers [62, 65, 66, 76-78], a general relativistic system consisting of gravity and classical matter can be determined by the action

$$A_{no-quantum} = \frac{1}{2k}\int d^4x\sqrt{-g}R + \int d^4x\sqrt{-g}\frac{\hbar^2}{m}\left(\frac{\rho}{\hbar^2}\partial_\mu S\partial^\mu S - \frac{m^2}{\hbar^2}\rho\right) \quad (37)$$

where $\rho = J^0$ is the ensemble density of the particles, $k = 8\pi G$ and hereafter we chose the units in which c=1. Since quantum effects are equivalent to the change of the space-time metric (determined by the quantum entropy) from $g_{\mu\nu}$ to $g_{\mu\nu} \rightarrow g^I_{\mu\nu} = \frac{g_{\mu\nu}}{\exp Q}$, the action with quantum effects can be written as:

$$A[\overline{g}_{\mu\nu},\Omega,S,\rho,\lambda]=\frac{1}{2k}\int d^4x\sqrt{-\overline{g}}\left(\overline{R}\Omega^2-6\overline{\nabla}_\mu\Omega\overline{\nabla}^\mu\Omega\right)+\int d^4x\sqrt{-\overline{g}}\left(\frac{\rho}{m}\Omega^2\overline{\nabla}_\mu S\overline{\nabla}^\mu S-m\rho\Omega^4\right)+$$
$$\int d^4x\sqrt{-\overline{g}}\lambda\left(\Omega^2-\left(1+\frac{\hbar^2\left(\nabla^2-\frac{\partial^2}{\partial t^2}\right)\sqrt{\rho}}{m^2\sqrt{\rho}}\right)\right) \tag{38}$$

where

$$\Omega^2=\exp Q$$

(namely $\Omega^2=\exp\left[-\frac{\hbar^2}{m^2c^2}\left(\nabla_\mu S_Q\right)^2+\frac{\hbar^2}{m^2c^2}\left(\left(\nabla^2-\frac{1}{c^2}\frac{\partial^2}{\partial t^2}\right)_g S_Q\right)\right]$ in the entropic picture) is the conformal factor, a bar over any quantity means that it corresponds to no-quantum regime and λ is a Lagrange multiplier introduced in order to identify the conformal factor with its Bohmian value.

By the variation of the above action with respect to $\overline{g}_{\mu\nu}$, Ω, ρ, S and λ one arrives at the following relations as quantum gravity equations of motion:

$$\overline{\nabla}_\mu\left(\rho\Omega^2\overline{\nabla}^\mu S\right)=0 \tag{39}$$

$$\overline{\nabla}_\mu S\overline{\nabla}^\mu S=m^2\Omega^2 \tag{40}$$

$$G_{\mu\nu}=-kT^{(m)}_{\mu\nu}-kT^{(\Omega)}_{\mu\nu} \tag{41}$$

where $T^{(m)}_{\mu\nu}$ is the matter energy-momentum tensor and

$$kT^{(\Omega)}_{\mu\nu}=\frac{\left[g_{\mu\nu}\left(\nabla^2-\frac{\partial^2}{\partial t^2}\right)-\nabla_\mu\nabla_\nu\right]\Omega^2}{\Omega^2}+6\frac{\nabla_\mu\Omega\nabla_\nu\Omega}{\omega^2}-3g_{\mu\nu}\frac{\nabla_\alpha\Omega\nabla^\alpha\Omega}{\Omega^2} \tag{42}$$

and

$$\Omega^2=1+\alpha\frac{\overline{\left(\nabla^2-\frac{\partial^2}{\partial t^2}\right)\sqrt{\rho}}}{\sqrt{\rho}} \tag{43}$$

and $\alpha = \frac{\hbar^2}{m^2}$. It can be noted that equation (40) is a Bohmian-type equation of motion, and if we write it in terms of the physical metric $g_{\mu\nu}$, it reads as

$$\nabla_{\mu} S \nabla^{\mu} S = m^2 c^2 \tag{44}$$

The equations of motion (39), (40), (41), (42) and (43) illustrate the quantum geometrodynamics of a general relativistic system with quantum effects: as a consequence of the link between the quantum potential and the quantum entropy, they tell us that there are back-reaction effects of the quantum factor on the background which are due to the quantum entropy. On the basis of the high-coupled five equations above listed, one can say that in the quantum gravity domain the quantum entropy is the fundamental entity, the ultimate visiting card which introduces the links (and thus the back-reaction terms) between the quantum effects and the background.

Finally, by considering the most general scalar-tensor action

$$A = \int d^4x \left\{ \phi R - \frac{\omega}{\phi} \nabla^{\mu}\phi \nabla_{\mu}\phi + 2\Lambda\phi + L_m \right\} \tag{45}$$

in which ω is a constant independent of the scalar field ϕ, Λ is the cosmological constant, and L_m is the matter lagrangian (which is assumed to be in the form

$$L_m = \frac{\rho}{m} \phi^a \nabla^{\mu} S \nabla_{\mu} S - m\rho\phi^b - \Lambda(1+Q)^d \tag{46}$$

in which *a*, *b*, and *d* are constants), using a perturbative expansion for the scalar field and the matter distribution density as

$$\phi = \phi_0 + \alpha\phi_1 + \ldots$$

$$\sqrt{\rho} = \sqrt{\rho_0} + \alpha\sqrt{\rho_1} + \ldots$$

(and imposing opportune physical constraints in order to determine the parameters *a*, *b*, and *d*), F. Shojai and A. Shojai found the following quantum gravity equations which make dynamical the conformal factor and the quantum potential:

$$\phi = 1 + Q - \frac{\alpha}{2}\left(\nabla^2 - \frac{\partial^2}{\partial t^2} \right) Q \tag{47}$$

$$\nabla^{\mu}S\nabla_{\mu}S = m^2\phi - \frac{2\Lambda m}{\rho}(1+Q)(Q-\tilde{Q}) + \frac{\alpha\Lambda m}{\rho}\left[\left(\nabla^2 - \frac{\partial^2}{\partial t^2}\right)Q - 2\nabla_{\mu}Q\frac{\nabla^{\mu}\sqrt{\rho}}{\sqrt{\rho}}\right] \quad (48)$$

$$\nabla_{\mu}\left(\rho\nabla^{\mu}S\right) = 0 \quad (49)$$

$$G^{\mu\nu} - \Lambda g^{\mu\nu} = -\frac{1}{\phi}T^{\mu\nu} - \frac{1}{\phi}\left[\nabla^{\mu}\nabla^{\nu} - g^{\mu\nu}\left(\nabla^2 - \frac{\partial^2}{\partial t^2}\right)\right]\phi + \frac{\omega}{\phi^2}\nabla^{\mu}\phi\nabla^{\nu}\phi - \frac{1}{2}\frac{\omega}{\phi^2}g^{\mu\nu}\nabla^{\alpha}\phi\nabla_{\alpha}\phi \quad (50)$$

where $\tilde{Q} = \alpha\frac{\nabla_{\mu}\sqrt{\rho}\nabla^{\mu}\sqrt{\rho}}{\sqrt{\rho}}$ and $T^{\mu\nu} = -\frac{1}{\sqrt{-g}}\frac{\delta}{\delta g_{\mu\nu}}\int d^4x\sqrt{-g}L_m$ is the energy-momentum tensor.

In the entropic view suggested by the author, the geometric quantum gravity equations of F. Shojai and A. Shojai allow us to draw some important conclusions:

- In this model equation (50) shows that the causal structure of the space-time metric $g^{\mu\nu}$ is determined by the gravitational effects of matter and thus by the quantum entropy, which must be considered as the ultimate entity which shows that the quantum effects and the gravitational effects are coupled (also in the quantum gravity domain). According to equation (47), quantum effects, and thus the quantum entropy, determine directly the scale factor of space-time andthe appearance of quantum mass justifies Mach's principle which leads to the existence of an interrelation between the global properties of the universe (space–time structure, the large scale structure of the universe) and its local properties (local curvature, motion in a local frame, etc.).A local variation of matter field distribution changes the quantum potential acting on the geometry. Thus the geometry is altered globally (in conformity with Mach's principle). In this sense the bohmian approach to quantum gravity is highly non–local as a consequence of the features of the quantum potential (and thus of the quantum entropy).
- The mass field given by the right-hand side of equation (48) consists of two parts. The first part, which is proportional to α, is a purely quantum effect, while the second part, which is proportional to $\alpha\Lambda$, is a mixture of the quantum effects and the large scale structure introduced via the cosmological constant.
- In this model, the scalar field produces the quantum force which appears on the right hand and violates the equivalence principle (just like, in Kaluza-Klein theory, the scalar field – dilaton – produces a fifth force leading to the violation of the equivalence principle [79]).

5. Towards an Entropic Bohmian Approach of the Holographic, Non-Local and Granular Space-time Foam

By assuming, on the basis of what we have seen in chapter 4, that the quantum entropy (32) is the fundamental physical entity which describes, in a bohmian picture, the non-local geometry of the background, one can explore the conditions which must be satisfied by the quantum entropy in order to provide the features of discreteness of the spacetime foam, analysed in chapter 3, in the presence of matter with quantum effects described by the action (38) (or (45)). In this chapter, some considerations are made on how one could construct an entropic bohmian approach of an holographic, non-local and granular spacetime foam.

In the quantum gravity domain, the spacetime background whose geometry is described by the quantum entropy (32), in the presence of matter with quantum effects described by the action (38) (or (45)), under opportune conditions is characterized by a discreteness at the Planck scale. By substituting equations (29) and (34) into equation (17) expressing the metric which characterizes loop quantum gravity, one obtains

$$g^{I}_{\mu\nu} = \frac{l_P^2}{\mu^2}\eta_{\mu\nu} = \exp\left[-\frac{\hbar^2}{m^2c^2}\left(\nabla_\mu S_Q\right)^2 + \frac{\hbar^2}{m^2c^2}\left(\left(\nabla^2 - \frac{1}{c^2}\frac{\partial^2}{\partial t^2}\right)_g S_Q\right)\right]\eta_{\mu\nu} \tag{51}$$

namely

$$\frac{l_P^2}{\mu^2} = \exp\left[-\frac{\hbar^2}{m^2c^2}\left(\nabla_\mu S_Q\right)^2 + \frac{\hbar^2}{m^2c^2}\left(\left(\nabla^2 - \frac{1}{c^2}\frac{\partial^2}{\partial t^2}\right)_g S_Q\right)\right] \tag{52}$$

This means that, in an entropic bohmian approach, the lattice size of the discrete spacetime foam corresponding to the metric (17) of loop quantum gravity may be written as

$$\mu = \left(\frac{l_P^2}{\exp\left[-\frac{\hbar^2}{m^2c^2}\left(\nabla_\mu S_Q\right)^2 + \frac{\hbar^2}{m^2c^2}\left(\left(\nabla^2 - \frac{1}{c^2}\frac{\partial^2}{\partial t^2}\right)_g S_Q\right)\right]}\right)^{1/2} \tag{53}$$

Therefore, equation (50) showing the link between the causal structure of the space-time metric and the gravitational effects of matter (and thus the quantum entropy), in the quantum gravity domain characterized by a discrete nature of the spacetime foam, becomes

$$G^{\mu\nu} - \Lambda\frac{l_P^2}{\mu^2}\eta^{\mu\nu} = -\frac{1}{\phi}T^{\mu\nu} - \frac{1}{\phi}\left[\nabla^\mu\nabla^\nu - \frac{l_P^2}{\mu^2}\eta^{\mu\nu}\left(\nabla^2 - \frac{\partial^2}{\partial t^2}\right)\right]\phi + \frac{\omega}{\phi^2}\nabla^\mu\phi\nabla^\nu\phi - \frac{1}{2}\frac{\omega}{\phi^2}\frac{l_P^2}{\mu^2}\eta^{\mu\nu}\nabla^\alpha\phi\nabla_\alpha\phi \tag{54}$$

where the energy-momentum tensor is

$$T^{\mu\nu} = -\frac{1}{\sqrt{-g}} \frac{\delta}{\delta \frac{l_P^2}{\mu^2} \eta_{\mu\nu}} \int d^4x \sqrt{-\eta} L_m \qquad (55).$$

Moreover, substituting equation (36) into equation (18) one obtains

$$8\pi\gamma\rho l_P (x + 2M) = \frac{1}{\left(\sqrt{(\nabla_\mu S_Q)^2 - \left(\nabla^2 - \frac{1}{c^2} \frac{\partial^2}{\partial t^2} \right)_g S_Q} \right)^3} \qquad (56)$$

namely

$$\rho = \frac{1}{8\pi\gamma l_P (x + 2M) \left(\sqrt{(\nabla_\mu S_Q)^2 - \left(\nabla^2 - \frac{1}{c^2} \frac{\partial^2}{\partial t^2} \right)_g S_Q} \right)^3} \qquad (57)$$

Which provides the condition which is satisfied by the quantum entropy in order to give origin to the holographic nature of the kinematical structure of loop quantum gravity in spherical symmetry of Gambini's and Pullin's approach.

In analogous way, the quantum entropy (32) describing the geometry of the background, under opportune conditions, gives rise to Ng's holographic spacetime foam characterized by quantum fluctuations in the form of the average minimum uncertainty (21). In this regard, by substituting equation (36) into equation (21), in an entropic bohmian approach the fundamental equation of Ng's holographic spacetime foam model becomes

$$(2\pi^2/3)^{1/3} l^{1/3} l_P^{2/3} = \frac{1}{\sqrt{(\nabla_\mu S_Q)^2 - \left(\nabla^2 - \frac{1}{c^2} \frac{\partial^2}{\partial t^2} \right)_g S_Q}} \qquad (58)$$

namely

$$l^{1/3} = \frac{1}{(2\pi^2/3)^{1/3} l_P^{2/3} \sqrt{(\nabla_\mu S_Q)^2 - \left(\nabla^2 - \frac{1}{c^2} \frac{\partial^2}{\partial t^2} \right)_g S_Q}} \qquad (59).$$

Equation (58) states that the average separation between neighbouring cells of Ng's quantized spacetime foam – and thus the accuracy in the measurement of the geometry of the spacetime foam – is associated with the quantum entropy. Conversely, according to equation (59), one can say that the measure of a distance l in the spacetime foam is linked with Planck's length as well as the quantum entropy.

As a consequence, in the entropic bohmian approach of the holographic spacetime foam, the cumulative factor $C=(l/\lambda)^{1/3}$ characterizing the cumulative effects of the spacetime fluctuations over the distance l becomes

$$C=\frac{1}{\left(2\pi^2\lambda/3\right)^{1/3}l_P^{2/3}\sqrt{\left(\nabla_\mu S_Q\right)^2-\left(\nabla^2-\frac{1}{c^2}\frac{\partial^2}{\partial t^2}\right)_g S_Q}} \tag{60}$$

while the maximum energy density (22) for a sphere of radius l that does not collapse into a black hole becomes

$$\rho=\frac{3}{8\pi}\left(8\pi^6/27\right)^2 l_P^2\left(\left(\nabla_\mu S_Q\right)^2-\left(\nabla^2-\frac{1}{c^2}\frac{\partial^2}{\partial t^2}\right)_g S_Q\right)^3 \tag{61}$$

namely

$$\rho=\frac{8\pi^{11}}{3^5}l_P^2\left(\left(\nabla_\mu S_Q\right)^2-\left(\nabla^2-\frac{1}{c^2}\frac{\partial^2}{\partial t^2}\right)_g S_Q\right)^3 \tag{62}$$

where thus the Hubble radius may be written as

$$R_H=\frac{1}{\left(2\pi^2/3\right)l_P^2\left(\sqrt{\left(\nabla_\mu S_Q\right)^2-\left(\nabla^2-\frac{1}{c^2}\frac{\partial^2}{\partial t^2}\right)_g S_Q}\right)^3} \tag{63}$$

According to the entropic bohmian approach of the spacetime foam based on equations (51)-(63), one can say that the holographic, non-local features of the spacetime foam at the Planck scale is determined by the quantum entropy. The quantum entropy may be considered as the ultimate physical entity describing the geometry of the background in the sense that it determines opportune mathematical conditions under which this geometry corresponds to a non-local holographic spacetime foam at the Planck scale. While on the basis of the original equations (17)-(18) of loop quantum gravity and (21)-(23) of Ng's model of holographic spacetime foam, one cannot provide a causal origin of the non-local and granular features of the spacetime foam, the equations (51)-(63) of the entropic bohmian approach here proposed suggest the possibility to explain causally the origin of the features of the spacetime foam.

The quantum entropy, satisfying appropriate mathematical conditions, can be considered as the ultimate entity responsible of the origin of the granular, holographic and non-local features of the spacetime foam at the Planck scale.

Conclusion

Relevant theoretical results suggest that the fundamental arena of processes has a quantized nature at the Planck scale. In loop quantum gravity the operators representing area, angle, length and volume have discrete spectra and a smooth geometry cannot be approximated at a physical scale lower than the Planck length. Another significant feature of the ultimate texture of the fundamental quantum geometry at the Planck scale is its holographic nature. In this regard, important results have been recently obtained by Ng who developed a model of a non-local, holographic quantum spacetime foam whose structure is determined by the accuracy with which one can measure its geometry.

On the other hand, in an entropic bohmian approach to quantum gravity, the non-local features of quantum gravity are associated with a quantum potential which is determined by a quantum entropy indicating the modification of the geometry in the presence of quantum effects. In this entropic picture, by taking account of A. Shojai's and F. Shojai's results, one can say that the structure of the space-time metric is determined by the quantum entropy which provides a link between gravity and quantum behaviour of matter. The quantum entropy may be considered as the ultimate entity which is responsible of the coupling between the effects of gravity on geometry and the quantum effects on the geometry of space-time. In this way, the conditions under which the spacetime background described by the quantum entropy is characterized by a discreteness at the Planck scale have been explored and an entropic bohmian approach of an holographic, non-local and granular spacetime foam has been introduced. According to this entropic approach, the holographic and non-local features of the spacetime foam at the Planck scale are determined by a quantum entropy which satisfies appropriate mathematical conditions.

References

[1] Licata, "Vision of Oneness. Spacetime Geometry and Quantum Physics", in *Vision of Oneness*, edited by I. Licata and A. Sakaji, Aracne Editrice, Roma (2011), pp. 1-12.

[2] C. Rovelli and L. Smolin, "Discreteness of area and volume in quantum gravity", *Nuclear Phys*. B, 442,593-619 (1995).

[3] J. Lewandowski, "Volume and quantizations", *Classical Quantum Gravity*, 14, 71-76, (1997); e-print arXiv:gr-qc/9602035.

[4] R. Loll, "Volume operator in discretized quantum gravity", *Phys. Rev. Lett*., 75, 3048-3051 (1995); e-print arXiv:gr-qc/9506014.

[5] E. Bianchi, "The length operator in loop quantum gravity", *Nuclear Phys*. B, 807, 591-624, (2009); e-print arXiv:0806.4710.

[6] Y. Ma, C. Soo and J. Yang, "New length operator for loop quantum gravity", *Phys. Rev*. D, 81, 124026, (2010); e-print arXiv:1004.1063.

[7] T. Thiemann, "A length operator for canonical quantum gravity", *J. Math. Phys.*, 39, 3372-3392 (1998); e-print arXiv:gr-qc/9606092.
[8] S. A. Major, "Operators for quantized directions", *Classical Quantum Gravity*, 16, 3859-3877 (1999); e-print arXiv:gr-qc/9905019.
[9] G. Amelino-Camelia, J. Ellis, N. E. Mavromatos, D. V. Nanopoulos and S. Sarkar, "Potential sensitivity of gammarayburster observations to wave dispersion in vacuo", Nature, 293, 763-765 (1998); e-print arXiv:astro-ph/9712103.
[10] G. Amelino-Camelia, "*Quantum gravity phenomenology*", arXiv:0806.0339.
[11] C. Rovelli, "A new look at loop quantum gravity", arXiv:1004.1780v1 [gr-qc] (2010).
[12] Ashtekar and J. Lewandowski, "Quantum theory of geometry. I. Area operators", *Classical Quantum Gravity,* 14, A55-A81 (1997); e-print arXiv:gr-qc/9602046.
[13] Rovelli, "A generally covariant quantum field theory and a prediction on quantum measurements of geometry", *Nuclear Phys*. B, 405, 797-815 (1993).
[14] Ashtekar, A. Corichi and J. A. Zapata, "Quantum theory of geometry. III. Non-commutativity of Riemannian structures", *Classical Quantum Gravity*, 15, 2955-2972 (1998); e-print arXiv:gr-qc/9806041.
[15] L. Freidel, M. Geiller and J. Ziprick, "Continuous formulation of the loop quantum gravity phase space", arXiv:1110.4833.
[16] Ashtekar and J. Lewandowski, "Quantum theory of geometry. II. Volume operators", *Adv. Theor. Math. Phys.* 1, 388-429 (1997); e-print arXiv:gr-qc/9711031.
[17] J. Brunnemann and D. Rideout, "Properties of the volume operator in loop quantum gravity. I. Results", *Classical Quantum Gravity*, 25, 065001 (2008); e-print arXiv:0706.0469.
[18] J. Brunnemann and D. Rideout, "Properties of the volume operator in loop quantum gravity. II. Detailed presentation", *Classical Quantum Gravity*, 25, 065002 (2008); e-print arXiv:0706.0382.
[19] J. Brunnemann and T. Thiemann, "Simplification of the spectral analysis of the volume operator in loop quantum gravity", *Classical Quantum Gravity*, 23, 1289-1346 (2006); e-print arXiv:gr-qc/0405060.
[20] R. De Pietri and C. Rovelli, "Geometry eigenvalues and the scalar product from recoupling theory in loopquantum gravity", *Phys. Rev*. D, 54, 2664-2690 (1996); e-print arXiv:gr-qc/9602023.
[21] K. A. Meissner, "Eigenvalues of the volume operator in loop quantum gravity", *Classical Quantum Gravity*, 23, 617-625 (2006); e-print arXiv:gr-qc/0509049.
[22] T. Thiemann, "Closed formula for the matrix elements of the volume operator in canonical quantum gravity", *J. Math. Phys*., 39, 3347-3371 (1998); e-print arXiv:gr-qc/9606091.
[23] K. Giesel and T. Thiemann, "Consistency check on volume and triad operator quantization in loop quantum gravity. I", *Classical Quantum Gravity*, 23, 5667-5691 (2006); e-print arXiv:gr-qc/0507036.
[24] K. Giesel and T. Thiemann, "Consistency check on volume and triad operator quantization in loop quantum gravity. II", *Classical Quantum Gravity*, 23, 5693-5771 (2006); e-print arXiv:gr-qc/0507037.
[25] L. Freidel and S. Speziale, "Twisted geometries: a geometric parametrisation of SU(2) phase space", *Phys. Rev. D*, 82, 084040 (2010); eprint arXiv:1001.2748.

[26] E. Bianchi, P. Donà and S. Speziale, "Polyhedra in loop quantum gravity", *Phys. Rev.* D, 83, 044035 (2011); e-print arXiv:1009.3402.
[27] E. Bianchi and H. Haggard, "Discreteness of the volume of space from Bohr-Sommerfeld quantization", *Phys. Rev. Lett.*, 107, 011301 (2011); e-print arXiv:1102.5439.
[28] Dittrich and T. Thiemann, "Are the spectra of geometrical operators in loop quantum gravity really discrete?", *J. Math. Phys.*, 50, 012503 (2009); e-print arXiv:0708.1721.
[29] Rovelli, "*Comment on Are the spectra of geometrical operators in loop quantum gravity really discrete? by B. Dittrich and T. Thiemann*", arXiv:0708.2481 [gr-qc] (2007).
[30] F. Girelli, F. Hinterleitner and S. A. Major, "Loop quantum gravity phenomenology: linking loops to observational physics", Symmetry, Integrability and Geometry: *Methods and Applications*, 8, 098 (2012); e-print arXiv:1210.1485v2 [gr-qc].
[31] Rovelli, *Quantum Gravity*, Cambridge University Press (2004).
[32] R. Gambini and J. Pullin, "*Holography in Spherically Symmetric Loop Quantum Gravity*", arXiv:0708.0250 [gr-qc].
[33] Y. J. Ng, "Holographic Foam, Dark Energy and Infinite Statistics", *Phys. Lett.* B, 657, 10-14 (2007).
[34] Y. Jack Ng, "Spacetime foam: from entropy and holography to infinite statistics and non-locality", *Entropy*, 10, 441-461 (2008).
[35] Y. Jack Ng, "*Holographic quantum foam*", arXiv:1001.0411v1 [gr-qc] (2010).
[36] Y. Jack Ng, "*Various facets of spacetime foam*", arXiv:1102.4109.v1 [gr-qc] (2011).
[37] N. Margolus and L. B. Levitin, "The Maximum Speed of Dynamical Evolution", *Physica D*, 120, 188-195 (1998).
[38] For similar results, the reader can find details also in X. Calmet, M. Graesser and S. D. Hsu, "Minimum Length from Quantum Mechanics and Classical General Relativity", *Phys. Rev. Lett.*, 93, 211101-1 - 211101-4 (2004).
[39] G. 'tHooft, in *Salamfestschrift*; Ali, A. et al., Ed., World Scientific, Singapore (1993).
[40] L. Susskind, "The World as a Hologram", *J. Math. Phys.*, 36, 6377-6396 (1995).
[41] J. D. Bekenstein, "Black Holes and Entropy", Phys. Rev. D, 7, 2333-2346 (1973).
[42] S. Hawking, "Particle Creation by Black Holes", *Comm. Math. Phys.*, 43, 199-220 (1975).
[43] S. B. Giddings, "Black Holes and Massive Remnants", *Phys. Rev.* D, 46, 1347-1352 (1992).
[44] R. Bousso, "The Holographic Principle", *Rev. Mod. Phys.*, 74, 825-874 (2002).
[45] M. Arzano, T. W. Kephart and Y. J. Ng, "From spacetime foam to holographic foam cosmology", *Phys. Lett.* B, 649, 243-246 (2007).
[46] O. W. Greenberg, "Example of Infinite Statistics", *Phys. Rev. Lett.*, 64, 705-708 (1990).
[47] K. Fredenhagen, "On the Existence of Antiparticles", *Commun. Math. Phys.*, 79, 141-151 (1981).
[48] M. Arzano, "*Quantum Fields, Non-Locality and Quantum Group Symmetries*", arXiv:0710.1083 [hep-th] (2007).
[49] P. Balachandran et al., "*S-Matrix on the Moyal Plane: Locality versus Lorentz Invariance*", arXiv:0708.1379 [hep-th] (2007).
[50] V. Jejjala, M. Kavic and D. Minic, "*Fine Structure of Dark Energy and New Physics*", arXiv:0705.4581 [hep-th] (2007).
[51] Benatti and R. Floreanini, "Non-Standard Neutral Kaons Dynamics from D-Brane Statistics", *Annals Physics*, 273, 58-71 (1999).

[52] J. M. Medved, "A comment or two on holographic dark energy", arXiv:0802.1753 [hep-th] (2008).
[53] S. B. Giddings, "*Black Holes, Information, and Locality*", arXiv:0705.2197 [hep-th] (2007).
[54] T. Horowitz, "*Black Holes, Entropies, and Information*", arXiv:0708.3680 [astro-ph] (2007).
[55] Hamma, F. Markopoulou, S. Lloyd, F. Caravelli, S. Severini, K. Markstrom, "A quantum Bose-Hubbard model with evolving graph as toy model for emergent spacetime", *Phys. Rev. D*, 81, 104032 (2010); e-print arXiv:0911.5075.
[56] Caravelli, A. Hamma, F. Markopoulou and A. Riera, "*Trapped surfaces and emergent curved space in the Bose-Hubbard model*", arXiv:1108.2013 (2011).
[57] F. Caravelli and F. Markopoulou, "*Disordered locality and Lorentz dispersion relations*", arXiv:1201.3206v3 [gr-qc] (2012).
[58] F. Cardone and R. Mignani, *Deformed spacetime. Geometrizing interactions in four and five dimensions*, Springer (2007).
[59] R. Mignani, F. Cardone and A. Petrucci, "Metric gauge fields in deformed special relativity", *Elec. Jour. Theor. Phys.*, 10, 29, 1-21 (2013).
[60] Shojai and F. Shojai, "About some problems raised by the relativistic form of de Broglie-Bohm theory of pilot wave", *Physica Scripta*, 64, 5, 413-416 (2001).
[61] F. Shojai and A. Shojai, "*Understanding quantum theory in terms of geometry*", arXiv:gr-qc/0404102 v1 (2004).
[62] Shojai, "Quantum, gravity and geometry", *Inter. Jour. Mod. Phys.* A, 15, 12, 1757-1771 (2000); e-print arXiv:gr-qc 0010013.
[63] F. Shojai and A. Shojai, "*Pure quantum solutions of bohmian quantum gravity*", arXiv:gr-qc 0105102 (2001).
[64] F. Shojai and A. Shojai, "Quantum Einstein's equations and constraints algebra", *Pramana*, 58, 1, 13-19 (2002); e-print: arXiv:gr-qc 0109052.
[65] F. Shojai and A. Shojai, "On the relation of Weyl geometry and Bohmian quantum gravity", *Gravitation and Cosmology*, 9, 3, 163-168 (2003); e-print arXiv:gr-qc 0306099.
[66] F. Shojai and A. Shojai, "Nonminimal scalar-tensor theories and quantum gravity", *Inter. Jour. Mod. Phys.* A, 15, 13, 1859-1868 (2000); e-print arXiv: gr-qc 0010012.
[67] Shojai and M. Golshani, "Direct particle interaction as the origin of the quantal behaviours", arXiv:quant-ph 9812019 (1998).
[68] Shojai and M. Golshani, "*On the relativistic quantum force*", arXiv:quant-ph 9612023 (1996).
[69] Shojai and M. Golshani, "*Some observable results of the retarded's Bohm's theory*", arXiv:quant-ph 9612020 (1996).
[70] Shojai and M. Golshani, "*Is superluminal motion in relativistic Bohm's theory observable*?", arXiv:quant-ph 9612021 (1996).
[71] F. Shojai and A. Shojai, "Constraint algebra and the equations of motion in the Bohmian interpretation of quantum gravity", *Class. Quant. Grav.*, 21, 1-9 (2004); e-print arXiv:gr-qc 0311076.
[72] F. Shojai and A. Shojai, "Causal loop quantum gravity and cosmological solutions", *Europhys. Lett.*, 71, 6, 886-892 (2005); e-print arXiv: gr-qc 0409020.

[73] F. Shojai and A. Shojai, "*Constraint algebra in causal loop quantum gravity*", arXiv:gr-qc 0409035 (2004).

[74] D. Fiscaletti, "The quantum entropy as an ultimate visiting card of the de Broglie-Bohm theory", *Ukrainian Journal of Physics*, 57, 9, 946-963 (2012).

[75] V. I. Sbitnev, "Bohmian split of the Schrödinger equation onto two equations describing evolution of real functions", *Kvantovaya Magiya*, 5, 1, 1101-1111 (2008). URL http://quantmagic.narod.ru/volumes/VOL512008/p1101.html.

[76] F. Shojai, A. Shojai and M. Golshani, "Conformal transformations and quantum gravity", *Mod. Phys. Lett.* A, 13, 34, 2725-2729 (1998).

[77] F. Shojai, A. Shojai and M. Golshani, "Scalar-tensor theories and quantum gravity", *Mod. Phys. Lett. A*, 13, 36, 2915-2922 (1998).

[78] Shojai, F. Shojai and M. Golshani, "Nonlocal effects in quantum gravity", *Mod. Phys. Lett.* A, 13, 37, 2965-2969 (1998).

[79] Y. M. Cho and D. H. Park, "Higher-dimensional unification and fifth force", *Il Nuovo Cimento B*, 105, 8-9, 817-829 (1990).

In: Space-Time Geometry and Quantum Events
Editor: Ignazio Licata, pp. 227-315
ISBN: 978-1-63117-455-1

Chapter 10

Structural Approach to the Elementary Particle Theory

Yuri A. Rylov*
Institute for Problems in Mechanics, Russian Academy of Sciences, Moscow, Russia

Abstract

There are two approaches to atomic physics: (1) structural approach and (2) empirical approach (chemistry). The structural approach uses methods of atomic physics and, in particular, quantum mechanics. The structural approach admits one to investigate the structure and arrangement of an atom (nucleus and electronic envelope). The empirical approach uses only experimental methods, in particular, the periodic system of chemical elements. It investigates the properties of chemical elements and their chemical reactions. It can predict new chemical elements and their properties (corresponding quantum numbers), but it cannot investigate the atom arrangement (nucleus and electronic envelope).

In the contemporary theory of elementary particles one has only the empirical approach. It admits one to obtain quantum numbers of elementary particles. It admits one to systematize elementary particles, to investigate their interactions, to predict new elementary particles, but it does not admit one to investigate an arrangement of elementary particles. Structural approach to the elementary particle theory does not exist now. The paper is devoted to development of the structural approach to the elementary particle theory. Being an axiomatic conception, the quantum theory cannot be used in the construction of the structural approach. Considering the quantum motion as statistical description of stochastically moving elementary particle, one succeeded to obtain some elements of the elementary particles arrangement. In particular, it appears, that a relativistic elementary particle generates some force field (κ-field), which is responsible for the pair production. Some properties of the κ-field are investigated in this paper. Stochastic motion of elementary particle can be freely explained by properties of the discrete space-time geometry, which admits one to construct the skeleton conception of elementary particles.

Keywords: structural approach, united formalism of dynamics, multivariant geometry, skeleton conception

*E-mail address: rylov@ipmnet.ru

1. Introduction

There are two different approaches to the elementary particle theory: (1) structural approach and (2) empirical approach. At the structural approach one attempts to investigate an arrangement of elementary particles and their structure. At the empirical approach one distinguishes between the different elementary particles by some "quantum numbers" ascribed to any elementary particle. These "quantum numbers" (parameters) are: mass, electric charge, spin, magnetic moment, baryon charge, isospin and so on. One can classify elementary particles by these parameters an predict new elementary particles on the basis of this classification. However, one cannot connect these parameters with the structure of elementary particles, because in the contemporary theory the structural approach is absent.

What is the structural approach one can understand in the example of the atomic theory, where there are both structural approach and empirical approach. At the structural approach one investigates the atom arrangement: its nucleus and electronic envelope. One uses the quantum mechanics, which admits one to calculate parameters of electronic envelopes of different atoms. At the empirical approach one classifies chemical elements by their properties, generated by parameters of electronic envelopes of their atoms. At the empirical approach one does not interested in structure and arrangement of atoms. At the empirical approach one uses periodical system of chemical elements which has been obtained from experiment. Empirical approach does not permit one to investigate the atomic structure. Investigation methods of empirical approach cannot be used at investigation of the atomic structure. The structural (physical) approach is more fundamental, than the empirical (chemical) approach.

In the contemporary theory of elementary particles the structural approach is absent. As a matter of fact the contemporary elementary particle theory is a chemistry (not physics) of elementary particles. Methods of the contemporary theory of elementary particles do not admit one to investigate structure (arrangement) of elementary particles. They admits only to ascribe quantum numbers to different elementary particles and distinguish between them by these quantum numbers. The reason of such a situation is a consideration of the quantum laws as fundamental laws of nature, whereas they describe only a mean motion of quantum particles. In the same way the laws of the gas dynamics describe only the mean motion of the gas molecules. Basing on the laws of the gas dynamics, one cannot investigate structure of gas molecules . In this paper the conceptual problems of the microcosm physics will be considered. The structural approach is based on a new conception of elementary particles.

It should note that we distinguish between a conception and a theory. A conception does not coincide with a theory. For instance, the skeleton conception of elementary particles [1] distinguishes from a theory of elementary particles. A conception investigates connections between concepts of a theory. For instance, the skeleton conception of elementary particles investigates the structure of a possible theory of elementary particles. It investigates, why an elementary particle is described by its skeleton (several space-time points), which contains all information on the elementary particle. The skeleton conceptions explains, why dynamic equations are coordinateless algebraic equations and why the dynamic equations a written in terms of the world functions. However, the skeleton conception does not answer the question, which skeleton corresponds to a concrete elementary particle and what is the world function of the real space-time. In other words, the skeleton conception deals with physical principles, but not with concrete elementary particles. The conception cannot be experimentally tested. However, if the world function of the real space-time geometry has been determined, and correspondence between a concrete elementary particle and its skeleton has been established, the skeleton conception turns to the elementary particle theory. The theory of elementary particle (but not a conception) can be tested experimentally.

In other words, it is useless to speak on experimental test of the skeleton conception, because it deals only with physical principles. Discussing properties of a conception, one should discuss only properties of the concepts and logical connection between them, but not to what extent they agree with experimental data. For instance, the statement, that dynamics of deterministic particles is described in terms of Lagrangians is a statement of the particle dynamics conception. It does not state what namely Lagrangian is used for some concrete particle. One obtains a theory of particle dynamics, when it is pointed which Lagrangian describes any particle. The Ptolemaic conception of the planet motion differs from the Newtonian conception of the planet motion, although experimentally both conceptions give the same result for the first six planets of the Solar system.

To use the quantum theory in the structural approach, one needs to replace the axiomatic conception of the quantum theory by the model conception. For instance, the wave function is the main object of quantum mechanics. But what is the wave function? From where did it appear? Nobody knows. The wave function is a method of description of any nondissipative continuous medium [2]. The fact that the Schrödinger equation describes a nonrotational flow of some "quantum" fluid was known from the beginning of the quantum mechanics [3, 4]. However, in these cases one started from the Schrödinger equation and quantum principles. One failed to start from hydrodynamics and to conclude quantum description in terms of the world function. To make this, one needs to integrate hydrodynamic equations and to present them in terms of hydrodynamic potentials (Clebsh potentials [5, 6]). Thereafter one can construct the wave function from hydrodynamic potentials and obtain description in terms of the wave function. Generally speaking, the dynamic equation in terms of the wave function is nonlinear. It becomes linear only for nonrotational flow. But linearity of dynamic equations written in terms of the world function is considered as a principle of the quantum mechanics.

But from where does the continuous medium appear at the description of quantum particles? Any quantum particle is stochastic particle, and there are no dynamic equation for description of a single stochastic particle. One can describe only a mean motion of a quantum particle. One needs to consider many independent identical stochastic particles. These particles form a statistical ensemble, which can be considered as a set of stochastic independent particles. The statistical ensemble is a dynamic system of the type of a continuous medium. This statistical ensemble may be considered as a gas of noninteracting stochastic particles. This gas (continuous medium) is described by the wave function. Such a description is convenient, because the dynamic equation is linear in terms of the wave function for nonrotational flows. Such a description explains, from where the wave function appears, and what it means.

Statistical ensemble $\mathcal{E}\left[\mathcal{S}_{\text{st}}\right]$ of stochastic particles $\mathcal{S}_{\text{st}}$ is a set of many independent identical stochastic particles $\mathcal{S}_{\text{st}}$. Stochastic particle $\mathcal{S}_{\text{st}}$ is not a dynamical system, and there are no dynamic equations for $\mathcal{S}_{\text{st}}$. However, the statistical ensemble $\mathcal{E}\left[\mathcal{S}_{\text{st}}\right]$ of stochastic particles $\mathcal{S}_{\text{st}}$ is a dynamic system, and there exist dynamic equations for $\mathcal{E}\left[\mathcal{S}_{\text{st}}\right]$. The dynamic system $\mathcal{E}\left[\mathcal{S}_{\mathrm{st}}\right]$ is a dynamic system of the type of a continuous medium (fluid). Dynamic equations for $\mathcal{E}\left[\mathcal{S}_{\mathrm{st}}\right]$ describe a mean motion of the stochastic particle $\mathcal{S}_{\mathrm{st}}$. Formally the statistical ensemble $\mathcal{E}\left[\mathcal{S}_{\mathrm{st}}\right]$ can be considered as a set (not statistical ensemble) of identical deterministic particles $\mathcal{S}_{\mathrm{d}}$ interacting between themselves by means of some force field. If this force field is considered as an attribute of the stochastic particle, then, investigating properties of this force field, one may investigate a structure of the stochastic particle.

For instance, the action for the statistical ensemble of stochastic particles $\mathcal{S}_{\mathrm{st}}$ has the form

$$\mathcal{A}_{\mathcal{E}\left[\mathcal{S}_{\mathrm{st}}\right]}\left[\mathbf{x},\mathbf{u}\right]=\int\int_{V_{\xi}}\left\{\frac{m}{2}\dot{\mathbf{x}}^{2}+\frac{m}{2}\mathbf{u}^{2}-\frac{\hbar}{2}\boldsymbol{\nabla}\mathbf{u}\right\}\rho_{1}\left(\boldsymbol{\xi}\right)dtd\boldsymbol{\xi},\qquad\dot{\mathbf{x}}\equiv\frac{d\mathbf{x}}{dt}\tag{1.1}$$

The variable $\mathbf{x}=\mathbf{x}\left(t,\boldsymbol{\xi}\right)$ describes the regular component of the particle motion. The independent variables $\boldsymbol{\xi}=\{\xi_{1},\xi_{2},\xi_{3}\}$ label elements (particles) of the statistical ensemble $\mathcal{E}\left[\mathcal{S}_{\mathrm{st}}\right]$. The variable $\mathbf{u}=\mathbf{u}\left(t,\mathbf{x}\right)$ describes the mean value of the stochastic velocity component, $\hbar$ is the quantum constant, $\rho_{1}\left(\boldsymbol{\xi}\right)$ is a weight function. One may set $\rho_{1}=1$. The second term in (1.1) describes the kinetic energy of the stochastic velocity component. The third term describes interaction between the stochastic component $\mathbf{u}\left(t,\mathbf{x}\right)$ and the regular component $\dot{\mathbf{x}}\left(t,\boldsymbol{\xi}\right)$. The operator

$$\boldsymbol{\nabla}=\left\{\frac{\partial}{\partial x^{1}},\frac{\partial}{\partial x^{2}},\frac{\partial}{\partial x^{3}}\right\}\tag{1.2}$$

is defined in the space of coordinates $\mathbf{x}$.

Formally the action (1.1) describes a set of deterministic particles $\mathcal{S}_{\mathrm{d}}$, interacting via the force field $\mathbf{u}$. The particles $\mathcal{S}_{\mathrm{d}}$ form a gas (or a fluid), described by the variables $\dot{\mathbf{x}}\left(t,\boldsymbol{\xi}\right)=\mathbf{v}\left(t,\boldsymbol{\xi}\right)$. Here this description is produced in the Lagrange representation. Hydrodynamic description is produced in terms of density ρ and velocity $\mathbf{v}$, where

$$\rho=\rho_{1}J,\qquad J\equiv\frac{\partial\left(\xi_{1},\xi_{2},\xi_{3}\right)}{\partial\left(x^{1},x^{2},x^{3}\right)}\tag{1.3}$$

Nonrotational flow of this gas is described by the Schrödinger equation [7].

The dynamic equation for the force field $\mathbf{u}$ is obtained as a result of variation of (1.1) with respect to $\mathbf{u}$. It has the form

$$\mathbf{u}=\mathbf{u}\left(t,\mathbf{x}\right)=-\frac{\hbar}{2m}\boldsymbol{\nabla}\ln\rho\tag{1.4}$$

The vector $\mathbf{u}$ describes the mean value of the stochastic velocity component of the stochastic particle $\mathcal{S}_{\mathrm{st}}$. In the nonrelativistic case the force field $\mathbf{u}$ is determined by its source: the fluid density ρ.

In terms of the wave function the action (1.1) takes the form [7]

$$\mathcal{A}[\psi,\psi^{*}]=\int\left\{\frac{i\hbar}{2}(\psi^{*}\partial_{0}\psi-\partial_{0}\psi^{*}\cdot\psi)-\frac{\hbar^{2}}{2m}\boldsymbol{\nabla}\psi^{*}\cdot\boldsymbol{\nabla}\psi+\frac{\hbar^{2}}{8m}\rho\boldsymbol{\nabla}s_{\alpha}\boldsymbol{\nabla}s_{\alpha}\right\}\mathrm{d}^{4}x\tag{1.5}$$

where the wave function $\psi = \left\{\begin{smallmatrix}\psi_1\\ \psi_2\end{smallmatrix}\right\}$ has two complex components.

$$\rho = \psi^*\psi, \qquad s_\alpha = \frac{\psi^*\sigma_\alpha\psi}{\rho}, \qquad \alpha = 1, 2, 3 \tag{1.6}$$

σ_α are 2×2 Pauli matrices

$$\sigma_1 = \begin{pmatrix} 0 & 1 \\ 1 & 0 \end{pmatrix}, \qquad \sigma_2 = \begin{pmatrix} 0 & -i \\ i & 0 \end{pmatrix}, \qquad \sigma_3 = \begin{pmatrix} 1 & 0 \\ 0 & -1 \end{pmatrix}, \tag{1.7}$$

Dynamic equation, generated by the action (1.5), has the form

$$i\hbar\partial_0\psi + \frac{\hbar^2}{2m}\nabla^2\psi + \frac{\hbar^2}{8m}\nabla^2 s_\alpha \cdot (s_\alpha - 2\sigma_\alpha)\,\psi - \frac{\hbar^2}{4m}\frac{\nabla\rho}{\rho}\nabla s_\alpha\sigma_\alpha\psi = 0 \tag{1.8}$$

In the case of one-component wave function ψ, when the flow is nonrotational and $\nabla s_\alpha = 0$, the dynamic equation has the form of the Schrödinger equation

$$i\hbar\partial_0\psi + \frac{\hbar^2}{2m}\nabla^2\psi = 0 \tag{1.9}$$

Thus, the Schrödinger equation is a special case of the dynamic equation, generated by the action (1.1) or (1.5).

There are several interpretations of the Schrödinger particle $\mathcal{S}_{\rm st}$: (1) quantum interpretation, (2) hydrodynamic interpretation, (3) dynamic interpretation.

At the conventional quantum interpretation the Schrödinger particle $\mathcal{S}_{\rm st}$ is a quantum particle, whose dynamics is described by the axiomatic Schrödinger equation (1.9). Any questions of the type: why the particle $\mathcal{S}_{\rm st}$ is quantum and what its parameters are responsible for its quantum behavior, are improper because of axiomatic character of description.

The hydrodynamic interpretation and dynamic interpretation are rather close. According to hydrodynamic interpretation the action (1.1) describes a set of deterministic particles interacting between themselves via the force field $\mathbf{u}$. One cannot consider a single particle, because in this case interaction between particles disappears. The hydrodynamic description does not admit a consideration of a single particle. When the action (1.1) describes a statistical ensemble (not a set of interacting identical deterministic particles), the dynamic interpretation admits one to consider experiments with a single stochastic particle. Experiments with the statistical ensemble can be realized as a set of experiments with identically prepared stochastic particles. Motion of single particles will be different, in general, in different experiments. However, result of the statistical handling of all experiments does not depend on the way of the experiments realization. Experiments with a single particle may be produced simultaneously at the same place or at different places in different time. The statistical averaging gives the same result in all these cases. For instance, the two-slit experiment can be produced with many electrons simultaneously, or with a single electron many times. Result of the statistical averaging will be the same in all cases. It shows, that the action (1.1) describes a statistical ensemble of stochastic particles, but not a gas of interacting deterministic particles.

Thus, the dynamic interpretations, when the action (1.1) describes the statistical ensemble of stochastic particles is the most true interpretation, which does not close the door for investigation of the stochastic behavior of quantum particles. The reason of such a stochastic behavior may be an interaction of a particle with the medium (vacuum, or ether), where the particle moves. The influence of the medium remains the same in all single experiments.

2. Relativistic Stochastic Particle

The pair production phenomenon takes place only for relativistic quantum particles. It is absent for classical particles. What is the reason of the pair production? Can it be described dynamically? A dynamic description is impossible in the framework of conventional axiomatic quantum theory. However, it is possible at the description of $\mathcal{S}_{\text{st}}$ in terms of the statistical ensemble. In the case of relativistic stochastic particle $\mathcal{S}_{\text{st}}$ the force field has its own degrees of freedom. It can escape from the source and travel in the space-time. In the relativistic case one obtains the action

$$\mathcal{A}[x,\kappa] = \int \left\{-mcK\sqrt{g_{ik}\dot{x}^i\dot{x}^k} - \frac{e}{c}A_k\dot{x}^k\right\} d^4\xi, \quad d^4\xi = d\xi_0 d\boldsymbol{\xi} \tag{2.1}$$

$$K = \sqrt{1+\lambda^2\left(\kappa_l\kappa^l + \partial_l\kappa^l\right)}, \qquad \lambda = \frac{\hbar}{mc}, \quad \tau = \boldsymbol{\xi}_0 \tag{2.2}$$

Here $x = \left\{x^i\left(\xi_0, \boldsymbol{\xi}\right)\right\}$, $i = 0,1,2,3$ are dependent variables, describing regular component of the particle motion. The variables $\xi = \{\xi_0, \boldsymbol{\xi}\} = \{\xi_k\}$, $k = 0,1,2,3$ are independent variables, labelling the particles of the statistical ensemble, and $\dot{x}^i \equiv dx^i/d\xi_0$. The quantities $\kappa^l = \left\{\kappa^l(x)\right\}$, $l = 0,1,2,3$ are dependent variables, describing stochastic component of the particle motion, $A_k = \{A_k(x)\}$, $k = 0,1,2,3$ is the potential of electromagnetic field. We shall refer to the dynamic system, described by the action (2.1), (2.2) as $\mathcal{S}_{\text{KG}}$, because irrotational flow of $\mathcal{S}_{\text{KG}}$ is described by the Klein-Gordon equation [8]. We present here this transformation to the Klein-Gordon form. Here and farther a summation is produced over repeated Latin indices $(0 \div 3)$ and over Greek indices $(1 \div 3)$. We present here transformation of (2.1), (2.2) to the Klein-Gordon form.

Dynamic equations generated by the action (2.1), (2.2) are equations of the hydrodynamical type. To present these equations in terms of the wave function, one needs to integrate them in general form. The problem of general integration of four hydrodynamic Euler equations

$$\partial_0\rho + \boldsymbol{\nabla}(\rho\mathbf{v}) = 0 \tag{2.3}$$

$$\partial_0\mathbf{v} + (\mathbf{v}\boldsymbol{\nabla})\mathbf{v} = -\frac{1}{\rho}\boldsymbol{\nabla}p, \qquad p = p(\rho, \boldsymbol{\nabla}\rho) \tag{2.4}$$

seems to be hopeless. It is really so, if the Euler system (2.3), (2.4) is considered to be a complete system of dynamic equations. In fact, the Euler equations (2.3), (2.4) do not form a complete system of dynamic equations, because it does not describe motion of fluid particles along their trajectories. To obtain the complete system of dynamic equations, we should add to the Euler system so called Lin constraints [9]

$$\partial_0\boldsymbol{\xi} + (\mathbf{v}\boldsymbol{\nabla})\boldsymbol{\xi} = 0 \tag{2.5}$$

where $\boldsymbol{\xi} = \boldsymbol{\xi}(t, \mathbf{x}) = \{\xi_1, \xi_2, \xi_3\}$ are three independent integrals of dynamic equations

$$\frac{d\mathbf{x}}{dt} = \mathbf{v}(t, \mathbf{x}),$$

describing motion of fluid particles in the given velocity field.

Seven equations (2.3) – (2.5) form the complete system of dynamic equations, whereas four Euler equations (2.3), (2.4) form only a closed subsystem of the complete system of dynamic equations. The wave function is expressed via hydrodynamic potentials $\boldsymbol{\xi} = \{\xi_1, \xi_2, \xi_3\}$, which are known also as Clebsch potentials [5, 6]. In general case of arbitrary fluid flow in three-dimensional space the complex wave function ψ has two complex components ψ_1, ψ_2 (or four independent real components)

$$\psi = \begin{pmatrix} \psi_1 \\ \psi_2 \end{pmatrix} = \begin{pmatrix} \sqrt{\rho} e^{i\varphi} u_1(\boldsymbol{\xi}) \\ \sqrt{\rho} e^{i\varphi} u_2(\boldsymbol{\xi}) \end{pmatrix}, \qquad |u_1|^2 + |u_2|^2 = 1 \tag{2.6}$$

It is impossible to obtain general solution of the Euler system (2.3), (2.4), but one can partially integrate the complete system (2.3) – (2.5), reducing its order to four dynamic equations for the wave function (2.6). Practically it means that one integrates dynamic equations (2.5), where the function $\mathbf{v}(t, \mathbf{x})$ is determined implicitly by equations (2.3), (2.4). Such an integration and reduction of the order of the complete system of dynamic equations appear to be possible, because the system (2.3) – (2.5) has the symmetry group, connected with transformations of the Clebsch potentials

$$\xi_\alpha \to \tilde{\xi}_\alpha = \tilde{\xi}_\alpha(\boldsymbol{\xi}), \qquad \alpha = 1, 2, 3, \qquad \frac{\partial\left(\tilde{\xi}_1, \tilde{\xi}_2, \tilde{\xi}_3\right)}{\partial(\xi_1, \xi_2, \xi_3)} \neq 0 \tag{2.7}$$

3. Transformation of the Action to Description in Terms of the Wave Function

Let us consider variables $\xi = \xi(x)$ in (2.1) as dependent variables and variables x as independent variables. Let the Jacobian

$$J = \frac{\partial(\xi_0, \xi_1, \xi_2, \xi_3)}{\partial(x^0, x^1, x^2, x^3)} = \det \left|\left|\xi_{i,k}\right|\right|, \qquad \xi_{i,k} \equiv \partial_k \xi_i, \qquad i, k = 0, 1, 2, 3 \tag{3.1}$$

be considered to be a multilinear function of $\xi_{i,k}$. Then

$$d^4\xi = J d^4 x, \qquad \dot{x}^i \equiv \frac{dx^i}{d\xi_0} \equiv \frac{\partial(x^i, \xi_1, \xi_2, \xi_3)}{\partial(\xi_0, \xi_1, \xi_2, \xi_3)} = J^{-1} \frac{\partial J}{\partial \xi_{0,i}} \tag{3.2}$$

After transformation to dependent variables ξ the action (2.1) takes the form

$$\mathcal{A}[\xi, \kappa] = \int \left\{ -mcK \sqrt{g_{ik} \frac{\partial J}{\partial \xi_{0,i}} \frac{\partial J}{\partial \xi_{0,k}}} - \frac{e}{c} A_k \frac{\partial J}{\partial \xi_{0,k}} \right\} d^4 x, \tag{3.3}$$

$$K = \sqrt{1 + \lambda^2 (\kappa_l \kappa^l + \partial_l \kappa^l)}, \qquad \lambda = \frac{\hbar}{mc}, \tag{3.4}$$

Let us introduce new variables

$$j^{k}=\frac{\partial J}{\partial \xi_{0,k}},\qquad k=0,1,2,3 \tag{3.5}$$

by means of Lagrange multipliers p_k

$$\mathcal{A}\left[\xi,\kappa,j,p\right]=\int\left\{-mcK\sqrt{g_{ik}j^{i}j^{k}}-\frac{e}{c}A_{k}j^{k}+p_{k}\left(\frac{\partial J}{\partial \xi_{0,k}}-j^{k}\right)\right\}d^{4}x, \tag{3.6}$$

Variation with respect to ξ_i gives

$$\frac{\delta \mathcal{A}}{\delta \xi_{i}}=-\partial_{l}\left(p_{k}\frac{\partial^{2}J}{\partial \xi_{0,k}\partial \xi_{i,l}}\right)=0,\qquad i=0,1,2,3 \tag{3.7}$$

Using identities

$$\frac{\partial^{2}J}{\partial \xi_{0,k}\partial \xi_{i,l}}\equiv J^{-1}\left(\frac{\partial J}{\partial \xi_{0,k}}\frac{\partial J}{\partial \xi_{i,l}}-\frac{\partial J}{\partial \xi_{0,l}}\frac{\partial J}{\partial \xi_{i,k}}\right) \tag{3.8}$$

$$\frac{\partial J}{\partial \xi_{i,l}}\xi_{k,l}\equiv J\delta_{k}^{i},\qquad \partial_{l}\frac{\partial^{2}J}{\partial \xi_{0,k}\partial \xi_{i,l}}\equiv 0 \tag{3.9}$$

one can test by direct substitution that the general solution of linear equations (3.7) has the form

$$p_{k}=b_{0}\left(\partial_{k}\varphi+g^{\alpha}\left(\boldsymbol{\xi}\right)\partial_{k}\xi_{\alpha}\right),\qquad k=0,1,2,3 \tag{3.10}$$

where $b_0\neq 0$ is a constant, $g^{\alpha}\left(\boldsymbol{\xi}\right),\ \ \alpha=1,2,3$ are arbitrary functions of $\boldsymbol{\xi}=\{\xi_1,\xi_2,\xi_3\}$, and φ is the dynamic variable ξ_0, which stops to be fictitious. Let us substitute (3.10) in (3.6). The term of the form $\partial J/\partial\xi_{0,k}\partial_k\varphi$ is reduced to Jacobian and does not contribute to dynamic equation. The terms of the form $\xi_{\alpha,k}\partial J/\partial\xi_{0,k}$ vanish due to identities (3.9). We obtain

$$\mathcal{A}\left[\varphi,\boldsymbol{\xi},\kappa,j\right]=\int\left\{-mcK\sqrt{g_{ik}j^{i}j^{k}}-j^{k}\pi_{k}\right\}d^{4}x, \tag{3.11}$$

where quantities π_k are determined by the relations

$$\pi_{k}=b_{0}\left(\partial_{k}\varphi+g^{\alpha}\left(\boldsymbol{\xi}\right)\partial_{k}\xi_{\alpha}\right)+\frac{e}{c}A_{k},\qquad k=0,1,2,3 \tag{3.12}$$

Integration of (3.7) in the form (3.10) is that integration which admits one to introduce a wave function. Note that coefficients in the system of equations (3.7) at derivatives of p_k are constructed of minors of the Jacobian (3.1). It is the circumstance that admits to produce a formal general integration.

Variation of (3.11) with respect to κ^l gives

$$\frac{\delta \mathcal{A}}{\delta \kappa^{l}}=-\frac{\lambda^{2}mc\sqrt{g_{ik}j^{i}j^{k}}}{K}\kappa_{l}+\partial_{l}\frac{\lambda^{2}mc\sqrt{g_{ik}j^{i}j^{k}}}{2K}=0,\qquad \lambda=\frac{\hbar}{mc} \tag{3.13}$$

It can be written in the form

$$\kappa_{l}=\partial_{l}\kappa=\frac{1}{2}\partial_{l}\ln\rho,\qquad e^{2\kappa}=\frac{\rho}{\rho_{0}}\equiv\frac{\sqrt{j_{s}j^{s}}}{\rho_{0}K}, \tag{3.14}$$

where ρ_0 =const is the integration constant. Substituting (3.4) in (3.14), we obtain dynamic equation for κ

$$\hbar^2\left(\partial_l\kappa\cdot\partial^l\kappa+\partial_l\partial^l\kappa\right)=m^2c^2\frac{e^{-4\kappa}j_sj^s}{\rho_0^2}-m^2c^2 \tag{3.15}$$

Variation of (3.11) with respect to j^k gives

$$\pi_k=-\frac{mcKj_k}{\sqrt{g_{ls}j^lj^s}} \tag{3.16}$$

or

$$\pi_kg^{kl}\pi_l=m^2c^2K^2 \tag{3.17}$$

Substituting $\sqrt{j_sj^s}/K$ from the second equation (3.14) in (3.16), we obtain

$$j_k=-\frac{\rho_0}{mc}e^{2\kappa}\pi_k, \tag{3.18}$$

Now we eliminate the variables j^k from the action (3.11), using relation (3.18) and (3.14). We obtain

$$\mathcal{A}\left[\varphi,\boldsymbol{\xi},\kappa\right]=\int\rho_0e^{2\kappa}\left\{-m^2c^2K^2+\pi^k\pi_k\right\}d^4x, \tag{3.19}$$

where π_k is determined by the relation (3.12). Using expression (2.2) for K, the first term of the action (3.19) can be transformed as follows.

$$\begin{aligned}-m^2c^2e^{2\kappa}K^2 &= -m^2c^2e^{2\kappa}\left(1+\lambda^2\left(\partial_l\kappa\partial^l\kappa+\partial_l\partial^l\kappa\right)\right)\\ &= -m^2c^2e^{2\kappa}+\hbar^2e^{2\kappa}\partial_l\kappa\partial^l\kappa-\frac{\hbar^2}{2}\partial_l\partial^le^{2\kappa}\end{aligned}$$

Let us take into account that the last term has the form of divergence. It does not contribute to dynamic equations and can be omitted. Omitting this term, we obtain

$$\mathcal{A}\left[\varphi,\boldsymbol{\xi},\kappa\right]=\int\rho_0e^{2\kappa}\left\{-m^2c^2+\hbar^2\partial_l\kappa\partial^l\kappa+\pi^k\pi_k\right\}d^4x, \tag{3.20}$$

Here π_k is defined by the relation (3.12), where the integration constant b_0 is chosen in the form $b_0=\hbar$

$$\pi_k=\hbar\left(\partial_k\varphi+g^\alpha\left(\boldsymbol{\xi}\right)\partial_k\xi_\alpha\right)+\frac{e}{c}A_k,\qquad k=0,1,2,3 \tag{3.21}$$

Instead of dynamic variables φ, $\boldsymbol{\xi}$, κ we introduce n-component complex function

$$\psi=\{\psi_\alpha\}=\left\{\sqrt{\rho}e^{i\varphi}u_\alpha\left(\boldsymbol{\xi}\right)\right\}=\left\{\sqrt{\rho_0}e^{\kappa+i\varphi}u_\alpha\left(\boldsymbol{\xi}\right)\right\},\qquad \alpha=1,2,...n \tag{3.22}$$

Here u_α are functions of only $\boldsymbol{\xi}=\{\xi_1,\xi_2,\xi_3\}$, having the following properties

$$\sum_{\alpha=1}^{\alpha=n}u_\alpha^*u_\alpha=1,\qquad -\frac{i}{2}\sum_{\alpha=1}^{\alpha=n}\left(u_\alpha^*\frac{\partial u_\alpha}{\partial\xi_\beta}-\frac{\partial u_\alpha^*}{\partial\xi_\beta}u_\alpha\right)=g^\beta\left(\boldsymbol{\xi}\right) \tag{3.23}$$

where (*) denotes the complex conjugation. The number n of components of the wave function ψ depends on the functions $g^{\beta}(\boldsymbol{\xi})$. The number n is chosen in such a way, that equations (3.23) have a solution. Then we obtain

$$\psi^*\psi \equiv \sum_{\alpha=1}^{\alpha=n} \psi_\alpha^*\psi_\alpha = \rho = \rho_0 e^{2\kappa}, \qquad \partial_l\kappa = \frac{\partial_l(\psi^*\psi)}{2\psi^*\psi} \tag{3.24}$$

$$\pi_k = -\frac{i\hbar(\psi^*\partial_k\psi - \partial_k\psi^*\cdot\psi)}{2\psi^*\psi} + \frac{e}{c}A_k, \qquad k = 0,1,2,3 \tag{3.25}$$

Substituting relations (3.24), (3.25) in (3.20), we obtain the action, written in terms of the wave function ψ

$$\mathcal{A}[\psi,\psi^*] = \int\left\{\left[\frac{i\hbar(\psi^*\partial_k\psi - \partial_k\psi^*\cdot\psi)}{2\psi^*\psi} - \frac{e}{c}A_k\right]\left[\frac{i\hbar(\psi^*\partial^k\psi - \partial^k\psi^*\cdot\psi)}{2\psi^*\psi} - \frac{e}{c}A^k\right]\right.$$
$$\left. + \hbar^2\frac{\partial_l(\psi^*\psi)\,\partial^l(\psi^*\psi)}{4(\psi^*\psi)^2} - m^2c^2\right\}\psi^*\psi d^4x \tag{3.26}$$

Let us use the identity

$$\frac{(\psi^*\partial_l\psi - \partial_l\psi^*\cdot\psi)(\psi^*\partial^l\psi - \partial^l\psi^*\cdot\psi)}{4\psi^*\psi} + \partial_l\psi^*\partial^l\psi$$
$$\equiv \frac{\partial_l(\psi^*\psi)\,\partial^l(\psi^*\psi)}{4\psi^*\psi} + \frac{g^{ls}}{2}\psi^*\psi\sum_{\alpha,\beta=1}^{\alpha,\beta=n} Q_{\alpha\beta,l}^* Q_{\alpha\beta,s} \tag{3.27}$$

where

$$Q_{\alpha\beta,l} = \frac{1}{\psi^*\psi}\begin{vmatrix}\psi_\alpha & \psi_\beta \\ \partial_l\psi_\alpha & \partial_l\psi_\beta\end{vmatrix}, \qquad Q_{\alpha\beta,l}^* = \frac{1}{\psi^*\psi}\begin{vmatrix}\psi_\alpha^* & \psi_\beta^* \\ \partial_l\psi_\alpha^* & \partial_l\psi_\beta^*\end{vmatrix} \tag{3.28}$$

Then we obtain

$$\mathcal{A}[\psi,\psi^*] = \int\left\{\left(i\hbar\partial_k + \frac{e}{c}A_k\right)\psi^*\left(-i\hbar\partial^k + \frac{e}{c}A^k\right)\psi - m^2c^2\psi^*\psi\right.$$
$$\left. + \frac{\hbar^2}{2}\sum_{\alpha,\beta=1}^{\alpha,\beta=n} g^{ls}Q_{\alpha\beta,l}Q_{\alpha\beta,s}^*\psi^*\psi\right\}d^4x \tag{3.29}$$

Let us consider the case of irrotational flow, when $g^{\alpha}(\boldsymbol{\xi}) = 0$ and the function ψ has only one component. It follows from (3.28), that $Q_{\alpha\beta,l} = 0$. Then we obtain instead of (3.29)

$$\mathcal{A}[\psi,\psi^*] = \int\left\{\left(i\hbar\partial_k + \frac{e}{c}A_k\right)\psi^*\left(-i\hbar\partial^k + \frac{e}{c}A^k\right)\psi - m^2c^2\psi^*\psi\right\}d^4x \tag{3.30}$$

Variation of the action (3.30) with respect to ψ^* generates the Klein-Gordon equation

$$\left(-i\hbar\partial_k + \frac{e}{c}A_k\right)\left(-i\hbar\partial^k + \frac{e}{c}A^k\right)\psi - m^2c^2\psi = 0 \tag{3.31}$$

Thus, description in terms of the Klein-Gordon equation is a special case of the stochastic particles description by means of the action (2.1), (2.2).

In the case, when the fluid flow is rotational, and the wave function ψ is two-component, the identity (3.27) takes the form

$$\frac{\left(\psi^*\partial_l\psi-\partial_l\psi^*\cdot\psi\right)\left(\psi^*\partial^l\psi-\partial^l\psi^*\cdot\psi\right)}{4\rho}-\frac{\left(\partial_l\rho\right)\left(\partial^l\rho\right)}{4\rho}$$
$$\equiv -\partial_l\psi^*\partial^l\psi+\frac{1}{4}\left(\partial_l s_\alpha\right)\left(\partial^l s_\alpha\right)\rho \qquad (3.32)$$

where 3-vector $\mathbf{s}=\{s_1,s_2,s_3,\}$ is defined by the relation

$$\rho=\psi^*\psi,\qquad s_\alpha=\frac{\psi^*\sigma_\alpha\psi}{\rho},\qquad \alpha=1,2,3 \qquad (3.33)$$

$$\psi=\binom{\psi_1}{\psi_2},\qquad \psi^*=\left(\psi_1^*,\psi_2^*\right), \qquad (3.34)$$

and Pauli matrices $\boldsymbol{\sigma}=\{\sigma_1,\sigma_2,\sigma_3\}$ have the form

$$\sigma_1=\left(\begin{array}{cc}0&1\\1&0\end{array}\right),\qquad \sigma_2=\left(\begin{array}{cc}0&-i\\i&0\end{array}\right),\qquad \sigma_3=\left(\begin{array}{cc}1&0\\0&-1\end{array}\right) \qquad (3.35)$$

Note that 3-vectors $\mathbf{s}$ and $\boldsymbol{\sigma}$ are vectors in the space V_ξ of the Clebsch potentials $\boldsymbol{\xi}=\{\xi_1,\xi_2,\xi_3\}$. They transform as vectors at the transformations (2.7)

In general, transformations of Clebsch potentials $\boldsymbol{\xi}$ and those of coordinates $\mathbf{x}$ are independent. However, the action (3.26) does not contain any reference to the Clebsch potentials $\boldsymbol{\xi}$ and transformations (2.7) of $\boldsymbol{\xi}$. If we consider only linear transformations of space coordinates $\mathbf{x}$

$$x^\alpha\rightarrow\tilde{x}^\alpha=b^\alpha+\omega^\alpha_{.\beta}x^\beta,\qquad \alpha=1,2,3 \qquad (3.36)$$

nothing prevents from accompanying any transformation (3.36) with the similar transformation

$$\xi_\alpha\rightarrow\tilde{\xi}_\alpha=b^\alpha+\omega^\alpha_{.\beta}\xi_\beta,\qquad \alpha=1,2,3 \qquad (3.37)$$

of Clebsch potentials $\boldsymbol{\xi}$. The formulas for linear transformation of vectors and spinors in V_x do not contain the coordinates $\mathbf{x}$ explicitly, and one can consider vectors and spinors in V_ξ as vectors and spinors in V_x, provided we consider linear transformations (3.36), (3.37) always together.

Using identity (3.32), we obtain from (3.26)

$$\mathcal{A}\left[\psi,\psi^*\right]=\int\left\{\left(i\hbar\partial_k+\frac{e}{c}A_k\right)\psi^*\left(-i\hbar\partial^k+\frac{e}{c}A^k\right)\psi-m^2c^2\rho\right.$$
$$\left.-\frac{\hbar^2}{4}\left(\partial_l s_\alpha\right)\left(\partial^l s_\alpha\right)\rho\right\}d^4x \qquad (3.38)$$

Dynamic equation, generated by the action (3.38), has the form

$$\left(-i\hbar\partial_k+\frac{e}{c}A_k\right)\left(-i\hbar\partial^k+\frac{e}{c}A^k\right)\psi-\left(m^2c^2+\frac{\hbar^2}{4}\left(\partial_l s_\alpha\right)\left(\partial^l s_\alpha\right)\right)\psi$$
$$=-\hbar^2\frac{\partial_l\left(\rho\partial^l s_\alpha\right)}{2\rho}\left(\sigma_\alpha-s_\alpha\right)\psi \qquad (3.39)$$

The gradient of the unit 3-vector $\mathbf{s} = \{s_1, s_2, s_3\}$ describes rotational component of the fluid flow. If $\mathbf{s} = \text{const}$, the dynamic equation (3.39) turns to the conventional Klein-Gordon equation (3.31). Curl of the vector field π_k, determined by the relation (3.25), is expressed only via derivatives of the unit 3-vector $\mathbf{s}$.

To show this, let us represent the wave function (3.22) in the form

$$\psi = \sqrt{\rho}e^{i\varphi}\left(\mathbf{n}\boldsymbol{\sigma}\right)\chi, \qquad \psi^* = \sqrt{\rho}e^{-i\varphi}\chi^*\left(\boldsymbol{\sigma}\mathbf{n}\right), \qquad \mathbf{n}^2 = 1, \qquad \chi^*\chi = 1 \tag{3.40}$$

where $\mathbf{n} = \{n_1, n_2, n_3\}$ is some unit 3-vector, $\chi = \binom{\chi_1}{\chi_2}$, $\chi^* = (\chi_1^*, \chi_2^*)$ are constant two-component quantities, and $\boldsymbol{\sigma} = \{\sigma_1, \sigma_2, \sigma_3\}$ are Pauli matrices (3.35). The unit vector $\mathbf{s}$ and the unit vector $\mathbf{n}$ are connected by means of the relations

$$\mathbf{s} = 2\mathbf{n}\left(\mathbf{n}\mathbf{z}\right) - \mathbf{z}, \qquad \mathbf{n} = \frac{\mathbf{s} + \mathbf{z}}{\sqrt{2\left(1 + (\mathbf{s}\mathbf{z})\right)}} \tag{3.41}$$

where $\mathbf{z}$ is a constant unit vector defined by the relation

$$\mathbf{z} = \chi^*\boldsymbol{\sigma}\chi, \qquad \mathbf{z}^2 = \chi^*\chi = 1 \tag{3.42}$$

All 3-vectors $\mathbf{n}$, $\mathbf{s}$, $\mathbf{z}$ are vectors in V_ξ. Let us substitute the relation (3.40) into expression $\partial_l\pi_k - \partial_k\pi_l$ for the curl of the vector field π_k defined by the relation (3.25). Then gradually reducing powers of σ by means of the identity

$$\sigma_\alpha\sigma_\beta \equiv \delta_{\alpha\beta} + i\varepsilon_{\alpha\beta\gamma}\sigma_\gamma, \qquad \alpha, \beta = 1, 2, 3 \tag{3.43}$$

where $\varepsilon_{\alpha\beta\gamma}$ is the Levi-Chivita pseudotensor ($\varepsilon_{123} = 1$), we obtain after calculations

$$\begin{aligned}\pi_k &= -\frac{i\hbar\left(\psi^*\partial_k\psi - \partial_k\psi^*\cdot\psi\right)}{2\psi^*\psi} + \frac{e}{c}A_k \\ &= \hbar\left(\partial_k\varphi + \varepsilon_{\alpha\beta\gamma}n_\alpha\partial_k n_\beta z_\gamma\right) + \frac{e}{c}A_k \qquad k = 0, 1, 2, 3\end{aligned} \tag{3.44}$$

$$\partial_k\pi_l - \partial_l\pi_k = -4\hbar\left[\partial_k\mathbf{n}\times\partial_l\mathbf{n}\right]\mathbf{z} + \frac{e}{c}\left(\partial_k A_l - \partial_l A_k\right), \qquad k, l = 0, 1, 2, 3 \tag{3.45}$$

The relation (3.45) may be expressed also via the 3-vector $\mathbf{s}$, provided we use the formulae (3.41).

Note that the two-component form of the wave function can describe irrotational flow. For instance, if $\psi = \binom{\psi_1}{\psi_1}$, $s_1 = 1$, $s_2 = s_3 = 0$, the dynamic equation (3.39) reduces to the form (3.31), and curl of π_k, defined by (3.45) reduces to

$$\partial_k\pi_l - \partial_l\pi_k = \frac{e}{c}\left(\partial_k A_l - \partial_l A_k\right), \qquad k, l = 0, 1, 2, 3 \tag{3.46}$$

4. κ-Field is Responsible for Pair Production

The nonrelativistic field $\mathbf{u}$ in the action (1.1) is an internal field of the nonrelativistic particle. It can act only on the motion of the nonrelativistic particle, making it stochastic. According to the action (2.1), (2.2) the κ-field looks also as an internal field of the particle. It seems that it may act only on the motion of the particle, and it cannot act on motion of other particles. However, it is not so. The κ-field (a relativistic version of nonrelativistic field $\mathbf{u}$) can produce pairs. In other words, the κ-field can turn the particle world line in the time direction. Formally, in such an action the κ-field acts as an internal field of the particle. But such a turn of the world line is possible only, if the κ-field is a given external field. Let us illustrate this in the example [10], when

$$K = \sqrt{1 + \lambda^2 \left(\kappa_l \kappa^l + \partial_l \kappa^l\right)} = \sqrt{1 + f(x)} \tag{4.1}$$

where $f(x)$ is some given function of coordinates x. The action (2.1), (2.2) takes the form

$$\mathcal{A}[q] = \int L(q, \dot{q})\, d\tau, \qquad L = -\sqrt{m^2 c^2 (1 + f(q))\, g_{ik} \dot{q}^i \dot{q}^k} - \frac{e}{c} A_k \dot{q}^k \tag{4.2}$$

where relations $x^i = q^i(\tau),\ i = 0,1,2,3$ describe the world line of the particle, and $\dot{q}^k \equiv dq^i/d\tau$. The quantities $A_k = A_k(q),\ k = 0,1,2,3$ are given electromagnetic potentials, and $f = f(q)$ is some given field, replacing the particle mass m by the effective particle mass $m_{\text{eff}} = m\sqrt{(1 + f(q))}$. The canonical momentum p_k is defined by the relation

$$p_k = \frac{\partial L}{\partial \dot{q}^k} = -\frac{mcK g_{ki} \dot{q}^i}{\sqrt{g_{ls} \dot{q}^l \dot{q}^s}} - \frac{e}{c} A_k, \qquad K = \sqrt{(1 + f(q))} \tag{4.3}$$

Dynamic equations have the form

$$\frac{dp_k}{d\tau} = -mc\sqrt{g_{ik} \dot{q}^i \dot{q}^k}\, \frac{\partial K}{\partial q^k} - \frac{e}{c} \frac{\partial A_i}{\partial q^k} \dot{q}^i \tag{4.4}$$

One can see from (4.3), that the vector

$$\dot{q}_k = \sqrt{\frac{g_{ls} \dot{q}^l \dot{q}^s}{1 + f(q)}}\, \frac{\left(p_k + \frac{e}{c} A_k\right)}{mc} \tag{4.5}$$

becomes to be spacelike $\left(g_{ls} \dot{q}^l \dot{q}^s < 0\right)$, if $f(q) < -1$, because only in this case the expression under radical in (4.5) is real.

The Hamilton-Jacobi equation for the action (4.2) has the form

$$g^{ik} \left(\frac{\partial S}{\partial q^i} + \frac{e}{c} A_i\right) \left(\frac{\partial S}{\partial q^k} + \frac{e}{c} A_k\right) = m^2 c^2 (1 + f(q)) \tag{4.6}$$

Let us consider solution of the Hamilton-Jacobi equation in the space-time, where $A_i = 0$, and $f = f(t)$ is a function of only time t. In this case the full integral $S(t, \mathbf{x}, \mathbf{p})$ of equation (4.6) has the form

$$S(t, \mathbf{x}, \mathbf{p}) = \mathbf{px} + \int_0^t c\sqrt{m^2 c^2 (1 + f(t)) + \mathbf{p}^2} dt + C, \qquad \mathbf{p}, C = \text{const} \tag{4.7}$$

where $\mathbf{p} = \{p_1, p_2, p_3\}$ are parameters. The equation of the world line is defined by the equation

$$\frac{\partial S(t, \mathbf{x}, \mathbf{p})}{\partial p_\alpha} = x^\alpha - x_0^\alpha, \quad x_0^\alpha = \text{const}, \quad \alpha = 1, 2, 3 \tag{4.8}$$

Substituting (4.7) in (4.8) and setting $p_2 = p_3 = 0$, one obtains

$$x^1 - x_0^1 + \int\limits_0^t \frac{p_1 c dt}{\sqrt{m^2c^2\left(1 + f\left(t\right)\right) + p_1^2}} = 0, \qquad x_0^1 = \text{const} \tag{4.9}$$

$$x^\alpha = x_0^\alpha = \text{const}, \quad \alpha = 2, 3 \tag{4.10}$$

Let for example

$$f\left(t\right) = \begin{cases} 0 & \text{if} \quad t < 0 \\ -\frac{V^2}{m^2c^4t_0^2} t\left(t - t_0\right) & \text{if} \quad 0 < t < t_0 \\ 0 & \text{if} \quad t_0 < t \end{cases}, \qquad t_0, V = \text{const} \tag{4.11}$$

The world line (4.9) takes the form

$$x^1 = \begin{cases} x_0^1 - \frac{p_1c^2}{E} t & \text{if} \quad t < 0 \\ x_0^1 - \int\limits_0^t \frac{p_1 c dt}{\sqrt{E^2 - V^2 t(t - t_0)/t_0^2}} & \text{if} \quad 0 < t < t_0 \\ x_1^1 + \alpha \frac{p_1c^2}{E}\left(t - t_0\right) & \text{if} \quad t_0 < t \end{cases}, \qquad E = c\sqrt{m^2c^2 + p_1^2} \tag{4.12}$$

where $\alpha = \pm 1$. Sign of α and the constant x_1^1 are determined from the continuity condition of the world line at $t = t_0$. The solution (4.12) has different form, depending on the sign of the constant $4E^2 - V^2$.

If $4E^2 > V^2$, the world line (4.12) takes the form

$$x^1 = \begin{cases} x_0^1 - \frac{p_1c^2}{E} t & \text{if } t < 0 \\ x_0^1 - \frac{p_1c^2t_0}{V} \arcsin \frac{2V\left(\sqrt{E^2t_0^2 - V^2 t(t-t_0)} - E(t_0 - 2t)\right)}{t_0(4E^2 + V^2)} & \text{if } 0 < t < t_0 \\ x_0^1 - p_1c^2 \frac{t_0}{V} \arcsin \frac{4EV}{4E^2+V^2} - \frac{p_1c^2}{E}\left(t - t_0\right) & \text{if } t_0 < t \end{cases}, \quad E^2 > V^2/4 \tag{4.13}$$

In the case, when $4E^2 < V^2$, the world line is reflected from the region Ω_{fb} of the space-time determined by the condition $0 < t < t_0$ in (4.11). In this case the coordinate x is not a single-valued function of the time t. We use a parametric representation for the world line (4.12). We have

$$x^1 = \begin{cases} x_0^1 - \frac{p_1c^2t_0}{2E}\left(1 - A\cosh\tau\right) & \text{if} \quad \tau < -\tau_0 \\ x_0^1 - \frac{p_1c^2t_0}{V}\left(\tau + \tau_0\right) & \text{if} \quad -\tau_0 < \tau < \tau_0 \\ x_0^1 - \frac{2p_1t_0}{V}\tau_0 + \frac{p_1c^2t_0}{2E}\left(1 - A\cosh\tau\right) & \text{if} \quad \tau_0 < \tau \end{cases} \tag{4.14}$$

$$t = \frac{t_0}{2}\left(1 - A\cosh\tau\right) \tag{4.15}$$

where

$$A = \sqrt{1 - \frac{4E^2}{V^2}}, \qquad \tau_0 = \operatorname{arccosh}\frac{1}{A} = \operatorname{arccosh}\frac{1}{\sqrt{1 - \frac{4E^2}{V^2}}} \tag{4.16}$$

The solution (4.14), (4.15) describes annihilation of particle and antiparticle with the energy $E < V/2$ in the region $0 < t < t_0$. The world line, describing the particle-antiparticle generation, has the form

$$x^1 = \begin{cases} x_0^1 - \frac{p_1 c^2 t_0}{2E}\left(A\cosh\tau - 1\right) & \text{if} \quad \tau < -\tau_0 \\ x_0^1 + \frac{p_1 c^2 t_0}{V}\left(\tau + \tau_0\right) & \text{if} \quad -\tau_0 < \tau < \tau_0 \\ x_0^1 + \frac{2p_1 t_0}{V}\tau_0 + \frac{p_1 c^2 t_0}{2E}\left(A\cosh\tau - 1\right) & \text{if} \quad \tau_0 < \tau \end{cases} \tag{4.17}$$

$$t = \frac{t_0}{2}\left(A\cosh\tau - 1\right) \tag{4.18}$$

where parameters A, τ_0 are defined by the relation (4.16), and the relation $2E < V$ takes place.

In both cases (4.14) and (4.17) at $|t| \to \infty$ the world line has two branches, which can be approximated by the relations

$$x^1 = x_0^1 + vt_1 \pm v\left(t - t_1\right), \qquad t_1 = t_0\frac{E}{V} \tag{4.19}$$

where $v = -\frac{p_1 c^2}{E}$ is the particle velocity, and $v = \frac{p_1 c^2}{E}$ is the antiparticle velocity.

The particle world line cannot turn its direction in time by means of its inner resources. It is possible only in some external field. Energy of the particle and of the antiparticle is absorbed by the external field $f\left(t\right)$. Thus, if it appears that the κ-field is not only internal field, it may be a force field which is responsible for pair production and pair annhilation, because both processes are connected with the turn of a world line in time.

5. Many Stochastic Relativistic Particles

Let us consider N identical stochastic relativistic particles, having electrical charge e and mass m. They interact via the electromagnetic field and via the force field κ. The action has the form

$$\mathcal{A}_{\mathcal{E}[\mathcal{S}_{\mathrm{st}}]}\left[X, \kappa, A\right] = \sum_{A=1}^{A=N} \int_{V_{\boldsymbol{\xi}}} L_{(A)}\left(x_{(A)}\left(\tau, \boldsymbol{\xi}\right)\right) d\tau d\boldsymbol{\xi} + \int_{V_x} L_{\mathrm{em}} d^4x \tag{5.1}$$

$$X = \left\{x_{(1)}, x_{(1)}, \ldots x_{(N)}\right\}, \quad x_{(A)} = \left\{x_{(A)}^0, x_{(A)}^1, x_{(A)}^2, x_{(A)}^3\right\}, \quad A = 1, 2, \ldots N \tag{5.2}$$

Here an index in brackets means the number of a particle.

$$L_{(A)}\left(x_{(A)}\left(\tau, \boldsymbol{\xi}\right)\right) = -mcK_{(A)}\left(x_{(A)}\right)\sqrt{g_{ik}\dot{x}_{(A)}^i \dot{x}_{(A)}^k} - \frac{e}{c}A_k\left(x_A\right)\dot{x}_{(A)}^k, \quad A = 1, 2, \ldots N \tag{5.3}$$

$$\dot{x}_{(A)}^i = \frac{dx_{(A)}^i}{d\tau}, \quad x_{(A)} = x_{(A)}\left(\tau, \boldsymbol{\xi}\right) \tag{5.4}$$

$$K_{(A)} = \sqrt{1 + \lambda^2\left(g_{kl}\kappa^k\left(x_A\right)\kappa^l\left(x_A\right) + \frac{\partial}{\partial x_{(A)}^k}\kappa^k\left(x_A\right)\right)}, \qquad \lambda = \frac{\hbar}{mc}, \quad A = 1, 2, \ldots N \tag{5.5}$$

$$L_{\text{em}} = \frac{1}{8\pi} g^{ik} \partial_i A_l(x) \partial_k A^l(x), \quad x = \{x^0, x^1, x^2, x^3\} \tag{5.6}$$

Variation with respect to $x^i_{(A)}$ gives

$$mc \frac{d}{d\tau} \left(K_{(A)}(x_{(A)}) \frac{x_{(A)i}}{\sqrt{\dot{x}_{(A)s} \dot{x}^s_{(A)}}} \right) - mc \frac{\partial \left(K_{(A)}(x_{(A)}) \sqrt{\dot{x}_{(A)s} \dot{x}^s_{(A)}} \right)}{\partial x^i_{(A)}}$$

$$+ \frac{e}{c} \frac{d}{d\tau} A_i(x_A) - \frac{e}{c} \dot{x}^k_{(A)} \frac{\partial}{\partial x^i_{(A)}} A_k(x_A) = 0, \quad A = 1, 2, ...N \tag{5.7}$$

Variation of (5.1) with respect to $\kappa^i(x_{(A)})$ gives

$$-mc \frac{\lambda^2 g_{ki} \kappa^k(x_{(A)}) J(x_{(A)}) \sqrt{\dot{x}_{(A)s} \dot{x}^s_{(A)}}}{K_{(A)}(x_{(A)})} + mc \frac{\partial}{\partial x^i_{(A)}} \frac{\lambda^2 J(x_{(A)}) \sqrt{\dot{x}_{(A)s} \dot{x}^s_{(A)}}}{2K_{(A)}(x_{(A)})} = 0 \tag{5.8}$$

$$A = 1, 2, ...N$$

$$J(x) = \frac{\partial(\tau, \xi_1, \xi_2, \xi_3)}{\partial(x^0, x^1, x^2, x^3)} \tag{5.9}$$

Jacobian $J(x_{(A)})$ appears in (5.8), because one needs before variation of (5.1) to go from integration over $d\tau d\boldsymbol{\xi}$ to integration over d^4x in (5.1).

Equations (5.8) can be written in the form

$$\kappa_i(x_{(A)}) = \frac{\partial}{\partial x^i_{(A)}} \kappa(x_{(A)}) = \frac{1}{2} \frac{\partial}{\partial x^i_{(A)}} \log \frac{J(x_{(A)}) \sqrt{\dot{x}_{(A)s} \dot{x}^s_{(A)}}}{K_{(A)}(x_{(A)})}, \quad A = 1, 2, ...N \tag{5.10}$$

where $\kappa(x_{(A)})$ is the potential of the κ-field κ^i.

Equations (5.10) can be integrated in the form

$$\kappa(x_{(A)}) = \frac{1}{2} \log \frac{J(x_{(A)}) \sqrt{\dot{x}_{(A)s} \dot{x}^s_{(A)}}}{K_{(A)}(x_{(A)})} + \frac{1}{2} \log C_{(A)}, \quad A = 1, 2, ...N \tag{5.11}$$

where $C_{(A)} = C_{(A)}(X)$, $A = 1, 2, ...N$ are functions of $X = \{x_{(1)}, x_{(2)}, ..x_{(N)}\}$. The functions $C_{(A)}$ satisfy the conditions

$$\frac{\partial C_{(A)}(X)}{\partial x^k_{(A)}} = 0, \quad A = 1, 2, ...N, \quad k = 0, 1, 2, 3 \tag{5.12}$$

Let us note that the flux $j^k_{(A)}$ of Ath particle in the statistical ensemble can be presented in the form

$$j^k_{(A)}(x_{(A)}) = \dot{x}^k_{(A)}(\tau, \boldsymbol{\xi}) J(x_{(A)}) \tag{5.13}$$

and equation (5.11) can be rewritten in the form

$$\kappa(x_{(A)}) = \frac{1}{2} \log \frac{\sqrt{j_{(A)s}(x_{(A)}) j^s_{(A)}(x_{(A)})}}{K_{(A)}(x_{(A)})} + \frac{1}{2} \log C_{(A)}, \quad A = 1, 2, ...N \tag{5.14}$$

Let chose $\log C_{(A)}$ in the form

$$\log C_{(A)} = \frac{1}{2}\sum_{B=1}^{B=N}\left(1-\delta_{AB}\right)\log\frac{\sqrt{j_{(B)s}\left(x_{(B)}\right)j_{(B)}^{s}\left(x_{(B)}\right)}}{K_{(B)}\left(x_{(B)}\right)} \tag{5.15}$$

According to (5.15) the κ-field at the point $x_{(A)}$ has the form

$$\kappa\left(x_{(A)}\right) = \kappa\left(x_{(A)}, X_{(A)}\right) = \frac{1}{2}\sum_{A=1}^{A=N}\log\frac{\sqrt{j_{(A)s}\left(x_{(A)}\right)j_{(A)}^{s}\left(x_{(A)}\right)}}{K_{(A)}\left(x_{(A)}\right)} \tag{5.16}$$

The second argument $X_{(A)}$

$$X_{(A)} = \left\{x_{(1)}, x_{(2)}, ..x_{(A-1)}, x_{(A+1)}, ...x_{(N)}\right\} \tag{5.17}$$

of κ shows that the κ-field at the point $x_{(A)}$ depends on all N particles of the statistical ensemble.

Expression (5.16) is symmetric with respect to transposition of any two particles of N considered identical particles.

Although the action (5.1) is a sum of actions for single particles, and the particles look as noninteracting particles, but actually the particles interact via the κ-field. The particles interact also via electromagnetic field. The electromagnetic interaction of particles arise because of the last term in (5.1), which contains time derivatives of A_k and describes the electromagnetic field as a dynamic system. Such a term, containing time derivatives of the κ-field, is present in any $L_{(A)}$. We have seen in the fourth section that external κ-field is responsible for pair production. In this section we have seen that any relativistic particle can generate the κ-field which is external with respect to other identical particles. Any single particle generate the κ-field, which acts on the particle motion. However, at the particle description in terms of the wave function the κ-field is incorporated in the definition of the wave function by formulas (3.24). And the κ-field is considered as an attribute of the wave function describing a free quantum particle (statistical ensemble of free stochastic particles).

6. κ-Field of a Single Particle

Let us consider an uniform statistical ensemble, whose state is described by the constant flux j^i of particles

$$j^0 = \text{const}, \quad j^\alpha = 0, \quad \alpha = 1,2,3 \tag{6.1}$$

For one particle ($N = 1$) the equation (5.14) takes the form

$$\exp(2\kappa) = \frac{\sqrt{j_s j^s}}{\sqrt{1+\lambda^2 e^{-\kappa} g^{kl}\frac{\partial^2}{\partial x^k \partial x^l}e^{\kappa}}} \tag{6.2}$$

Or

$$1+\lambda^2 e^{-\kappa} g^{kl}\frac{\partial^2}{\partial x^k \partial x^l}e^{\kappa} = \frac{j_s j^s}{\exp(4\kappa)} \tag{6.3}$$

Introducing designation

$$w = e^{\kappa} \tag{6.4}$$

one obtains dynamic equation for w

$$w + \lambda^2 \frac{\partial^2 w}{c^2 \partial t^2} - \lambda^2 \Delta w = \frac{j_s j^s}{w^3} \tag{6.5}$$

We consider the simplest case, when the flux j^k is taken in the form

$$j^{\alpha} = 0, \quad \alpha = 1, 2, 3, \quad j^0 = \left\{ \begin{array}{ll} \rho, \text{ if} & r > r_0 \\ 0, \text{ if} & r < r_0 \end{array} \right., \quad \rho > 0, \quad r_0 \leq \lambda \tag{6.6}$$

We search for stationary spherically symmetric solution, which is an analog of the Coulomb solution for electromagnetic field. Neglecting the term with the timelike derivatives, we shall solve the equation (6.5) taken in the form

$$w - \lambda^2 \frac{1}{r^2} \frac{\partial}{\partial r} \left(r^2 \frac{\partial w}{\partial r} \right) = \frac{f}{w^3}, \quad f = j_s j^s \tag{6.7}$$

We shall search solution in the form

$$w = \frac{A}{r} + B \frac{e^{-r/\lambda}}{r} \tag{6.8}$$

where the constants A and B are to be determined. Substituting (6.8) in (6.7), one obtains

$$\frac{f}{w_0^3} - \frac{A}{r} = 0, \quad w_0 = w(0) \tag{6.9}$$

The quantity B is a constant, which is determined from the continuity condition. One obtains

$$w = \left\{ \begin{array}{c} f^{1/4}, \text{ if } r < r_0 \\ f^{1/4} \frac{e^{-r/\lambda}}{r}, \text{ if } r > r_0 \end{array} \right. \tag{6.10}$$

$$\kappa = \log w = -\frac{r}{\lambda} + \frac{1}{4} \log \frac{f}{r^4}, \quad r > r_0 \tag{6.11}$$

Of course, there is also a solution of linear equation

$$w + \lambda^2 \frac{\partial^2 w}{c^2 \partial t^2} - \Delta w = 0 \tag{6.12}$$

which takes place in the region, where $j_k = 0$.

Thus, we have investigated the case, when the external κ-field produces pairs, and the case, when the κ-field is generated by a statistical ensemble of stochastic (quantum) relativistic particles. Unfortunately, a self-consistent conception of pair production can be hardly formulated in terms of the described formalism, because this formalism does not distinguish between particles and antiparticles. In the fourth section the particle and antiparticle are distinguished by their orientation $\varepsilon = \pm 1$. But the orientation is a discrete quantity, and there is no dynamic equation for ε. We hope that one will succeed to modify the statistical ensemble formalism in such a way, that it will distinguish formally between particle and antiparticle.

7. Multivariant Space-Time Geometry

Explanation of quantum effects by stochastic motion of elementary particles admits one to remove quantum principles as the primary laws of nature. But simultaneously the stochastic motion of free particles arises the question on reasons of the stochasticity. This stochasticity may be explained as a result of interaction with some medium (ether, vacuum) distributed in the space-time. Another reason may be an interaction of free elementary particles with the space-time directly. In other words, space-time geometry may be such one that a free elementary particle moves stochastically in this space-time geometry. World line of the stochastically moving particle wobbles. This wobbling is conditioned by a multivariance of the real space-time geometry.

Geometrical vector (g-vector) $\mathbf{AB}$ is defined as a the ordered set $\mathbf{AB} = \{A, B\}$ of two points $A, B \in \Omega$. Here Ω is the set of points (events) of the space-time, where the geometry is given. We use the term g-vector (vector), because there are linear vectors (linvectors) u, which are defined as elements of the linear vector space $\mathcal{L}_n$. Linvectors $u \in \mathcal{L}_n$ are abstract quantities, whose properties are defined by a system of axioms. In particular, operations of summation of linvectors and multiplication of a linvector by a real number are defined in $\mathcal{L}_n$. Under some conditions the operation on linvectors may be applied to g-vectors.

Linvectors and g-vectors have different properties. Any linvector exists in one copy, whereas there are many g-vectors $\mathbf{CD}$ which are equivalent to the g-vector $\mathbf{AB}$. Geometric vector $\mathbf{CD}$ is equivalent (equal) to g-vector $\mathbf{AB}$ ($\mathbf{CD}$eqv$\mathbf{AB}$), if

$$(\mathbf{CD}\text{eqv}\mathbf{AB}): \quad (\mathbf{AB.CD}) = |\mathbf{CD}| \cdot |\mathbf{AB}| \wedge |\mathbf{CD}| = |\mathbf{AB}| \tag{7.1}$$

where $(\mathbf{CD.AB})$ is the scalar product of two vectors $\mathbf{CD}$ and $\mathbf{AB}$, and $|\mathbf{AB}| = \sqrt{(\mathbf{AB.AB})}$ is the length of the vector $\mathbf{AB}$. The two g-vectors equivalence is defined by the relation (7.1) in the proper Euclidean geometry, where

$$(\mathbf{AB.CD}) = \sigma(A, D) + \sigma(B, C) - \sigma(A, C) - \sigma(B, D) \tag{7.2}$$

$$|\mathbf{AB}| = \sqrt{2\sigma(A, B)} \tag{7.3}$$

Here $\sigma(A, B)$ is the world function $\sigma(A, B) = \frac{1}{2}\rho^2(A, B)$, where $\rho(A, B)$ is the distance between the points A and B. Definition (7.1) - (7.3) of two g-vectors equivalence depends only on the world function. It does depend neither on dimension, nor on the coordinate system. Definition (7.1) - (7.3) of two g-vectors equivalence can be used in any geometry which is described completely by its world function and only by its world function. Such a geometry is called the physical geometry. If the world function is restricted by some conditions (the triangle axiom, nonnegativity of the distance ρ), such a geometry is known as metric geometry. Metric geometry is a special case of the physical geometry. The metric geometry as well as the distance geometry [11] (restricted only by the condition of nonnegativity of the distance ρ) cannot be used for description of the space-time, because in the space-time geometry the space-time distance ρ may be imaginary.

In the proper Euclidean geometry there is only one g-vector $\mathbf{CD}$ at the point C which is equivalent to the g-vector $\mathbf{AB}$ at the point A. It means that there exist only one point $D \in \Omega$ which is a solution of two equations

$$(\mathbf{AB.CD}) = |\mathbf{CD}| \cdot |\mathbf{AB}|, \quad |\mathbf{CD}| = |\mathbf{AB}| \tag{7.4}$$

at fixed points $A, B, C \in \Omega$.

In a physical geometry, generally speaking, there are many g-vectors $\mathbf{CD}$, $\mathbf{CD}'$, $\mathbf{CD}''$, ...which are equivalent to g-vector $\mathbf{AB}$. Such a geometry is considered as a multivariant geometry. The multivariance is a reason of the world line wobbling of the free particle motion. The world line is described as a set $\mathcal{C}$ of points ... $P_0, P_1, ...P_s, ...$divided by a constant distance $\rho(P_s, P_{s+1}) = \mu$, $s = ...0, 1, ...$

$$\mathcal{C} = \bigcup_s P_s, \quad \rho(P_s, P_{s+1}) = \mu = \text{const}, \quad s = ...0, 1, ... \tag{7.5}$$

If the limit at $\mu \to 0$ exists, the set $\mathcal{C}$ tends to a smooth world line of the particle. For free particle $(\mathbf{P}_s\mathbf{P}_{s+1}\text{eqv}\mathbf{P}_{s+1}\mathbf{P}_{s+2})$, $s = ...0, 1, ...$If there is an unique solution of two equations

$$(\mathbf{P}_s\mathbf{P}_{s+1}.\mathbf{P}_{s+1}\mathbf{P}_{s+2}) = |\mathbf{P}_s\mathbf{P}_{s+1}|\cdot|\mathbf{P}_{s+1}\mathbf{P}_{s+2}|, \quad |\mathbf{P}_s\mathbf{P}_{s+1}| = |\mathbf{P}_{s+1}\mathbf{P}_{s+2}|, s = ...0, 1, .. \tag{7.6}$$

for P_{s+2} at any given P_s, P_{s+1}, then the world line does not wobble. In this case the space-time geometry is single-variant, the limit $\mu \to 0$ exists and the point set $\mathcal{C}$ forms a smooth world line L. If the space-time geometry is multivariant, there are several point P_{s+2} determined by the points P_s, P_{s+1}. The set $\mathcal{C}$ does not form a smooth world line. The set $\mathcal{C}$ forms a wobbling broken line, consisting of connected segments of the straight line.

Even the space-time geometry of Minkowski is multivariant with respect to spacelike g-vectors. For instance, spacelike g-vectors $\mathbf{P}_{s+1}\mathbf{P}_{s+2} = \left\{\sqrt{r^2 + z^2}, r\cos\phi, r\sin\phi, z\right\}$ and $\mathbf{P}_{s+1}\mathbf{P}'_{s+2} = \left\{\sqrt{r_1^2 + z^2}, r_1\cos\phi_1, r_1\sin\phi_1, z\right\}$ are equivalent to the spacelike g-vector $\mathbf{P}_s\mathbf{P}_{s+1} = \{0, 0, 0, z\}$ at arbitrary values of quantities r, r_1, ϕ, ϕ_1. But g-vectors $\mathbf{P}_{s+1}\mathbf{P}_{s+2}$ and $\mathbf{P}_{s+1}\mathbf{P}'_{s+2}$ are not equivalent between themselves, generally speaking. Amplitude of this difference is infinite in the sense that the value of $\left|\mathbf{P}_{s+2}\mathbf{P}'_{s+2}\right|$

$$\left|\mathbf{P}_{s+2}\mathbf{P}'_{s+2}\right|^2 = \sqrt{(r^2 + z^2)\left(r_1^2 + z^2\right)} - rr_1\cos(\phi - \phi_1) - z^2 \tag{7.7}$$

has neither minimum, no maximum. The particle with spacelike world line is called tachyon. Absence of supremum of (7.7) means that the world line of a tachyon wobbles with infinite amplitude, and tachyon cannot be detected, even if it exists. As far as a free tachyon cannot be detected, the contemporary scientists prefer to think that tachyons do not exist. They prefer not to consider the wobbling spacelike world lines. However, although a single tachyon cannot be detected, the tachyon gas can be detected by its gravitational field. Existence of so-called dark matter may be freely explained by a presence of the tachyon gas in cosmos [12, 13].

Tardions (i.e. particles with timelike world line) have a smooth world line in the space-time geometry of Minkowski $\mathcal{G}_{\mathrm{M}}$, because $\mathcal{G}_{\mathrm{M}}$ is single-variant with respect to any timelike g- vectors. However, if the space-time geometry $\mathcal{G}$ differs from $\mathcal{G}_{\mathrm{M}}$, the space-time geometry $\mathcal{G}$ may be multivariant with respect to timelike g-vectors. In this case the world line of a free tardion wobbles. In particular, if the space-time geometry $\mathcal{G}_{\mathrm{d}}$ is discrete, and world function σ_{d} of this geometry $\mathcal{G}_{\mathrm{d}}$ has the form

$$\sigma_{\mathrm{d}} = \sigma_{\mathrm{M}} + \frac{\lambda_0^2}{2}\mathrm{sgn}(\sigma_{\mathrm{M}}) \tag{7.8}$$

where λ_0 is the elementary length, and σ_{M} is the world function of the geometry of Minkowski, the world lines of tardions wobble also. The discrete space-time geometry $\mathcal{G}_{\mathrm{d}}$ is given on the same manifold Ω_{M}, where the geometry of Minkowski $\mathcal{G}_{\mathrm{M}}$ is given. But any distance ρ_{d} in the geometry $\mathcal{G}_{\mathrm{d}}$ has the property

$$|\rho_{\mathrm{d}}(P,Q)| \notin (0,\lambda_0), \quad \forall P,Q \in \Omega_{\mathrm{M}} \tag{7.9}$$

which means that any distance $|\rho_{\mathrm{d}}(P,Q)|$ is not less, than elementary length λ_0. It is easy to verify that the distance $\rho_{\mathrm{d}} = \sqrt{2\sigma_{\mathrm{d}}}$ of geometry (7.8) satisfies the condition (7.9). It follows from (7.9) that λ_0 is minimal distance in $\mathcal{G}_{\mathrm{d}}$ (but $\rho_{\mathrm{d}}(P,Q) = 0$ is possible). The discrete geometry $\mathcal{G}_{\mathrm{d}}$ is multivariant with respect to all g-vectors. But wobbling of timelike world lines is restricted by the elementary length λ_0. This wobbling is responsible for quantum effects. If $\lambda_0^2 = \hbar/(bc)$, then statistical description of wobbling world lines is equivalent to description in terms of the Schrödinger equation [14]. Here $\hbar$ and c are respectively the quantum constant and the speed of the light. The quantity b is an universal constant which connects the geometric mass μ, defined in (7.5) with the particle mass m by means of the relation

$$m = b\mu \tag{7.10}$$

The real space-time geometry may distinguish from (7.8), but in any case the space-time geometry is multivariant, and the multivariance of the space-time geometry is a reason of quantum effects.

8. Fluidity of Boundary Between the Particle Dynamics and Space-Time Geometry

The particle motion occurs in the space-time, and properties of the space-time are essential for description of the particle motion. The boundary between the properties of the space-time and properties of laws of motion (dynamics) is indefinite. One may choose simple properties of the space-time geometry and obtain complicated laws of dynamics. On the contrary, one may choose a simple dynamics (free particle motion) and obtain a complicated space-time geometry. It is possible intermediate version, when dynamics and space-time geometry are not very simple. Historically the boundary between physics and space-time geometry moved towards space-time geometry. This process may be qualified as the physics geometrization. One can see several steps of the physics geometrization: (1) conservation laws as a corollaries of the space-time geometry symmetry, (2) spacial relativity, (3) general relativity, (4) five-dimensional geometry of Kaluza-Klein, where motion of a charged particle in the given electromagnetic and gravitational fields is described as a free particle motion in the Kaluza-Klein space-time geometry [15].

In the classical physics, where gravitational field and electromagnetic field are the only possible force fields, the Kaluza-Klein representation realizes the complete physics geometrization. But this geometrization is not complete one in microcosm, where the quantum effects are essential. Besides, the Riemannian geometry which is used in the Kaluza-Klein description is rather complicated. The Riemannian geometry is founded on several basic concepts: (1) concepts of topology, (2) concepts of local geometry such as dimension, coordinate system, metric tensor and parallel transport. Work with concepts of the Riemannian geometry is not simpler, than the work with numerous concepts of dynamics. As a result one prefers to work with customary concepts of dynamics.

At the metric approach to geometry, when the space-time geometry is described in terms of only distance ρ or in terms of only world function $\sigma = \rho^2/2$, any modification of the space-time geometry looks very simple. To obtain a modification of a geometry, one replaces world function and obtains a modified geometry described by the new world function. If the geometry is described by means of several fundamental concepts, any modification of the geometry needs a modification of all fundamental concepts. This modification of different fundamental concepts is to be concerted, in order the modified geometry be consistent. The more number of the basic concepts the difficult agreement between the modified concepts. The monistic conception of a geometry, when there is only one fundamental quantity is the best conception, because the problem of agreement of different basic modified concepts is absent. From this viewpoint the metric approach to the space-time geometry is the best approach.

9. Metric Approach to Geometry and Multivariance of Geometry

The proper Euclidean geometry can be presented in terms and only in terms of its world function. However, attempt of generalization of the proper Euclidean geometry [11] failed in the sense, that Blumental was forced to introduce concept of continuous mapping in addition to concept of distance. The condition of the continuous mapping cannot be expressed in terms of only distance. But the continuous mapping was necessary to construct one-dimensional continuous curve in the distance geometry of Blumental. As a result Blumental failed to realize a consistent metric approach to geometry, when the geometry is discribed in terms and only in terms of a distance.

What was a reason of failure? During two thousand years we knew only proper Euclidean geometry $\mathcal{G}_{\mathrm{E}}$. All statements of $\mathcal{G}_{\mathrm{E}}$ are derived logically from several basic statements (axioms). In all presentations of $\mathcal{G}_{\mathrm{E}}$ one considers the ways of derivation (theorems) of different statements of $\mathcal{G}_{\mathrm{E}}$ from axioms of $\mathcal{G}_{\mathrm{E}}$. The impression arises that these theorems form the content of $\mathcal{G}_{\mathrm{E}}$, whereas these theorems form only the way of the proper Euclidean geometry construction. The proper Euclidean geometry itself is a set $\mathcal{P}_{\mathrm{E}}$ of statements of $\mathcal{G}_{\mathrm{E}}$. At the modification $\mathcal{G}$ of the proper Euclidean geometry $\mathcal{G}_{\mathrm{E}}$ the set $\mathcal{P}$ of statements of the geometry $\mathcal{G}$ is obtained from the set $\mathcal{P}_{\mathrm{E}}$ of statements of $\mathcal{G}_{\mathrm{E}}$. Such a derivation of $\mathcal{P}$ from $\mathcal{P}_{\mathrm{E}}$ may differ from the way of the proper Euclidean geometry construction. It is possible such a situation that the modified (generalized) geometry $\mathcal{G}$ cannot be derived from a system of axioms. In other words, the geometry $\mathcal{G}$ may be nonaxiomatizable. Unfortunately, the nonaxiomatizablity of a geometry is perceived as something impossible, and this perception is a result of identification of geometry $\mathcal{G}_{\mathrm{E}}$ with the way of derivation of $\mathcal{G}_{\mathrm{E}}$.

In general, at the metric approach to geometry the modified (generalized) geometry $\mathcal{G}$ is obtained from $\mathcal{G}_{\mathrm{E}}$ by means of a deformation, when the world function σ_{E} is replaced by the world function σ of the modified geometry $\mathcal{G}$ in all definitions and all general geometric statements containing only σ_{E}. Such a construction of a physical geometry will be referred to as the deformation principle [16, 17]. The nonaxiomatizability of a physical geometry is connected with its multivariance. Indeed, in order the logical Euclidean method of the geometry construction could work, the equivalence relation (7.1) is to be transitive, when from $(\mathbf{AB}\mathrm{eqv}\mathbf{CD})$ and $(\mathbf{AB}\mathrm{eqv}\mathbf{FH})$ it follows that $(\mathbf{CD}\mathrm{eqv}\mathbf{FH})$. If the equivalence relation (7.1) is intransitive, and it does not follow from $(\mathbf{AB}\mathrm{eqv}\mathbf{CD}) \wedge (\mathbf{AB}\mathrm{eqv}\mathbf{FH})$ that $(\mathbf{CD}\mathrm{eqv}\mathbf{FH})$, a logical construction is impossible. But impossibility of derivation of multivariant physical geometry $\mathcal{G}$ by means of the Euclidean logical method does not mean that the set $\mathcal{P}$ of statements of a multivariant geometry $\mathcal{G}$ cannot be constructed. It can be constructed by means of the deformation principle.

Note that the Riemannian space-time geometry is multivariant with respect to remote vectors. But in the Riemannian geometry one removes fernparallelism (equivalency of remote vectors). Instead in the Riemannian geometry one introduces the parallel transport of a vector. In the Riemannian geometry the finite distance, defined as an integral along a geodesic, appears to be many-valued in many cases. Many-valued distance seems to be inadmissible from physical viewpoint.

Summation of linvectors and multiplication of a linvector by a real number are operation which are defined in the linear vector space $\mathcal{L}_n$. These operations are not adequate in application to g-vectors of multivariant geometry, although the are adequate in application to g-vectors of $\mathcal{G}_{\mathrm{E}}$, because the proper Euclidean geometry $\mathcal{G}_{\mathrm{E}}$ is single-variant.

Let $S_{\mathbf{AB}}$ be a set of g-vectors $\mathbf{CD}$, which are equivalent to g-vector $\mathbf{AB}$. If the equivalence relation is transitive, the set $S_{\mathbf{AB}}$ is a equivalence class $[\mathbf{AB}]$ of the g-vector $\mathbf{AB}$. It contains only g-vectors which are equivalent between themselves. In this case any equivalence class $[\mathbf{AB}]$ may be corresponded by some linvector $u \in \mathcal{L}_n$, and this correspondence will be one-to-one, because any equivalence class exist only in one copy. If the equivalence relation is intransitive and the set $S_{\mathbf{AB}}$ does not form an equivalence class, the correspondence between the linvectors and g-vectors cannot be established. As a result operation of the linear vector space $\mathcal{L}_n$ are not adequate in the multivariant geometry, where the equivalence relation is intransitive.

Formally one may introduce summation of g-vectors in multivariant geometry, but this summation will be many-valued. Let one needs to sum g-vectors $\mathbf{AB}$ and $\mathbf{CD}$, and $B \neq C$. Let g-vector $\mathbf{PQ} = \mathbf{AB} + \mathbf{CD}$, where the point P is given, and the point Q should be determined. One obtains

$$\mathbf{PQ} = \mathbf{PF} + \mathbf{FQ} \tag{9.1}$$

where points F and Q are determined from the relations

$$(\mathbf{PF}\mathrm{eqv}\mathbf{AB}) \wedge (\mathbf{FQ}\mathrm{eqv}\mathbf{CD}) \tag{9.2}$$

In the multivariant geometry the equations (9.2) has many solutions for the points F and Q, and the operation of summation appears to be many-valued. In the single-variant geometry the relations (9.2) have unique solution for points F and Q and the summation (9.1) is defined one-to-one.

Multiplication of a g-vector by a real number is also many-valued in the multivariant geometry, because definition of multiplication contains a reference to a relation of equivalence, which is many-valued in the multivariant geometry. Let $\mathbf{PQ}' = a\mathbf{PQ}$, where the points P, Q and the number a are given, then the point Q' is to be determined from the relations

$$(\mathbf{PQ}'.\mathbf{PQ}) = |\mathbf{PQ}'| \cdot |\mathbf{PQ}|, \quad |\mathbf{PQ}'| = a\,|\mathbf{PQ}| \tag{9.3}$$

Solution of the two equations is many-valued in the multivariant geometry, generally speaking. Thus, methods of differential geometry, developed for the proper Euclidean geometry $\mathcal{G}_{\rm E}$ are inadequate in the multivariant geometry. *However, inadequacy of the differential geometry methods in the multivariant geometry does not mean that multivariant geometries do not exist.*

The physics geometrization in the classical physics, when the space-time geometry is a Riemannian geometry, is not effective, because for determination of the Kaluza-Klein geometry one needs to determine the metric tensor g_{ik} and electromagnetic potential A_k, $k = 0, 1, 2, 3$. However, if these quantities are known, one may write dynamic equations for the particle motion in the space-time geometry of Minkowski and determine the particle world line. A use of the Kaluza-Klein geometry appears to be needless.

In the physics geometrization inside microcosm the force fields acting on a particle are not known. They are different for different elementary particles. One supposes, that in the proper (true) space-time geometry the elementary particle motion is free. Writing dynamic equations for the free particle motion in the true space-time geometry, one may rewrite the dynamic equations in the case of the space-geometry of Minkowski. In this case the dynamic equations cease to be free dynamic equations. Dynamic equations will contain force fields, arising as a result of deflection of the Minkowski geometry $\mathcal{G}_{\rm M}$ from the true space-time geometry, where the particle motion is free. The microcosm dynamic equations in the space-time geometry of Minkowski are not known primarily. They arise as a result of transformation of free dynamic equations, written in a true space-time geometry. In the microcosm the fluidity of boundary between the particle dynamics and the space-time geometry admits one to reduce determination of the particle dynamics laws to the determination of the world function of the true space-time geometry, where the elementary particles move free.

The number of variants of the dynamics laws for indefinite number of different sorts of elementary particles is more, than the number of variants of the world functions $\sigma(P, Q)$ of two space-time points P, Q. As a result a use of the hypothesis on the boundary fluidity for any elementary particles seems to be more effective, than suppositions on dynamics of any single elementary particle, which are extracted from complicated experiments with elementary particles. Of course, the hypothesis on the boundary fluidity for any elementary particles should be tested by experiment. However, in the case of classical physics this hypothesis is true distinctly. Besides, primarily it is not clear what is responsible for peculiar properties of particle motion: the space-time geometry or the laws of the particle dynamics.

Usage of the hypothesis on the boundary fluidity between the dynamics and the space-time geometry generates a conception of the elementary particle dynamics. In other words, a connection arises between the concepts of dynamics and those of the space-time geometry. This connection is a logical connection. It arises on the logical basis, but not on basis of a single experiment or on the basis of several single experiments. It concerns all elementary particles. This conception may appear to be valid or wrong, but it is a conception.

A like conception is absent in the contemporary elementary particle theory, where one invents suppositions on dynamics and interaction of different sorts of elementary particles, which are labelled by some quantum numbers. Absence of a conception in the contemporary theory generates numerous variants of a theory. These variants contain numerous interaction constants, which are to be determined from experiment. To understand, why it is bad, let us imagine that we have not a conception of classical particle dynamics, which states that any deterministic classical particle is a dynamic system, and its motion is described by a Lagrange function. In absence of such a conception one needs to invent dynamic equations for any particle, depending on its mass, color, temperature, shape and so on. In absence of the dynamics conception one cannot distinguish between essential parameters (mass) and unessential ones (color, temperature). As a result any investigation of dynamics becomes to be complicated.

10. Description of Geometrical Objects in Multivariant Geometry

A geometrical object is a geometrical image of a physical body. Any geometrical object is some subset of points in the space-time. However, a geometrical object is not an arbitrary set of points. In the physical geometry a geometrical object is to be defined in such a way, that similar geometrical objects (which are images of similar physical bodies) could be recognized in different space-time geometries.

Definition 1: A geometrical object $g_{\mathcal{P}_n,\sigma}$ of the geometry $\mathcal{G} = \{\sigma, \Omega\}$ is a subset $g_{\mathcal{P}_n,\sigma} \subset \Omega$ of the point set Ω. This geometrical object $g_{\mathcal{P}_n,\sigma}$ is a set of roots $R \in \Omega$ of the function $F_{\mathcal{P}_n,\sigma}$

$$g_{\mathcal{P}_n,\sigma} = \{R | F_{\mathcal{P}_n,\sigma}(R) = 0\}, \quad F_{\mathcal{P}_n,\sigma}: \quad \Omega \to \mathbb{R} \tag{10.1}$$

where $F_{\mathcal{P}_n,\sigma}$ depends on the point R via world functions of arguments $\{\mathcal{P}_n, R\} = \{P_0, P_1, ...P_n, R\}$

$$F_{\mathcal{P}_n,\sigma} \quad : \quad F_{\mathcal{P}_n,\sigma}(R) = G_{\mathcal{P}_n,\sigma}(u_1, u_2, ...u_s), \qquad s = \frac{1}{2}(n+1)(n+2) \tag{10.2}$$

$$u_l \quad = \quad \sigma(w_i, w_k), \quad i, k = 0, 1, ...n+1, \quad l = 1, 2, ...\frac{1}{2}(n+1)(n+2) \tag{10.3}$$

$$w_k \quad = \quad P_k \in \Omega, \qquad k = 0, 1, ...n, \qquad w_{n+1} = R \in \Omega \tag{10.4}$$

Here $\mathcal{P}_n = \{P_0, P_1, ..., P_n\} \subset \Omega$ are $n+1$ points which are parameters, determining the geometrical object $g_{\mathcal{P}_n,\sigma}$

$$g_{\mathcal{P}_n,\sigma} = \{R | F_{\mathcal{P}_n,\sigma}(R) = 0\}, \qquad R \in \Omega, \qquad \mathcal{P}_n \in \Omega^{n+1} \tag{10.5}$$

$F_{\mathcal{P}_n,\sigma}(R) = G_{\mathcal{P}_n,\sigma}(u_1, u_2, ...u_s)$ is a function of $\frac{1}{2}(n+1)(n+2)$ arguments u_k and of $n+1$ parameters $\mathcal{P}_n$. The set $\mathcal{P}_n = \{P_0, P_1, ...P_n\} \in \Omega^{n+1}$ of the geometric object parameters will be referred to as the skeleton of the geometrical object. The subset $g_{\mathcal{P}_n,\sigma} \subset \Omega$ will be referred to as the envelope of the skeleton. The skeleton is an analog of a frame of reference, attached rigidly to a physical body. Tracing the skeleton motion, one can trace the motion of the physical body. When a particle is considered as a geometrical object, its motion in the space-time is described by the motion of skeleton $\mathcal{P}_n$. At such an approach (the rigid body approximation) the shape of the envelope is of no importance.

Remark: An arbitrary subset Ω' of the point set Ω is not a geometrical object, generally speaking. It is supposed, that physical bodies may have only a shape of a geometrical object, because only in this case one can identify identical physical bodies (geometrical objects) in different space-time geometries.

Existence of the same geometrical objects in different space-time regions, having different geometries, brings up the question on equivalence of geometrical objects in different space-time geometries. Such a question did not arise before, because one does not consider such a situation, when a physical body moves from one space-time region to another space-time region, having another space-time geometry. In general, mathematical technique of the conventional space-time geometry (differential geometry) is not applicable for simultaneous consideration of several different geometries of different space-time regions.

We can perceive the space-time geometry only via motion of physical bodies in the space-time, or via construction of geometrical objects corresponding to these physical bodies. As it follows from the *definition 1* of the geometrical object, the function $G_{\mathcal{P}_n,\sigma}$ as a function of its arguments u_k, $k = 1, 2, ...n(n+1)/2$ (of world functions of different points) is the same in all physical geometries. It means, that a geometrical object $\mathcal{O}_1$ in the geometry $\mathcal{G}_1 = \{\sigma_1, \Omega_1\}$ is obtained from the same geometrical object $\mathcal{O}_2$ in the geometry $\mathcal{G}_2 = \{\sigma_2, \Omega_2\}$ by means of the replacement $\sigma_2 \to \sigma_1$ in the definition of this geometrical object.

Definition 2: Geometrical object $g_{\mathcal{P}'_n,\sigma'}$ ($\mathcal{P}'_n = \{P'_0, P'_1, ..P'_n\}$) in the geometry $\mathcal{G}' = \{\sigma', \Omega'\}$ and the geometrical object $g_{\mathcal{P}_n,\sigma}$ ($\mathcal{P}_n = \{P_0, P_1, ..P_n\}$) in the geometry $\mathcal{G} = \{\sigma, \Omega\}$ are similar geometrical objects, if

$$\sigma'\left(P'_i, P'_k\right) = \sigma\left(P_i, P_k\right), \qquad i, k = 0, 1, ..n \tag{10.6}$$

and the functions $G'_{\mathcal{P}'_n,\sigma'}$ for $g_{\mathcal{P}'_n,\sigma'}$ and $G_{\mathcal{P}_n,\sigma}$ for $g_{\mathcal{P}_n,\sigma}$ in the formula (10.2) are the same functions of arguments $u_1, u_2, ...u_s$

$$G'_{\mathcal{P}'_n,\sigma'}\left(u_1, u_2, ...u_s\right) = G_{\mathcal{P}_n,\sigma}\left(u_1, u_2, ...u_s\right) \tag{10.7}$$

In this case

$$u_l \equiv \sigma\left(P_i, P_k\right) = u'_l \equiv \sigma'\left(P'_i, P'_k\right), \qquad i, k = 0, 1, ...n, \qquad l = 1, 2, ..n(n+1)/2 \tag{10.8}$$

The functions $F'_{\mathcal{P}'_n,\sigma'}$ for $g_{\mathcal{P}'_n,\sigma'}$ and $F_{\mathcal{P}_n,\sigma}$ for $g_{\mathcal{P}_n,\sigma}$ in the formula (10.2) have the same roots, if the relation (10.7) is fulfilled. As a result one-to-one connection between the geometrical objects $g_{\mathcal{P}'_n,\sigma'}$ and $g_{\mathcal{P}_n,\sigma}$ arises.

As far as the physical geometry is determined by its geometrical objects construction, a physical geometry $\mathcal{G} = \{\sigma, \Omega\}$ can be obtained from some known standard physical geometry $\mathcal{G}_{\text{st}} = \{\sigma_{\text{st}}, \Omega\}$ by means of a deformation of the standard geometry $\mathcal{G}_{\text{st}}$. Deformation of the standard geometry $\mathcal{G}_{\text{st}}$ is realized by the replacement of σ_{st} by σ in all definitions of the geometrical objects in the standard geometry. The proper Euclidean geometry $\mathcal{G}_{\text{E}}$ is an axiomatizable geometry. It has been constructed by means of the Euclidean method as a logical construction. Using Euclidean method, one obtains $\mathcal{G}_{\text{E}}$ in the vector representation [19]. Simultaneously the proper Euclidean geometry is a physical geometry. In this case one obtains $\mathcal{G}_{\text{E}}$ in terms of the world function σ_{E}, i.e. in the σ-representation [19]. It may be used as a standard geometry $\mathcal{G}_{\text{st}}$. Construction of a physical geometry as a deformation of the proper Euclidean geometry $\mathcal{G}_{\text{E}}$ will be referred to as the deformation principle [17]. The most physical geometries are nonaxiomatizable geometries. They can be constructed only by means of the deformation principle.

11. General Geometric Relations

Describing a physical geometry in terms of the world function, one should distinguish between general geometric relations and specific geometric relations. The general geometric relations are the relations, which are written only in terms of the world function. The general geometric relations are valid for any physical geometry.

The first general geometric definition is the definition of the scalar product of two vectors (7.2). Definition of the two vectors equivalence (7.1) - (7.3) is also a general geometric relation.

Linear dependence of n g-vectors $\mathbf{P}_0\mathbf{P}_1, \mathbf{P}_0\mathbf{P}_2, ...\mathbf{P}_0\mathbf{P}_n$ is defined by the relation,

$$F_n(\mathcal{P}_n) = 0, \quad F_n(\mathcal{P}_n) \equiv \det ||(\mathbf{P}_0\mathbf{P}_i.\mathbf{P}_0\mathbf{P}_k)||, \qquad i,k = 1,2,...n \tag{11.1}$$

where $\mathcal{P}_n = \{P_0, P_1, ...P_n\}$ and $F_n(\mathcal{P}_n)$ is the Gram's determinant. Vanishing of the Gram's determinant is the necessary and sufficient condition of the linear dependence of n g-vectors. Condition of linear dependence relates usually to the properties of the linear vector space. It seems rather meaningless to use it, if the linear vector space cannot be introduced. Nevertheless, the relation (11.1) written as a general geometric relation describes some general geometric properties of g-vectors, which in the proper Euclidean geometry transform to the property of linear dependence. In particular, the dimension of the proper Euclidean geometry is defined in terms of the world function by means of the relations of the type (11.1) as a maximal number of linear independent vectors, which is possible in the Euclidean space. This circumstance seems to be rather unexpected, because in conventional presentation (vector representation [19]) of the Euclidean geometry $\mathcal{G}_{\mathrm{E}}$ the geometry dimension is postulated in the beginning of the geometry construction.

The general geometric relations describe general geometric properties of g-vectors, which are used at construction of geometrical objects. General geometric relations are essentially definitions of the scalar product, equivalence of g-vectors and their linear dependence. As we have seen, a definition of geometrical objects in the form of general geometric relations (i.e. in terms of the world function) is necessary to recognize the same physical body (and corresponding geometrical object) in different space-time geometries.

The general geometric relations are parametrized by the form of the world function. Changing the form of the world function, one obtains the general geometric relations at a new value of the parameter σ (new form of the world function).

12. Specific Properties of the n-Dimensional Euclidean space

Along of general geometric properties, connecting mainly with the properties of the linear vector space, there are special geometric relations, describing properties of the world function. For instance, there are relations, which are necessary and sufficient conditions of the fact, that the world function σ_{E} is the world function of n-dimensional Euclidean space. They have the form [18]:

I. Definition of the dimension:

$$\exists \mathcal{P}^n \equiv \{P_0, P_1, ...P_n\} \subset \Omega, \qquad F_n(\mathcal{P}^n) \neq 0, \qquad F_k\left(\Omega^{k+1}\right) = 0, \qquad k > n \tag{12.1}$$

where $F_n(\mathcal{P}^n)$ is the n-th order Gram's determinant (11.1). Geometric vectors $\mathbf{P}_0\mathbf{P}_i$, $i = 1, 2, ...n$ are basic g-vectors of the rectilinear coordinate system K_n with the origin at the point P_0. The metric tensors $g_{ik}(\mathcal{P}^n)$, $g^{ik}(\mathcal{P}^n)$, $i, k = 1, 2, ...n$ in K_n are defined by the relations

$$\sum_{k=1}^{k=n} g^{ik}(\mathcal{P}^n)\, g_{lk}(\mathcal{P}^n) = \delta_l^i, \qquad g_{il}(\mathcal{P}^n) = (\mathbf{P}_0\mathbf{P}_i.\mathbf{P}_0\mathbf{P}_l), \qquad i, l = 1, 2, ...n \quad (12.2)$$

$$F_n(\mathcal{P}^n) = \det ||g_{ik}(\mathcal{P}^n)|| \neq 0, \qquad i, k = 1, 2, ...n \quad (12.3)$$

II. Linear structure of the Euclidean space:

$$\sigma_{\mathrm{E}}(P, Q) = \frac{1}{2}\sum_{i,k=1}^{i,k=n} g^{ik}(\mathcal{P}^n)\,(x_i(P) - x_i(Q))\,(x_k(P) - x_k(Q)), \qquad \forall P, Q \in \Omega \quad (12.4)$$

where coordinates $x_i(P), x_i(Q), i = 1, 2, ...n$ of the points P and Q are covariant coordinates of the g-vectors $\mathbf{P}_0\mathbf{P}$, $\mathbf{P}_0\mathbf{Q}$ respectively in the coordinate system K_n. The covariant coordinates are defined by the relation

$$x_i(P) = (\mathbf{P}_0\mathbf{P}_i.\mathbf{P}_0\mathbf{P}), \qquad i = 1, 2, ...n \quad (12.5)$$

III: The metric tensor matrix $g_{lk}(\mathcal{P}^n)$ has only positive eigenvalues g_k

$$g_k > 0, \qquad k = 1, 2, ..., n \quad (12.6)$$

IV. The continuity condition: the system of equations

$$(\mathbf{P}_0\mathbf{P}_i.\mathbf{P}_0\mathbf{P}) = y_i \in \mathbb{R}, \qquad i = 1, 2, ...n \quad (12.7)$$

considered to be equations for determination of the point P as a function of coordinates $y = \{y_i\}$, $i = 1, 2, ...n$ has always one and only one solution. Conditions I – IV contain a reference to the dimension n of the Euclidean space, which is defined by the relations (12.1).

All relations I – IV are written in terms of the world function. They are constraints on the form of the world function of the proper Euclidean geometry $\mathcal{G}_{\mathrm{E}}$. Constraints (12.1), determining the dimension via the form of the world function, look rather unexpected. They contain a lot of constraints imposed on the world function of the proper Euclidean geometry $\mathcal{G}_{\mathrm{E}}$, and they are necessary. At the conventional approach to geometry one uses a very simple supposition: "Let the dimension of the Euclidean space be n." instead of numerous constraints (12.1).

At the vector representation of the proper Euclidean geometry, which is based on a use of the linear vector space, the dimension is considered as a primordial property of the linear vector space and as a primordial property of the Euclidean geometry $\mathcal{G}_{\mathrm{E}}$. Situation, when the geometry dimension is different at different points of the space Ω, or when it is indefinite, is not considered. At the vector representation of the Euclidean geometry $\mathcal{G}_{\mathrm{E}}$ one does not distinguish between the general geometric relations and the specific relations of the geometry.

Instead of constraints (12.1) – (12.7) one may use an explicit form of the world function

$$\sigma_{\mathrm{E}}\left(x,x'\right)=\frac{1}{2}\sum_{k=1}^{k=n}\left(x^{k}-x'^{k}\right)^{2} \tag{12.8}$$

where x^k, $x'^k \in \mathbb{R}$, $k = 1, 2, ...n$ are Cartesian coordinates of points P and P' respectively. The relation (12.8) satisfies all constraints (12.1) – (12.7). It uses concepts of dimension and of coordinates as primordial concepts of geometry. Using the world function only in such an explicit form, one cannot imagine a generalized geometry without such concepts as a dimension and a coordinate system, although these concepts are only means of a geometry $\mathcal{G}_{\mathrm{E}}$ description.

In general, after the logical reloading to σ-representation, when such base concepts of $\mathcal{G}_{\mathrm{E}}$ as dimension and coordinate system are replaced by the only base concept (world function), the proper Euclidean geometry $\mathcal{G}_{\mathrm{E}}$ looks rather unexpected. Some concepts look very simple in the vector representation. The same concepts look complicated in the σ-representation and vice versa. As a result the proper Euclidean geometry in the σ-representation is perceived hardly. In the vector representation one has several fundamental quantities: dimension, coordinate system, linear dependence, whereas in the σ-representation there is only one fundamental quantity: world function. The dimension, the coordinate system and the linear dependence are derivative concepts. Agreement between these quantities is achieved in any physical geometry automatically, because they are defined as some attributes of a world function. But this agreement looks very strange for researchers, who learned the Euclidean geometry in its conventional presentation and believe that any properties of $\mathcal{G}_{\mathrm{E}}$ take place in any generalized geometry.

In reality $\mathcal{G}_{\mathrm{E}}$ is a degenerate geometry, where the equality relation is transitive and the property of multivariance is absent in $\mathcal{G}_{\mathrm{E}}$. According to its properties $\mathcal{G}_{\mathrm{E}}$ can be constructed as a logical construction. Most researchers believe that any space-time geometry can be derived as a logical construction. They can imagine no other method of the geometry construction. They cannot imagine that the equivalence relation may be intransitive. They assume that the equivalence relation is transitive by definition. (How can one construct a geometry, if the equivalence relation is intransitive!?). In reality such a viewpoint is a corollary of the fact that researchers have been working only with $\mathcal{G}_{\mathrm{E}}$ which is a degenerate single-variant geometry. In $\mathcal{G}_{\mathrm{E}}$ some natural geometric properties (intransitivity of the equivalence relation and multivariance) are absent. How can one accept a geometry, where customary operations: (1) summation of g-vectors, (2) multiplication of a g-vector by a number and (3) decomposition of a g-vector are inadequate?

If $\mathbf{P}_0\mathbf{P}_i$, $i = 1, 2, ...n$ are basic g-vectors in some coordinate system K_n, one can determine projections of g-vector $\mathbf{P}_0\mathbf{P}$ on the basic g-vectors

$$\Pr\left(\mathbf{P}_0\mathbf{P}\right)_{\mathbf{P}_0\mathbf{P}_i}=\frac{\left(\mathbf{P}_0\mathbf{P}.\mathbf{P}_0\mathbf{P}_i\right)}{\left|\mathbf{P}_0\mathbf{P}_i\right|} \tag{12.9}$$

However, the g-vector $\mathbf{P}_0\mathbf{P}$ cannot be represented as a sum of its projections, because the summation of g-vectors is inadequate operation in the multivariant geometry. Thus, coordinates may be used for labelling of space-time points, but they cannot be used for realization of the differential geometry operations.

13. Equivalence of Physical Geometries

Generalization of general geometric expressions (7.1) – (7.3) on the case of the discrete geometry $\mathcal{G}_\mathrm{d}$ is obtained by means of the replacement of σ_E by σ_d, where σ_d is the world function (7.8) of the discrete geometry $\mathcal{G}_\mathrm{d}$. We are to be ready, that properties of concepts of dimension, linear dependence of g-vectors and segment of the straight line in $\mathcal{G}_\mathrm{d}$ differ strongly from their properties in $\mathcal{G}_\mathrm{E}$. However, we have no alternative to these relations for definition of these geometrical quantities in a discrete geometry $\mathcal{G}_\mathrm{d}$.

Definition 4: The physical geometry $\mathcal{G} = \{\sigma, \Omega\}$ is a point set Ω with the single-valued function σ on it

$$\sigma: \qquad \Omega \times \Omega \to \mathbb{R}, \qquad \sigma(P,P) = 0, \qquad \sigma(P,Q) = \sigma(Q,P), \qquad \forall P,Q \in \Omega \tag{13.1}$$

Definition 5: Two physical geometries $\mathcal{G}_1 = \{\sigma_1, \Omega_1\}$ and $\mathcal{G}_2 = \{\sigma_2, \Omega_2\}$ are equivalent $(\mathcal{G}_1 \mathrm{eqv} \mathcal{G}_2)$, if the point set $\Omega_1 \subseteq \Omega_2 \wedge \sigma_1(P,Q) = \sigma_2(P,Q)$, $\forall P,Q \in \Omega_1$, or $\Omega_2 \subseteq \Omega_1 \wedge \sigma_2(P,Q) = \sigma_1(P,Q)$, $\forall P,Q \in \Omega_2$

Remark: Coincidence of point sets Ω_1 and Ω_2 is not necessary for equivalence of geometries $\mathcal{G}_1$ and $\mathcal{G}_2$. If one demands coincidence of Ω_1 and Ω_2 in the case of equivalence of $\mathcal{G}_1$ and $\mathcal{G}_2$, then an elimination of one point P from the point set Ω_1 turns the geometry $\mathcal{G}_1 = \{\sigma_1, \Omega_1\}$ into geometry $\mathcal{G}_2 = \{\sigma_1, \Omega_1 \backslash P\}$, which appears to be not equivalent to the geometry $\mathcal{G}_1$. Such a situation seems to be inadmissible, because a geometry on a part $\omega \subset \Omega_1$ of the point set Ω_1 appears to be not equivalent to the geometry on the whole point set Ω_1.

According to definition the geometries $\mathcal{G}_1 = \{\sigma, \omega_1\}$ and $\mathcal{G}_2 = \{\sigma, \omega_2\}$ on parts $\omega_1 \subset \Omega$ and $\omega_2 \subset \Omega$ of Ω are equivalent $(\mathcal{G}_1 \mathrm{eqv} \mathcal{G})$, $(\mathcal{G}_2 \mathrm{eqv} \mathcal{G})$ to the geometry $\mathcal{G} = \{\sigma, \Omega\}$, whereas the geometries $\mathcal{G}_1 = \{\sigma, \omega_1\}$ and $\mathcal{G}_2 = \{\sigma, \omega_2\}$ are not equivalent, generally speaking, if $\omega_1 \nsubseteq \omega_2$ and $\omega_2 \nsubseteq \omega_1$. Thus, the relation of the geometries equivalence is intransitive, in general. The space-time geometry may vary in different regions of the space-time. It means, that a physical body, described as a geometrical object, may evolve in such a way, that it appears in regions with different space-time geometry.

The space-time geometry of Minkowski as well as the Euclidean geometry are continuous geometries. It is true for usual scales of distances. However, one cannot be sure, that the space-time geometry is continuous in microcosm. The space-time geometry may appear to be discrete in microcosm. We consider a discrete space-time geometry and discuss the corollaries of the suggested discreteness.

14. Discreteness and its Manifestations

The simplest discrete space-time geometry $\mathcal{G}_\mathrm{d}$ is described by the world function (7.8). Density of points in $\mathcal{G}_\mathrm{d}$ with respect to point density in $\mathcal{G}_\mathrm{M}$ is described by the relation

$$\frac{d\sigma_\mathrm{M}}{d\sigma_\mathrm{d}} = \begin{cases} 0 & \text{if} \quad |\sigma_\mathrm{d}| < \frac{1}{2}\lambda_0^2 \\ 1 & \text{if} \quad |\sigma_\mathrm{d}| > \frac{1}{2}\lambda_0^2 \end{cases} \tag{14.1}$$

If the world function has the form

$$\sigma_\mathrm{g} = \sigma_\mathrm{M} + \frac{\lambda_0^2}{2} \begin{cases} \mathrm{sgn}(\sigma_\mathrm{M}) & \text{if} \quad |\sigma_\mathrm{M}| > \sigma_0 \\ \frac{\sigma_\mathrm{M}}{\sigma_0} & \text{if} \quad |\sigma_\mathrm{M}| \leq \sigma_0 \end{cases} \tag{14.2}$$

where $\sigma_0 = \text{const}$, $\sigma_0 \geq 0$, the relative density of points has the form

$$\frac{d\sigma_{\mathrm{M}}}{d\sigma_{\mathrm{g}}} = \begin{cases} \frac{2\sigma_0}{2\sigma_0+\lambda_0^2} & \text{if} \quad |\sigma_{\mathrm{g}}| < \sigma_0 + \frac{1}{2}\lambda_0^2 \\ 1 & \text{if} \quad |\sigma_{\mathrm{g}}| > \sigma_0 + \frac{1}{2}\lambda_0^2 \end{cases} \tag{14.3}$$

If the parameter $\sigma_0 \to 0$, the world function $\sigma_{\mathrm{g}} \to \sigma_{\mathrm{d}}$ and the point density (14.3) tends to the point density (14.1). The space-time geometry $\mathcal{G}_{\mathrm{g}}$, described by the world function (14.2) is a geometry, which is a partly discrete geometry, because it is intermediate between the discrete geometry $\mathcal{G}_{\mathrm{d}}$ and the continuous geometry $\mathcal{G}_{\mathrm{M}}$. We shall refer to the geometry $\mathcal{G}_{\mathrm{g}}$ as a granular geometry.

Deflection of the discrete space-time geometry from the continuous geometry of Minkowski generates special properties of the geometry, which are corollaries of impossibility of the linear vector space introduction.

Let $\mathbf{P}_0\mathbf{P}_1$ be a timelike g-vector in $\mathcal{G}_{\mathrm{d}}$ ($\sigma_{\mathrm{d}}(P_0, P_1) > 0$). We try to determine a g-vector $\mathbf{P}_1\mathbf{P}_2$ at the point P_1, which is equivalent to g-vector $\mathbf{P}_0\mathbf{P}_1$. Geometrical vectors $\mathbf{P}_0\mathbf{P}_1$ and $\mathbf{P}_1\mathbf{P}_2$ may be considered as two adjacent links of a broken world line, describing a pointlike particle.

Let for simplicity coordinates have the form

$$P_0 = \{0, 0, 0, 0\}, \qquad P_1 = \{\mu, 0, 0, 0\}, \qquad P_2 = \left\{x^0, \mathbf{x}\right\} = \left\{x^0, x^1, x^2, x^3\right\} \tag{14.4}$$

In this coordinate system the world function of geometry Minkowski has the form

$$\sigma_{\mathrm{M}}\left(x, x'\right) = \frac{1}{2}\left(\left(x^0 - x^{0\prime}\right)^2 - \left(\mathbf{x} - \mathbf{x}'\right)^2\right) \tag{14.5}$$

and σ_{d} is determined by the relation (7.8). We are to determine coordinates x of the point P_1 from two equations (7.1), which can be written in the form

$$\sigma_{\mathrm{d}}\left(P_0, P_1\right) = \sigma_{\mathrm{d}}\left(P_1, P_2\right), \quad \sigma_{\mathrm{d}}\left(P_0, P_2\right) = 4\sigma_{\mathrm{d}}\left(P_0, P_1\right) \tag{14.6}$$

After substitution of world function (7.8) one obtains

$$\frac{1}{2}\left(\left(x^0 - \mu\right)^2 - \mathbf{x}^2 + \lambda_0^2\right) = \frac{1}{2}\left(\mu^2 + \lambda_0^2\right) \tag{14.7}$$

$$\frac{1}{2}\left(\left(x^0\right)^2 - \mathbf{x}^2 + \lambda_0^2\right) = 2\left(\left(x^0 - \mu\right)^2 - \mathbf{x}^2 + \lambda_0^2\right) \tag{14.8}$$

Solution of these equations has the form

$$x^0 = 2\mu + \frac{3}{2}\frac{\lambda_0^2}{\mu}, \qquad \mathbf{x}^2 = 3\lambda_0^2\left(1 + \frac{3\lambda_0^2}{4\mu^2}\right) \tag{14.9}$$

As a result the point P_2 has coordinates

$$P_2 = \left\{2\mu + \frac{3}{2}\frac{\lambda_0^2}{\mu}, r\sin\theta\cos\varphi, r\sin\theta\sin\varphi, r\cos\theta\right\}, \quad r = \lambda_0\sqrt{3 + \frac{9}{4}\frac{\lambda_0^2}{\mu^2}} \tag{14.10}$$

where θ and φ are arbitrary quantities. Thus, spatial coordinates of the point P_2 are determined to within $\sqrt{3}\lambda_0$. In the limit $\lambda_0 \to 0$ the point P_2 is determined uniquely. Two solutions

$$P_2' = \left\{2\mu + \frac{3}{2}\frac{\lambda_0^2}{\mu}, 0, 0, r\right\}, \quad P_2'' = \left\{2\mu + \frac{3}{2}\frac{\lambda_0^2}{\mu}, 0, 0, -r\right\}$$

are divided by spatial distance $i\,|\mathbf{P}_2'\mathbf{P}_2''| = \sqrt{4r^2 + \lambda_0^2} \approx \sqrt{13}\lambda_0\ (\lambda_0 \ll \mu)$. It is a maximal distance between two solutions $\mathbf{P}_2'$ and $\mathbf{P}_2''$.

If $\lambda_0 = 0$, then the discrete geometry turns to the geometry of Minkowski, and $P_2 = \{2\mu, 0, 0, 0\}$. The relations

$$x^0 = 2\mu, \quad x^1 = 0, \quad x^2 = 0, \quad x^3 = 0 \tag{14.11}$$

follow from one equation $\mathbf{x}^2 = 0$. It means, that the geometry of Minkowski is a degenerate geometry, because different solutions of the discrete geometry merge into one solution of the geometry of Minkowski.

Let us consider the same problem for spacelike g-vectors $\mathbf{P}_0\mathbf{P}_1$, $\mathbf{P}_1\mathbf{P}_2$, when

$$P_0 = \{0, 0, 0, 0\}, \quad P_1 = \{0, l, 0, 0\}, \quad P_2 = \{ct, x, y, z\} \tag{14.12}$$

We have the same equations (14.6), but now we have another solution

$$x = 2l + \frac{3\lambda_0^2}{2l}, \quad y = a_2, \quad z = a_3, \quad c^2t^2 = r^2 = 3\lambda_0^2 + \frac{9}{4}\frac{\lambda_0^4}{l^2} \tag{14.13}$$

where a_2 and a_3 are arbitrary numbers. The point P_2 has coordinates

$$P_2 = \left\{\sqrt{a_2^2 + a_3^2 + r^2}, 2l + \frac{3\lambda_0^2}{2l}, a_2, a_3\right\}, \quad r^2 = 3\lambda_0^2\left(1 + \frac{3\lambda_0^2}{4l^2}\right) \tag{14.14}$$

The difference between two solutions P_2' and P_2''

$$P_2' = \left\{\sqrt{a_2^2 + a_3^2 + r^2}, 2l + \frac{3\lambda_0^2}{2l}, a_2, a_3\right\}, \quad P_2'' = \left\{\sqrt{b_2^2 + b_3^2 + r^2}, 2l + \frac{3\lambda_0^2}{2l}, b_2, b_3\right\}$$

may be infinitely large

$$|\mathbf{P}_2'\mathbf{P}_2''| = \sqrt{2a_2b_2 + 2a_3b_3 - 2\sqrt{r^2 + a_2^2 + a_3^2}\sqrt{r^2 + b_2^2 + b_3^2} + 2r^2 - \lambda_0^2}$$

This difference remains very large, even if $\lambda_0 \to 0$.

Thus, both the discrete geometry and the geometry of Minkowski are multivariant with respect to spacelike g-vectors. However, this circumstance remains to be unnoticed in the conventional relativistic particle dynamics, because the spacelike g-vectors do not used there.

Multivariance of the discrete geometry leads to intransitivity of the equivalence relation of two vectors. Indeed, if $(\mathbf{Q}_0\mathbf{Q}_1 \text{eqv} \mathbf{P}_0\mathbf{P}_1)$ and $(\mathbf{Q}_0\mathbf{Q}_1 \text{eqv} \mathbf{P}_0\mathbf{P}_1')$, but g-vector $(\mathbf{P}_0\mathbf{P}_1 \overline{\text{eqv}} \mathbf{P}_0\mathbf{P}_1')$, generally speaking. It means intransitivity of the equivalence relation.

Besides, it means that the discrete geometry is nonaxiomatizable, because in any logical construction the equivalence relation is transitive.

Transitivity of the equivalence relation in the case of the proper Euclidean geometry is a corollary of the special conditions (12.1) – (12.7). In the case of the arbitrary physical geometry they are not satisfied, generally speaking.

Parallel transport of a g-vector $\mathbf{P}_0\mathbf{P}_1$ to some point Q_0 leads to some indeterminacy of the result of this transport, because at the point Q_0 there are many g-vectors $\mathbf{Q}_0\mathbf{Q}_1$, $\mathbf{Q}_0\mathbf{Q}_1'$,..., which are equivalent to the g-vector $\mathbf{P}_0\mathbf{P}_1$.

According to (9.1) - (9.3) results of g-vectors summation and of a multiplication of a g-vector by a real number are not unique, in general, in the discrete geometry. It means, that one cannot introduce a linear vector space in the discrete geometry.

Let the discrete geometry be described by n coordinates. Let the skeleton $\mathcal{P}_n = \{P_0, P_1, ...P_n\}$ determine n g-vectors $\mathbf{P}_0\mathbf{P}_k$, $k = 1, 2, ...n$, which are linear independent in the sense

$$F_n(\mathcal{P}_n) = \det \|(\mathbf{P}_0\mathbf{P}_i.\mathbf{P}_0\mathbf{P}_k)\| \neq 0 \qquad i, k = 1, 2, ...n \tag{14.15}$$

One can determine uniquely projections of a g-vector $\mathbf{Q}_0\mathbf{Q}_1$ onto g-vectors $\mathbf{P}_0\mathbf{P}_k$, $k = 1, 2, ...n$ by means of relations (12.9). However, one cannot reestablish the g-vector $\mathbf{Q}_0\mathbf{Q}_1$, using its projections onto g-vectors $\mathbf{P}_0\mathbf{P}_k$, $k = 1, 2, ...n$, because a summation of the g-vector components is many-valued. Thus, all operations of the linear vector space are not unique in the discrete geometry.

Mathematical technique of differential geometry is not adequate for application in a discrete geometry, because it is too special and it is adapted for a continuous (differential) geometry. This circumstance is especially important in a description of the elementary particle dynamics. The state of a particle cannot be described by its position and its momentum, because the limit $\mu \to 0$ in (14.4) does not exist in a discrete geometry. Besides, dynamic equations cannot be differential equations.

15. Skeleton Conception of Particle Dynamics

An elementary particle is a physical body. In the discrete space-time geometry a position of a physical body is described by its skeleton $\mathcal{P}_n = \{P_0, P_1, ..P_n\}$. Of course, such a description of a physical body position may be used in any space-time geometry. The skeleton is an analog of the frame of reference attached rigidly to the particle (physical body). Tracing the skeleton motion, one traces the physical body motion. Direction of the skeleton displacement is described by the leading vector $\mathbf{P}_0\mathbf{P}_1$.

The skeleton motion is described by a world chain $\mathcal{C}$ of connected skeletons

$$\mathcal{C} = \bigcup_{s=-\infty}^{s=+\infty} \mathcal{P}_n^{(s)} \tag{15.1}$$

Skeletons $\mathcal{P}_n^{(s)}$ of the world chain are connected in the sense, that the point P_1 of a skeleton is a point P_0 of the adjacent skeleton. It means

$$P_1^{(s)} = P_0^{(s+1)}, \qquad s = ...0, 1, ... \tag{15.2}$$

The geometric vector $\mathbf{P}_0^{(s)}\mathbf{P}_1^{(s)} = \mathbf{P}_0^{(s)}\mathbf{P}_0^{(s+1)}$ is the leading g-vector, which determines the direction of the world chain.

If the particle motion is free, the adjacent skeletons are equivalent

$$\mathcal{P}_n^{(s)}\text{eqv}\mathcal{P}_n^{(s+1)}: \qquad \mathbf{P}_i^{(s)}\mathbf{P}_k^{(s)}\text{eqv}\mathbf{P}_i^{(s+1)}\mathbf{P}_k^{(s+1)}, \qquad i,k=0,1,...n, \qquad s=..0,1,.. \tag{15.3}$$

If the particle is described by the skeleton $\mathcal{P}_n^{(s)}$, the world chain (15.1) has $n(n+1)/2$ invariants

$$\mu_{ik}^2 = \left|\mathbf{P}_i^{(s)}\mathbf{P}_k^{(s)}\right|^2 = 2\sigma\left(P_i^{(s)}, P_k^{(s)}\right), \qquad i,k=0,1,...n, \qquad s=...0,1,... \tag{15.4}$$

which are constant along the whole world chain.

Equations (15.3) form a system of $n(n+1)$ difference equations for determination of nD coordinates of n skeleton points $\{P_1, P_2, ..P_n\}$, where D is the coordinate dimension of the space-time. The number of dynamical variables, liable for determination distinguishes, generally speaking, from the number of dynamic equations. It is the main difference between the skeleton conception of particle dynamics and the conventional conception of particle dynamics, where the number of dynamic variables coincides with the number of dynamic equations.

In the case of pointlike particle, when $n=1$, $D=4$, the number of equations $n_e = 2$, whereas the number of variables $n_v = 4$. The number of equations is less, than the number of dynamic variables. In the discrete space-time geometry (7.8) the position of the adjacent skeleton is not uniquely determined. As a result the world chain wobbles. In the nonrelativistic approximation a statistical description of the stochastic world chains leads to the Schrödinger equations [14], if the elementary length λ_0 has the form

$$\lambda_0^2 = \frac{\hbar}{bc} \tag{15.5}$$

where $\hbar$ is the quantum constant, c is the speed of the light and b is a universal constant, connecting the particle mass m with the length μ of the world chain link by the relation (7.10).

Dynamic equations (15.3) are difference equations. At the large scale, when one may go to the limit $\lambda_0 = 0$, the dynamic equations (15.3) turn to the differential dynamic equations. In the case of pointlike particle $(n=1)$ and of the Kaluza-Klein five-dimensional space-time geometry these equations describe the motion of a charged particle in the given electromagnetic field. One can see in this example, that the space-time geometry "assimilates" the electromagnetic field. It means that one may consider only a free particle motion, keeping in mind, that the space-time geometry can "assimilate" all force fields.

Dynamic equations (15.3) realize the skeleton conception of particle dynamics in the microcosm. The skeleton conception of dynamics distinguishes from the conventional conception of particle dynamics in the relation, that the number of dynamic equations may differ from the number of dynamic variables, which are to be determined. In the conventional conception of particle dynamics the number of dynamic equations (of the first order) coincides always with the number of dynamic variables, which are to be determined. As a result the motion of a particle (or of an averaged particle) appears to be deterministic. In the case of quantum particles, whose motion is stochastic (indeterministic), the dynamic equations are written for a statistical ensemble of indeterministic particles (or for the statistically averaged particle).

In the conventional conception of the particle dynamics one can obtain dynamic equation for the statistically averaged particle (i.e. statistical ensemble normalized to one particle), but there are no dynamic equations for a single stochastic particle. In the skeleton conception of the particle dynamics there are dynamic equations for a single particle. These equations are many-valued (multivariant), but they do exist. In the conventional conception of the particle dynamics one can derive dynamic equations for the statistically averaged particle, which are a kind of equations for a fluid (continuous medium). But one cannot obtain dynamic equations for a single indeterministic particle [7].

The skeleton conception of the particle dynamics realizes a more detailed description of elementary particle. One may hope to obtain some information on the elementary particle structure.

We have now only two examples of the skeleton conception application. Considering compactification in the 5-dimensional discrete space-time geometry of Kaluza-Klein, and imposing condition of uniqueness of the world function, one obtains that the value of the electric charge of a stable elementary particle is restricted by the elementary charge [20]. This result has been known from experiments, but it could not be explained theoretically, because in the continuous space-time geometry nobody considers the world function as a fundamental quantity, and one does not demand its uniqueness.

Another example concerns structure of Dirac particles (fermions). Writing the Dirac equation as dynamic equation for an ensemble of a stochastic particle [21, 22, 23] one obtains that the mean world line of this particle is a helix with timelike axis. Spin and magnetic moment of the Dirac particle are conditioned by the particle rotation in its motion along the helical world line. Thus, statistical description provides some information on the Dirac particle structure, whereas the quantum approach cannot give such information, although in both cases one investigates the same dynamic equation.

Consideration in the framework of skeleton conception [24] shows, that a world chain of a fermion is a (spacelike or timelike) helix with timelike axis. The averaged world chain of a free fermion is a timelike straight line. The helical motion of a skeleton generates an angular moment (spin) and magnetic moment. Such a result looks rather reasonable. In the conventional conception of the particle dynamics the spin and magnetic moment of a fermion are postulated without a reference to its structure. Helical world chain of the Dirac particle is connected with the fact that the skeleton of the Dirac particle contains three, or more points and it is described by three (or more) invariants $\mu_{ik} = \mu_{ki}$, $i, k = 1, 2, 3$ which are defined by the relation (15.4). In the case of the two-point skeleton describing the pointlike particle there is only one parameter μ which describes the particle mass. At the quantum approach the parameter μ is absent and the particle mass is considered as some external (not geometric) parameter of a particle. In the skeleton conception all particle parameteres are geometric quantities. In the case of the Dirac particle its mass and spin are expressed via geometical invariants μ_{ik}. (this connection is not yet obtained).

16. Concluding Remarks

Structural approach to the elementary particle theory appears as a result of a skeleton conception, where the particle state is described by means of the particle skeleton. Such a description of the particle state is relativistic. Besides, this description is coordinateless, and it can be produced in any space-time geometry (continuous or discrete). Relativistic concept of the particle state admits one to replace the quantum description by statistical description of stochastic world lines of elementary particles. As a result the quantum principles and quantum essences appeared to be unnecessary. Such a replacement of the quantum description by the statistical description appears to be possible because of a logical reloading in the particle dynamics, when a single particle as the basic object of dynamics is replaced by a statistical ensemble of particles, and dynamics of stochastic and deterministic particles is described in the same terms.

Further development of the skeleton conception arises after the logical reloading in space-time geometry, when such basic geometric concepts as dimension, coordinate system, infinitesimal distance are replaced by the unique basic concept: finite distance or world function. As a result a monistic conception of the space-time geometry appears. Capacities of this conception increases essentially because of monistic character of the description. In particular, one succeeds to overcome the degenerate character of $\mathcal{G}_{\rm E}$ and to construct multivariant space-time geometry. Multivariance of the space-time geometry explains freely the elementary particles stochasticity (quantum effects).

Thus, the supposition on the space-time geometry discreteness seems to be more natural and reasonable, than the supposition on quantum nature of physical phenomena in microcosm. Discreteness is simply a property of the space-time, whereas quantum principles assume introduction of new essences.

Formalism of the discrete geometry is very simple. It does not contain theorems with complicated proofs. Nevertheless the discrete geometry and its formalism is perceived hardly. The discrete geometry was not developed in the twentieth century, although the discrete space-time was necessary for description of physical phenomena in microcosm. It was rather probably, that the space-time is discrete in microcosm. What is a reason of the discrete geometry disregard? We try to answer this important question.

The discrete geometry was not developed, because it could be obtained only as a generalization of the proper Euclidean geometry. Almost all concepts and quantities of the proper Euclidean geometry use essentially concepts of the continuous geometry. They could not be used for construction of a discrete geometry. Only world function (or distance) does not use a reference to the geometry continuity. Only coordinateless expressions (7.1) –(7.4) of the Euclidean geometry presented in terms of world function admit one to construct a discrete geometry and other physical geometries.

Assurance, that any geometry is to be axiomatizable, was the second obstacle on the way of the discrete geometry construction. The fact, that the proper Euclidean geometry is a degenerate geometry, was the third obstacle. In particular, being a physical geometry, the proper Euclidean geometry is an axiomatizable geometry, and this circumstance is an evidence of its degeneracy. It is very difficult to obtain a general conception as a generalization of a degenerate conception, because some different quantities of the general conception coincide in the degenerate conception. It is rather difficult to disjoint them. For instance, a physical geometry is multivariant, generally speaking. Single-variant physical geometry is a degenerate geometry. In the physical geometry the straight segment $\mathcal{T}_{[P_0P_1]}$

$$\mathcal{T}_{[P_0P_1]} = \left\{R | \sqrt{2\sigma(P_0, R)} + \sqrt{2\sigma(R, P_1)} = \sqrt{2\sigma(P_0, P_1)}\right\} \tag{16.1}$$

is a surface (tube), generally speaking. In the degenerate physical geometry (the proper Euclidean geometry $\mathcal{G}_E$) the straight segment is a one-dimensional set. How can one guess, that a straight segment is a surface, generally speaking? Besides, multivariance of the equivalence relation leads to nonaxiomatizability of geometry. But we learn only axiomatizable geometries in the last two thousand years. How can we guess, that nonaxiomatizable geometries exist? *The multivariance is a natural property of a geometry. Non-acceptance of this concept is the main reason of the discrete physical geometry disregard.* The straight way from the Euclidean geometry to physical geometries was very difficult, and the physical geometry has been derived on an oblique way.

J.L.Synge [25, 26] has introduced the world function for description of the Riemannian geometry. I was a student. I did not know the papers of Synge, and I introduced the world function for description of the Riemannian space-time in general relativity. My approach differed slightly from the approach of Synge. In particular, I had obtained an equation for the world function of Riemannian geometry [27], which contains only the world function and their derivatives,

$$\frac{\partial\sigma(x, x')}{\partial x^i} G^{ik'}(x, x') \frac{\partial\sigma(x, x')}{\partial x'^k} = 2\sigma(x, x'), \qquad G^{ik'}(x, x') G_{lk'}(x, x') = \delta^i_l, \tag{16.2}$$

where

$$G_{lk'}(x, x') \equiv \frac{\partial^2\sigma(x, x')}{\partial x^l \partial x'^k}, \quad l, k = 0, 1, 2, 3 \tag{16.3}$$

The metric tensor is expressed via world function σ by the relation

$$g_{ik}(x) = -G_{lk'}(x, x) = -\left[G_{lk'}(x, x')\right]_{x'=x} \tag{16.4}$$

but it is used at the determination of the world function $\sigma(x, x')$ only as a initial (or boundary) condition. Equation (16.2) was obtained as a corollary of the world function definition as an integral along the geodesic, connecting points x and x'. This equation contains only world function and its derivatives, but it does not contain a metric tensor.

This equation arose the question. Let a world function σ do not satisfy the equation (16.2). Does this world function describe a non-Riemannian geometry or it does describe no geometry at all? It was very difficult to answer this question. On one hand, the formalism, based on the world function, is a more developed formalism, than formalism based on a usage of metric tensor, because a geodesic between points P_0, P_1 is described in terms of the world function by algebraic equation (16.1), whereas the same geodesic is described by differential equations in terms the metric tensor.

On the other hand, the geodesic described by (16.1) is one-dimensional only in the Riemannian geometry. In n-dimensional space the equation (16.1) describes a $(n-1)$-dimensional surface. I did not know, whether the surface is a generalization of a geodesic in any geometry. I was not sure, if it is possible, because in the Euclidean geometry a straight line segment is one-dimensional by definition. I left this question unsolved and returned to it almost thirty years later, in the beginning of ninetieth.

When the string theory of elementary particles appeared, it became clear for me, that the particle may be described by means of a world surface (tube) but not only by a world line. As far as the particle world line associates with a geodesic, I decided, that a world tube may describe a particle. It meant that there exist space-time geometries, where straights (geodesics) are described by world tubes. The question on possibility of the physical space-time geometry has been solved for me finally, when the quantum description appeared to be a corollary of the space-time multivariance [14].

References

[1] Yu. A.Rylov, Discrete space-time geometry and skeleton conception of particle dynamics. *Int. J. Theor. Phys.* {51, *iss. 6 1847-1865, (2012)*, see also *e-print 1110.3399v1*

[2] Yu.A. Rylov, Spin and wave function as attributes of ideal fluid. (*Journ. Math. Phys.* {40, *pp. 256 - 278, (1999)*

[3] E. Madelung, Quanten theorie in hydrodynamischer Form, Z. Physik, {40, 322-326, (1926).

[4] D. Bohm, On interpretation of quantum mechanics on the basis of the "hidden" variable conception. *Phys.Rev.* {85, 166, 180, (1952).

[5] A. Clebsch, *J. reine angew. Math.* {54 , 293, (1857).

[6] A. Clebsch, *J. reine angew. Math.* {56 , 1, (1859).

[7] Yu. A.Rylov, Uniform formalism for description of dynamic, quantum and stochastic systems. *e-print /physics/0603237v6*

[8] Yu.A.Rylov, Quantum mechanics as a dynamic construction. *Found. Phys.* {28, No.2, 245-271, (1998).

[9] C.C. Lin, *Proc. International School of Physics "Enrico Fermi"*. Course XXI, Liquid Helium , New York, Academic. 1963, pp. 93-146.

[10] Yu.A.Rylov, Classical description of pair production, *e-print /abs/physics/0301020)*

[11] L.M. Blumenthal, *Theory and Applications of Distance Geometry*, Oxford, Clarendon Press, 1953

[12] Yu. A. Rylov, Dynamic equations for tachyon gas. *Int. J. Theor. Phys.* (2012)

[13] Yu. A. Rylov, Tachyon gas as a candidate for dark matter. *Vestnik RUDN, mathematics, informatics, physics (2013) iss 2 pp.159-173.*

[14] Yu.A.Rylov, "Non-Riemannian model of the space-time responsible for quantum effects. *Journ. Math. Phys.* {32(8), *2092-2098, (1991)*

[15] Yu.S.Wladimirov, *Geometrodynamics*, chpt. 8, Moscow, Binom, 2005 (in Russian)

[16] Yu. A.Rylov, Deformation principle and further geometrization of physics.*e-print* /0704.3003

[17] Yu.A. Rylov, Non-Euclidean method of the generalized geometry construction and its application to space-time geometry in *Pure and Applied Differential geometry* pp.238-246. eds. Franki Dillen and Ignace Van de Woestyne. Shaker Verlag, Aachen, 2007. See also *e-print Math.GM/0702552*

[18] Yu.A. Rylov, Geometry without topology as a new conception of geometry. *Int. Jour. Mat. & Mat. Sci.* {30, *iss. 12, 733-760, (2002).*

[19] Yu. A. Rylov, Different conceptions of Euclidean geometry and their application to the space-time geometry. *e-print /0709.2755v4*

[20] Yu. A. Rylov, Discriminating properties of compactification in discrete uniform isotropic space-time. *e-print 0809.2516v2*

[21] Yu.A.Rylov, Dirac equation in terms of hydrodynamic variables. *Advances in Applied Clifford Algebras*, {5, *pp 1-40, (1995)*). See also *e-print* {/1101.5868.

[22] Yu. A.Rylov, Is the Dirac particle composite? *e-print* /physics/0410045.

[23] Yu. A. Rylov (2004), Is the Dirac particle completely relativistic? *e-print* /physics/0412032.

[24] Yu. A. Rylov, Geometrical dynamics: spin as a result of rotation with superluminal speed. *e-print 0801.1913.*

[25] J.L. Synge, A characteristic function in Riemannian space and its applications to the soution of geodesic triangles. *Proc. London Math. Soc.* {32, *241, (1931).*

[26] J.L.Synge, *Relativity: the General Theory.* Amsterdam, North-Holland Publishing Company, 1960.

[27] Yu.A.Rylov, On a possibility of the Riemannian space description in terms of a finite interval. *Izvestiya Vysshikh Uchebnych Zavedenii, Matematika. No.3(28), 131-142.* (1962). (in Russian).

In: Space-Time Geometry and Quantum Events
Editor: Ignazio Licata, pp. 317-336
ISBN: 978-1-63117-455-1

Chapter 11

FORMING PHYSICAL FIELDS AND PSEUDOMETRIC AND METRIC MANIFOLDS. NONCOMMUTATIVITY AND DISCRETE STRUCTURES IN CLASSICAL AND QUANTUM PHYSICS

L. I. Petrova*
Moscow State University, Moscow, Russia

Abstract

It is shown that from the equations of conservation laws for energy, linear momentum, angular momentum, and mass, which are conservation laws for material systems (such as thermodynamic, gas dynamical, cosmological systems, systems of charged particles, systems of elementary particles and others), it follows the evolutionary relation for functionals the examples of which are the action functional, entropy, the Pointing vector, the Einstein tensor, and so on. This relation, which appears to be nonidentical due to the noncommutativity of conservation laws, describes the evolutionary processes proceeded in material systems and accompanied by the origination of physical structures. The physical fields and relevant manifolds are formed by such physical structures. In material system the origination of physical structures is accompanied by the emergence of observable formations such as waves, vortices, turbulent pulsations and so on. Such a duality discloses the connection of physical fields with material systems and allows one to understand the properties and peculiarities of physical fields and relevant manifolds.

The generation of physical fields and pseudometric and metric manifolds by material systems points to the fact that the space-time is not an originally unchanged substance.

1. Introduction

As it will be shown, the process of forming physical fields and manifolds is controlled by the conservation laws. This is due to the fact that there exist two types of conservation

*E-mail address: ptr@cs.msu.su

laws, namely, the conservation laws for physical fields and the conservation laws laws for material systems (such as thermodynamic, gas dynamical, cosmological systems, systems of charged particles and others).

The conservation laws for physical fields are those that claim an existence of conservative physical quantities or objects. Such conservation laws are described by closed exterior skew-symmetric forms (the Noether theorem is an example). It is well known that the closed exterior forms and relevant dual forms made up a differential-geometric structure. The physical structures that made up physical fields and relevant manifolds are such physical structures.

In the paper it is shown that closed exterior forms, which correspond to conservation laws for physical fields and describe physical structures, are obtained from the equations of conservation laws for energy, linear momentum, angular momentum, and mass, which are equations of conservation laws for material systems.

The conservation laws for material systems establish the balance between the variance of physical quantities and external actions. Such conservation laws, which can be named as balance ones, are described by differential equations.

From the analysis of conservation law equations for material systems it follows that the balance conservation laws are noncommutative ones. The noncommutativity of the conservation laws is a driving force of evolutionary processes proceeded in material systems. Such processes are accompanied by origination of physical structures.

The process of origination of physical structures is described by the evolutionary relation that is obtained from the conservation law equations for material systems and appears to be nonidentical due to the noncommutativity of balance conservation laws.

The evolutionary relation obtained from the equations for material systems is a relation for such functionals as the action functional, entropy, the Pointing vector, the Einstein tensor, and so on. This proves the fact that physical fields and relevant manifolds are generated by material systems. The connection between physical fields and material systems enables one to understand the properties of physical fields, their characteristics and to introduce the classification of physical fields.

2. The Mechanism of Generation of Physical Structures by Material Systems

It will be shown below that the physical structures, the examples of which are structures from which physical fields and relevant manifolds are made up, are generated by material systems such as the thermodynamic, gas-dynamic and cosmological systems, the systems of charged particles and so on. This follows from the analysis of the properties and peculiarities of differential equations that describe material systems.

2.1. The Functional Properties of the Mathematical Physics Equations, which Describe Material Systems

The equations of material systems are equations of conservation laws for energy, linear momentum, angular momentum, and mass [1-3]. These equations, as it is well known, are used for description of physical quantities, which specify the behavior of material systems.

But, it appears that the equations of material systems define not only the variation of physical quantities. Their role is much wider. They control the evolutionary processes in material systems that are accompanied by the origination of physical structures.

The functions for equations of material media sought are usually functions which relate to such physical quantities like a particle velocity (of elements), temperature or energy, pressure and density. Since the functions desired relate to a one material system, it has to exist a connection between them. This connection is described by state functional that specifies the material system state. The action functional, entropy, the Pointing vector, Einstein tensor, wave function and others can be regarded as examples of such functional [4].

From equations of material system it follows the evolutionary relation for these functionals. Such a relation, which appears to be evolutionary and nonidentical one, just describes evolutionary processes in material systems and the origination of physical structures.

The evolutionary relation is obtained in the analysis of the integrability of the material system equations, which depends on the consistency of derivatives along different directions and on the consistency of equations in the set of equations. (In particular, the consistency of equations was analyzed in paper [5]. In that paper the consistency conditions were referred to as dynamical conditions.)

From the analysis of consistency of derivatives along different directions it follows that the differential equation, which describes any processes, has the solutions of two types, namely, the solutions that are not functions and the generalized solutions that are functions but are realized discretely. The transition from the solutions of the first type to generalized one, as it has been noted in the author's paper (see, for example, [6]), points to the origination of some structure.

In the analysis of the consistency of equations in the set of equations for material systems (which will be presented below) the evolutionary relation obtained not only discloses the properties of the solutions to differential equations, but it also describes the process of physical structure originations.

2.2. Study the Consistency of the Conservation Law Equations for Material Systems. Evolutionary Relation

The equations are consistent if they can be contracted into identical relations for the differentials, i.e. for closed exterior skew-symmetric forms. (Some properties of skew-symmetric forms are presented in the Appendix 1.)

Let us now analyze the consistency of the equations for energy and linear momentum.

We introduce two frames of reference: the first is an inertial one and the second is an accompanying one (this system is connected with the manifold built by the trajectories of material system elements).

In the inertial frame of reference the energy equation can be reduced to the form:

$$\frac{D\psi}{Dt} = A_1 \tag{1}$$

where D/Dt is the total derivative with respect to time, ψ is the functional of the state that specifies the material system, coefficient A_1 is a quantity which depends on the specific features of the system and on the external energy actions onto the system. [The action functional, entropy and wave function can be regarded as examples of the functional ψ [4]. Thus, the equation for energy expressed in terms of the action functional S has the following form: $DS/Dt = L$, where $\psi = S$ and $A_1 = L$ is the Lagrange function. The equation for energy of ideal gas can be presented in the form: $Ds/Dt = 0$, where s is entropy [4].]

In the accompanying frame of reference the total derivative with respect to time converts into the derivative along trajectory. Equation (1) is now written in the form

$$\frac{\partial\psi}{\partial\xi^1} = A_1 \tag{2}$$

here ξ^1 is the coordinate along trajectory.

In a similar manner, in the accompanying frame of reference the equation for linear momentum appears to be reduced to the equation of the form

$$\frac{\partial\psi}{\partial\xi^\nu} = A_\nu, \quad \nu = 2, \ldots \tag{3}$$

where ξ^ν are the coordinates in the direction normal to trajectory, the A_ν are quantities that depend on specific features of the system and external (with respect to local domain) force actions.

Eqs. (2) and (3) can be convoluted into the relation

$$d\psi = A_\mu\, d\xi^\mu, \quad (\mu = 1, \nu) \tag{4}$$

where $d\psi$ is the differential expression $d\psi = (\partial\psi/\partial\xi^\mu)d\xi^\mu$.

Relation (4) can be written as

$$d\psi = \omega \tag{5}$$

here $\omega = A_\mu\, d\xi^\mu$ is the skew-symmetric differential form of the first degree. (The summation over repeated indices is implied)

Since the equations of conservation laws are evolutionary ones, the relation obtained is also an evolutionary relation.

[It should be noted that skew-symmetric differential forms, which are obtained from differential equations, are not exterior skew-symmetric forms because, in contrast to exterior forms, they are defined on accompanying manifolds, which are not integrable. Such skew-symmetric differential forms are evolutionary ones since they are obtained from evolutionary equations [7]. Below it will be shown that the properties of such evolutionary forms enable one to study specific features of solutions to the equations of mathematical physics. (About some properties of such skew-symmetric differential forms it is written in Appendix 1.)]

Relation (5) has been obtained from the equation of conservation laws for energy and linear momentum. In this relation the form ω is that of the first degree. If the conservation law equation for angular momentum be added to the equations for energy and linear momentum, this form in the evolutionary relation will be a form of the second degree. And in combination with the equation of the balance conservation law for mass this form will be a form of degree 3. Thus, in the general case, the evolutionary relation can be written as

$$d\psi = \omega^p \tag{6}$$

where the form degree p takes the values $p = 0, 1, 2, 3$.

[A concrete form of relation (5) and its properties in the case of the Euler and Navier-Stokes equations were considered in papers [8, 9]. In this case the functional ψ is the entropy s. A concrete form of relation (6) for $p = 2$ were considered for electromagnetic field in Appendix 3 of paper [10] and in paper http://arxiv.org/pdf/math-ph/0310050v1.pdf]

2.3. Nonidentity of Evolutionary Relation. Inconsistency of the Balance Conservation Law Equations

Evolutionary relation obtained from the equation of the balance conservation laws possesses some peculiarity. This relation proves to be nonidentical since the differential form in the right-hand side of this relation is not a closed form, and, hence, this form can't be a differential like the left-hand side.

To justify this, we shall analyze the relation (5).

The evolutionary form $\omega = A_\mu d\xi^\mu$ is not a closed form since its differential $d\omega$ is nonzero. The differential $d\omega$ can be written as $K_{\alpha\beta}d\xi^\alpha d\xi^\beta$, where $K_{\alpha\beta} = \partial A_\beta/\partial\xi^\alpha - \partial A_\alpha/\partial\xi^\beta$ are the components of the differential form commutator built of the mixed derivatives (here the term connected with the nonintegrability of the manifold has not yet been taken into account).

The coefficients A_μ of the form ω can be obtained either from the equation of the balance conservation law for energy or from that for linear momentum. This means that in the first case the coefficients depend on the energetic action and in the second case they depend on the force action. In actual processes energetic and force actions have different nature and appear to be inconsistent. The commutator of the form ω constructed from the derivatives of such coefficients is nonzero. The differential of the form ω is nonzero as well. Thus, the form ω proves to be unclosed and cannot be a differential. **This means that the evolutionary relation cannot be an identical one.** (In the left-hand side of this relation it stands a differential, whereas in the right-hand side it stands an unclosed form that is not a differential.)

Hence, without a knowledge of the particular expression for the form ω, one can argue that for actual processes the evolutionary relation proves to be nonidentical. [Nonidentical relation was analyzed in paper J.L.Synge "Tensorial Methods in Dynamics" (1936). And yet it was allowed a possibility to use the sign of equality in nonidentical relation.]

Similarly it can be shown that general relation (6) is also nonidentical. (The analysis of some particular equations of balance conservation laws and relevant evolutionary relations are presented in paper [7]). [The nonidentity of evolutionary relation are connected with the peculiarities of evolutionary skew-symmetric form ω^p that enters into this relation. The skew-symmetric form in evolutionary relation is defined on the manifold made up by trajectories of the material system elements. Such a manifold is a deforming nonintegrable manifold [11]. The differential, and hence the skew-symmetric form commutator defined on such a manifold, (as it was noted in Appendix 1)) includes an additional term related to the differential of basis. This term specifies the manifold deformation and hence is nonzero. Both terms of the commutator (obtained by differentiating the basis and the form coefficients) have a different nature and, therefore, cannot compensate one another. This fact once more emphasize that the evolutionary form commutator, and, hence, its differential, are nonzero. And this means that the evolutionary skew-symmetric form, which enters into evolutionary relation, cannot be closed. Nonclosure of evolutionary form and the properties of commutator of such form define properties and peculiarities of the relation obtained from the conservation law equations for material systems.]

The nonidentity of the evolutionary relation means that the balance conservation law equations turn out to be inconsistent.

2.4. Noncommutativity of the Balance Conservation Laws. Evolutionary Processes in Material System

Inconsistency of the balance conservation law equations points to the fact that the balance conservation laws are noncommutative [8].

2.5. Nonequilibrium State of Material System

Noncommutativity of the balance conservation laws reflects the state of material system.

Since the evolutionary relation is nonidentical, from this relation one cannot get the differential $d\psi$ of the state functional. This means that the functional ψ is not a state function. And this points to the fact that the material system is in nonequilibrium state. It is evident that the internal force producing such nonequilibrium state is described by the evolutionary form commutator. (If the evolutionary form commutator be zero, the evolutionary relation would be identical, and this would point to the equilibrium state, i.e. the absence of internal forces.) Everything that makes contribution to the commutator of the form ω^p leads to the emergence of internal force.

Nonidentical evolutionary relation also describes how the state of material system changes. This is due to that the evolutionary nonidentical relation is a selfvarying one. [This relation includes two objects one of which appears to be unmeasurable. The variation of any object of the relation in any process leads to the variation of another object and, in turn, the variation of the latter leads to the variation of the former. Since one of the objects is an unmeasurable quantity, the other cannot be compared with the first one, and hence, the process of mutual variation cannot terminate. This process is governed by the evolutionary form commutator, that is, by interaction between the commutator made up by derivatives of the form itself and by metric form commutator of deforming manifold made up by trajectories of elements of material system.]

The process of selfvariation of the evolutionary relation points to a change of the material system state. But the material system state remains nonequilibrium in this process because the internal forces do not vanish due to the evolutionary form commutator remains to be nonzero.

Such selfvariation of the material system state proceeds under the action of internal (rather than external) forces. That will continue even in the absence of external forces. (Here it should be noted that in an actual physical process the internal forces can be increasing, and this can lead to development of instability in the material system.)

2.6. Transition of Material System into a Locally Equilibrium State. Origination of Observable Formations

During selfvariation of evolutionary relation the conditions when the closed inexact (closed only *on pseudostructure*) exterior form is obtained from evolutionary form can be realized. This leads to the fact that from nonidentical evolutionary relation it will be obtained an identical relation, from which one can get the state functional, and this will point to the transition of material system from nonequilibrium state to locally equilibrium one.

The transition from unclosed evolutionary form (which differential is nonzero) to closed exterior form (which differential equals to zero) is possible only as a degenerate transformation, namely, a transformation that does not conserve the differential. [The availability of some symmetries is a condition of the degenerate transformation existence. Such conditions can be due to degrees of freedom of material system (like, for example, translation, rotation, oscillation and so on) that are realized while selfvarying of the material system nonequilibrium state.]

The conditions of degenerate transformation are those that determine the direction on which interior (only along a given direction) differential of evolutionary form vanishes. These are conditions that define the pseudostructure, i.e. the closure conditions of dual form, and lead to realization of the exterior form closed on pseudostructure.

As it has been already mentioned, the differential of the evolutionary form ω^p involved into nonidentical relation (6) is nonzero. That is, $d\omega^p \neq 0$. If the conditions of degenerate transformation are realized, it will take place the transition

$d\omega^p \neq 0 \to$ (degenerate transformation) $\to d_\pi \omega^p = 0$, $d_\pi{}^* \omega^p = 0$

The relations obtained

$$d_\pi \omega^p = 0, d_\pi{}^* \omega^p = 0 \tag{7}$$

are closure conditions for exterior inexact form and for relevant dual form. This means that it is realized the exterior form closed on pseudostructure.

In this case, on the pseudostructure π evolutionary relation (6) converts into the relation

$$d_\pi \psi = \omega_\pi^p \tag{8}$$

which proves to be an identical relation. Since the form ω_π^p is a closed one, on the pseudostructure this form turns out to be a differential. There are differentials in the left-hand and right-hand sides of this relation. This means that the relation obtained is an identical one.

From identical relation one can obtain the differential $d_\pi \psi$ of the state functional and find the state function. This points to that the material system state is an equilibrium state. But this state is realized only locally since the state differential is interior one defined exclusively on pseudostructure. (*The total state of material system turns out to be nonequilibrium* because the evolutionary relation itself remains to be nonidentical one.)

The transition of material system from nonequilibrium state into a locally equilibrium one means that the unmeasurable quantity described by the nonzero commutator of the unclosed evolutionary differential form ω^p, that acted as an internal force, transforms into the measurable quantity. In material system this reveals as an **occurrence of certain observable formations**, which develop spontaneously. Such formations and their manifestations are fluctuations, turbulent pulsations, waves, vortices, and others. The intensity of such formations is controlled by a quantity accumulated by the evolutionary form commutator. (In paper [9] the process of production of vorticity and turbulence was described.)

Below it will be shown that the process of origination of observable formations relates to generation of physical structures.

2.7. Origination of Physical Structures

In subsection 1.2 it has been shown that the transition from nonidentical evolutionary relation to identical one (which describes the transition of material system from nonequilibrium state to equilibrium one) relates to realization of closed dual form and exterior inexact skew-symmetric form.

Closed inexact exterior form and relevant closed dual form describe the differential-geometrical structure: the pseudostructure (dual form) and the conservative quantity (closed exterior form). [The dual form is a metric form of manifold]. It is evident that the pseudostructures with conservative quantity are structures, on which exact conservation laws are satisfied.

The realization of the closed dual form and the exterior inexact skew-symmetric form in transition from nonidentical evolutionary relation to identical one points to an occurrence of pseudostructure (dual form) with conservative quantity (closed exterior form), i.e. an occurrence of structure on which the exact conservation law is fulfilled.

Below it will be shown that such structures have a physical meaning, and therefore, they can be named as physical structures. [It should be noted that the differential-geometrical structures have also a mathematical meaning [6]. The integral structures and relevant integrable manifolds such as the characteristics, singular points, characteristic and potential surfaces, which are obtained when solving the mathematical physics equations, are such differential-geometrical structures.]

As it has been shown, the transition from nonidentical evolutionary relation to identical one and the realization of differential $d_\pi \psi$ of the state functional describe the transition of material system from nonequilibrium state to locally-equilibrium one and the origination of observable formations.

On the other hand, the transition from nonidentical evolutionary relation to identical one and the realization of the closed closed dual form and the exterior inexact skew-symmetric form describe the origination of physical structure.

It appears that the process of origination of observable formation relates to the occurrence of physical structure. This fact is also fixed by identical relation (8), which possesses the duality. The existence of the state differential (left-hand side of relation (8)) points to the transition of material system from nonequilibrium state to the locally-equilibrium state and the origination of observable formations. And the emergency of the closed (on pseudostructure) inexact exterior form (right-hand side of relation (8)) points to the origination of the physical structure.

Physical structures and the formations of material system observed are a manifestation of the same phenomena. On the other hand, the observed formation and the physical structure are not identical objects. [The light is an example of such a duality. The light manifests itself in the form of a massless particle (photon) and as a wave. If the wave be such a formation, the element of wave front made up the physical structure in its motion.]

(The formations observed can be considered as certain structures obtained from the equations of mathematical physics which describe material systems. It will be shown below that the physical structures from which physical fields are made up and which are generated by material systems are described by the field-theory equations. Therefore one may say that the formation observed are structures of classic physics, whereas the physical structures are structures of quantum physics. And, as it is shown, there is the connection between those.)

Thus, the analysis of the balance conservation law equations for material systems shows that material systems generate structures that are described by closed exterior forms. As it will be said below, the closed exterior forms correspond to the conservation laws for physical fields.

Below it will be presented the substantiation that the structures forming physical fields and relevant manifolds are those that are generated by material systems. This follows from the fact that the field-theory equations, which describe physical fields, are connected with the equations for material systems.

3. Connection of the Field-theory Equations with the Equations of Conservation Law for Material Systems

3.1. Closed Inexact Exterior Forms as the Basis of Field Theories

As it has been shown, from the equations for material system it follows the evolutionary relation from which the closed exterior forms are obtained.

The closed (inexact) exterior forms, which correspond to the conservation laws for physical fields and describe relevant physical structures, are solutions to the field-theory equations. And there is the following correspondence:

-Closed exterior forms of zero degree correspond to quantum mechanics.

-The Hamilton formalism bases on the properties of closed exterior and dual forms of first degree.

-The properties of closed exterior and dual forms of second degree are at the basis of the electromagnetic field equations.

-The closure conditions of exterior and dual forms of third degree form the basis of equations for gravitational field.

[One can see that field theories equations are connected with closed exterior forms of a certain degree. This enables one to introduce a classification of physical fields. Such a classification shows that there exists an internal connection between field theories, which describe physical fields of various types. It is evident that the degree of closed exterior forms is a parameter that integrates field theories into unified field theory. This can serve as a step to constructing the unified field theory.]

Below it will be shown that the closed inexact exterior forms, on which the field theories are based, are closed exterior forms obtained from the equations for material systems. This follows from the properties of field-theory equations.

3.2. Specific Features of the Field-theory Equations

The field-theory equations differ from the equations for material systems by their functional properties. The equations for material systems are differential equations, its solutions are functions (which describe physical quantities such as a velocity, pressure and density). And the solutions to the field-theory equations are closed inexact exterior forms, i.e. they are differentials. Only the equations that have the form of relations (nonidentical) may have the solutions which are differentials rather then functions.

One can verify that all equations of existing field theories have the form of nonidentical relations in differential forms or in the forms of their tensor or differential (i.e. expressed in terms of derivatives) analogs.

The Einstein equation is a relation in differential forms.

The Dirac equation relates Dirac's *bra*- and *cket*- vectors, which made up the differential forms of zero degree.

The Maxwell equations have the form of tensor relations.

The field equation and Schrődinger's one have the form of relations expressed in terms of derivatives and their analogs.

Another specific feature of the field-theory equations consists in the fact that all field-theory equations are nonidentical relations for functionals such as a wave function, action functional, the Pointing vector, Einstein's tensor and so on [4]. (Entropy is such a functional for the fields generated by thermodynamical and gas-dynamical systems.)

The evolutionary relation obtained from the equations for material systems is a nonidentical relation for all these functionals. That is, all field-theory equations are an analog to the evolutionary relation obtained from the equations for material system.

This points out to the connection of the field-theory equations, which describe physical fields, with the equations for material system.

Connection of the field-theory equations with the equations for material system points to the fact that physical structures, from which physical fields are made up, are generated by material systems. This means that there exists the connection between physical fields and material systems. This connection enables one to understand the mechanism of forming the physical fields and relevant manifolds and their properties.

4. Properties of Physical Structures that Made Up Physical Fields. The Mechanism of Forming the Pseudometric and Metric Manifolds

In section 1 the mechanism of evolutionary processes in material media was described. In section 2 it was shown that the structures, which made up physical fields and relevant manifolds, are structures that are realized in the evolutionary processes proceeded on material systems.

Below the characteristics of physical structures and the classification of relevant physical fields will be presented, and it will be described the mechanism of forming physical fields and relevant manifolds.

4.1. Characteristics of Physical Structures that Made Up Physical Fields

The physical structure is an object obtained by conjugating the conserved physical quantity, which is described by closed inexact exterior form, and the pseudostructure, which is described by relevant dual form. (The dual form corresponds to the metric form of the manifold on which the skew-symmetric form is defined).

Since the closed inexact exterior forms corresponding to physical structures are generated by the evolutionary forms obtained from the equations of balance conservation laws for the material system, it is evident that the characteristics of physical structures are determined by the characteristics of material system and the characteristics of evolutionary form. Moreover, the characteristics of physical structures are related to the degrees of freedom of material system, i.e. they depend on the material system construction. This is explained by the fact that the pseudostructure is obtained from the conditions of degenerate transformation, which are due to the degrees of freedom of material system. Vanishing certain functional expressions (like Jacobians, determinants, the Poisson brackets, residues and so on) must correspond to a degenerate transformation. The pseudostructure is obtained from these functional expressions that can be realized when any symmetries (conditioned by the degrees of freedom of material system) are obtained from the coefficients of the metric form commutators of accompanying manifold.

The connection between physical structures and closed exterior and dual forms allows to disclose the properties of the characteristics of physical fields.

Conserved physical quantity (closed exterior form) describes a certain charge.

Under transition from some structure to another, the conserved on pseudostructure quantity, which corresponds to the closed exterior form, changes discretely, and the pseudostructure changes discretely as well.

The discrete changes of the conserved quantity and pseudostructure are determined by the value of the evolutionary form commutator, which is a commutator at the time when the physical structure emerges. The first term of the evolutionary form commutator obtained from the derivatives of the evolutionary form coefficients controls the discrete change of the conserved quantity. The second one obtained from the derivatives of the metric form coefficients of the initial manifold controls the pseudostructure change.

Spin is the example of the second characteristic. Spin is a characteristic that determines the character of the manifold deformation before the emergence of physical structure. (The spin value depends on the form degree.)

The discontinuities of the derivatives normal to potential surface, the breaks of the derivative in passing throughout the characteristic surfaces and in passing throughout the wave front, and others are the examples of discrete change of the conserved quantity.

A discrete change of the conserved quantity and that of the pseudostructure produce the quantum that is obtained while going from one structure to another. The evolutionary form commutator formed at the instant of the structure origination determine characteristics of this quantum.

Here it should be called attention to one more fact. The material system can generate physical structures only if the system is in nonequilibrium state, that is, if it experiences the influence of any actions. As it has been shown, the internal forces arisen are described by the evolutionary form commutator. Therefore, the characteristics of the physical structures arisen will also depend on the characteristics of the evolutionary form commutator describing any actions onto a material system.

[As it was pointed before, the physical structure emerged reveals in material system as a observed formation. It should be noted some accordance between the characteristics of the formations emerged and the characteristics of the evolutionary forms, the evolutionary form commutators and the material system that allows one to clarify some properties of physical structures.

1) the intensity of the formation (a potential force) $\leftrightarrow$ *the value of the first term in the evolutionary form commutator* at the instant when the formation is created;

2) the vorticity $\leftrightarrow$ *the second term in the commutator that is connected with the dual (metric) form commutator*;

3) the absolute propagation speed of the formation created (a speed in the inertial frame of reference) $\leftrightarrow$ *additional conditions connected with degrees of freedom of the material system*;

4) the speed of the formation propagation relative to the material system $\leftrightarrow$ *additional conditions connected with degrees of freedom of the material system and the velocity of the local domain elements*.

The pseudostructures are connected with the front of the formation translation. Since a certain physical quantity is conserved on the pseudostructure (the closed form), the pseudostructure is a level surface. The equation of the pseudostructure is the equation of the eikonal surface. (The eikonal is an example of physical structure.)]

4.2. Classification of Physical Structures and Physical Fields

The connection of physical structures with the skew-symmetric differential forms allows to introduce a classification of these structures and corresponding physical fields in dependence on parameters that specify the skew-symmetric differential forms.

The closed forms corresponding to physical structures are generated by the evolutionary relation that contains the evolutionary form of the degree p where the degree p is connected with the number of interacting balance conservation laws. It is evident that physical structures may be classified by the parameter p.

The order parameter is the degree of closed forms generated by the evolutionary relation.

To determine this parameter, one has to consider the problem of integration of the nonidentical evolutionary relation.

If a relation is identical, this relation can be integrated directly since it involves closed forms that are differentials. In the integration process the transition to the relation with differential forms of lesser degrees takes place.

The integration and transitions with lowering the form degree are also allowed in nonidentical relations (with evolutionary forms) but only in the case of degenerate transformations. Under the degenerate transformation on the pseudostructure it can be obtained the identical relation that can be integrable, and this enables one to get a relation with the forms of degree less by one. The relation obtained turns out to be nonidentical as well. After further realization of degenerate transformation from this relation it can be once again obtained the identical relation that can be integrated, and the closed exterior forms will be obtained.

Under the degenerate transformation, the identical relation with closed form of degree $k = p$ is obtained from the nonidentical evolutionary relation with the evolutionary form of degree p. The identical relation obtained can be integrated. As the result, a new nonidentical relation with the evolutionary form of the degree $p-1$ will be obtained, from which, under realization of degenerate transformation, a new identical relation but with the closed form of lower degree $k = p-1$ can be again obtained, and this relation can be integrated once again.

Thus, under integrating the nonidentical evolutionary relation with the forms of degree p, one can successively obtain the closed forms on the pseudostructure of sequential degrees $k = p$, $k = p-1$, ..., $k = 0$. This points out to the realization of physical structures of relevant type.

The degree k of closed form can serve as another (in addition to the evolutionary form degree p in the evolutionary relation) parameter that classifies physical fields.

Closed exterior forms of the same degree realized in spaces of different dimensions prove to be distinguishable because the dimension of the pseudostructures, on which the closed forms are defined, depends on the space dimension. When generating closed forms of sequential degrees $k = p$, $k = p-1$, ..., $k = 0$ the pseudostructures of dimensions $(n+1-k)$: $1, \ldots, n+1$ are obtained.

As a result, the space dimension n also specifies physical fields.

The degree k of the closed forms realized and the number p connected with the number of balance conservation laws determine the type of physical fields.

The properties of physical structures are governed by the space dimension. The physical structures with closed forms of degree k realized in spaces of different dimensions are distinctive in their properties.

Thus, from the analysis of the evolutionary relation one can see that the type and the properties of the physical structures and, accordingly, of physical fields for a given material system depend on a number of balance conservation laws p, the degree of realized closed forms k, and the space dimension n. By introducing the classification by numbers p, k, n one can understand the internal connection between various physical fields. Since the physical fields are the carriers of interactions, such classification enables one to see the connection between interactions. One can see the correspondence between the degree k of the closed forms realized and the type of interactions. Thus, $k = 0$ corresponds to strong interaction, $k = 1$ corresponds to weak interaction, $k = 2$ corresponds to electromagnetic interaction, and $k = 3$ corresponds to gravitational interaction. In Appendix 2 it is presented the table of elementary particles, where physical fields and interactions in their dependence on the parameters p, k, n of evolutionary and closed exterior forms are demonstrated [12].

4.3. Formation of Pseudometric and Metric Manifolds

It was shown that there exists the connection between physical fields and material systems. The physical structures generated by material systems produce physical fields and relevant manifolds. The process of generation of physical structures by material systems leads to forming the pseudometric and metric manifolds.

Since the physical structures, i.e. the pseudostructures with conservative quantities, are described by closed inexact and relevant dual forms, it is evident that the process of forming the manifolds can be understood if one knows how the closed inexact exterior forms are realized. As it was shown above, such exterior forms are generated by the evolutionary form that enters into the evolutionary relation obtained from the conservation law equations for material systems. The above described process of generation of closed exterior forms and realization of physical structures discloses the process of forming the pseudometric and metric manifolds. It should be emphasized that the dual form realized in this process is a metric form of the pseudometric manifold.

Here it should be noted that the realization of pseudostructures and forming the pseudometric manifolds proceed as the result of the fact that there are two spatial objects.

When deriving the evolutionary relation there were used two spatial objects: the accompanying manifold (manifold made up by the trajectories of material system elements), which is not a metric one, and the inertial space which is a metric manifold. As the generator of the forming the metric space it serves the material system, and as the base it serves the accompanying manifold. The metric space formed becomes an inertial space of the dimension greater by one as compared to the initial space. The process of generation of physical structures is connected with the evolutionary process proceeded in material system under the action of internal forces arised due to the noncommutativity of the conservation laws for material systems. It appears to be possible to describe such a process due to the topologic properties of the commutator of the skew-symmetric evolutionary form (obtained from the conservation law equations for material systems). If the skew-symmetric differential forms are defined on the deforming nonintegrable manifolds with unclosed metric forms, the commutators of metric forms enter into the commutators of differential forms defined on such manifolds. The evolutionary form is defined on the accompanying deforming manifolds, i.e. on the manifold with unclosed metric forms (see, Appendix 2). For this reason the evolutionary form commutator contains two terms, one of which depends on the evolutionary form coefficients and the second term depends on the commutator of the evolutionary unclosed metric form. Due to such specific feature the evolutionary form commutators possess the topological properties, namely, they can execute the interplay between the evolutionary form and the basis, i.e. the metric form of the manifold. Such properties of the evolutionary form commutators and the peculiarities of the evolutionary relation obtained from the balance conservation laws for material systems just allows to disclose the mechanism of forming the pseudometric and metric spaces.

From the evolutionary relation, which describes the mechanism of physical structure emergence, it follows that the accompanying nonmetric manifold generates the pseudometric and metric manifold.

It was shown above that the evolutionary relation of degree p can generate (at the availability of degenerate transformations) closed forms of the degree $0 \leq k \leq p$ on the pseudostructures. When generating closed forms of sequential degrees $k = p$, $k = p - 1, \ldots,$ $k = 0$, as it was already noted, the pseudostructures of dimensions $(n+1-k)$: $1, \ldots, n+1$ are obtained. As a result of transition to the exact closed form of zero degree the metric structure of the dimension $n + 1$ is obtained.

The realization of metric structures points to the fact that the metric space is also formed by the material system itself. Under the influence of an external action (and with the availability of the degrees of freedom) the material system can transfer the initial inertial space into the space of the dimension $n + 1$ (see, Appendix 2). [It is known that the skew-symmetric tensors of the rank k correspond to the closed exterior differential forms, and the pseudotensors of the rank $(N-k)$, where N is the space dimension, correspond to the relevant dual forms. The pseudostructures correspond to such tensors, but on the space formed with the dimension $n+1$. That is, $N = n+1$.] The mechanism of forming the pseudostructures and the metric structures can explain, in particular, how the internal structure of elements of the material system is formed.

So it can be seen that the inertial spaces are not absolute spaces where actions are developed, they are spaces generated by the material systems.

Here the following should be pointed out. Physical structures are generated by local domains of material system. These are elementary physical structures. By combining with one another they can form large-scale structures making up pseudomanifolds and physical fields.

Sections of the cotangent bundles (Yang-Mills fields), cohomologies by de Rham, singular cohomologies, the pseudo-Riemann and pseudo-Euclidean spaces and others can be regarded as examples of the pseudostructures and spaces that are formed in a similar manner. Riemannian space is the example of metric manifold obtained in passing to exact forms [7].

Something can be said about the Pseudo-Riemannian manifold and Riemannian space.

The distinctive property of the Riemannian manifold is an availability of the curvature. This means that the metric form commutator of the third degree is nonzero. Hence, the commutator of the evolutionary form of third degree ($p = 3$), which enters into the proper metric form commutator, is not equal to zero. That is, the evolutionary form that enters into the evolutionary relation is unclosed, and the relation is nonidentical one.

When realizing pseudostructures of the dimensions $1, 2, 3, 4$ and obtaining the closed inexact forms of the degrees $k = 3, k = 2, k = 1, k = 0$ the pseudo-Riemannian space is formed, and the transition to the exact form of zero degree corresponds to the transition to the Riemannian space.

It is well known that when obtaining the Einstein equations it was assumed that the following conditions are satisfied [11,13]: 1) the Bianchi identity is satisfied, 2) the coefficients of connectedness are symmetric, 3) the requirement that the coefficients of connectedness are the Christoffel symbols, and 4) an existence of the transformation under which the coefficients of connectedness vanish. These conditions are the conditions of realization of degenerate transformations for nonidentical relations obtained from the evolutionary nonidentical relation with evolutionary form of the degree $p = 3$ and after going to identical relations. In this case the identical relations with forms of the first degree are assigned to the Einstein equation.

Conclusion

From the description of evolutionary processes in material system one can see that physical fields and relevant manifolds are generated by material system. (And thus the causality of physical processes and phenomena is explained.)

Here it should be emphasized that the conservation laws for material system, i.e. the balance conservation laws for energy, linear momentum, angular momentum, and mass, which are noncommutative ones, play a controlling role in these processes. This is precisely the noncommutativity of the balance conservation laws produced by external actions onto material system, which is a moving force of evolutionary processes leading to the emergence of physical structures (to which the exact conservation laws are assigned).

Noncommutativity of balance conservation laws for material system and their controlling role in evolutionary processes accompanied by emerging physical structures practically have not been taken into account in the explicit form anywhere. The mathematical apparatus of evolutionary differential forms enables one to take into account and to describe these points.

The above described mechanism of forming the manifolds elucidates the connection between the space and the material objects. In his work [3] S.Weinberg gives more than one historical concepts of the space. He wrote that the connection of the space with the material objects was pointed out by Leibnitz who believed that there is no philosophical necessity in any concept of space apart from that following from the connections of the material objects. In addition S.Weinberg cites another similar conception, namely, the Mach principle, which claims that in the definition of inertial system the masses of Earth and celestial bodies play a role. The idea of the physical space as a continuum whose properties are governed by the matter was realized by Einstein. The above described mechanism of forming manifolds is one more substantiation of the connection between the space and material objects.

Appendix 1

Some Properties of Skew-Symmetric forms Corresponding to the Conservation Laws

Closed Inexact Exterior Forms: Differential-Geometrical Structures

The exterior differential form of the degree p (p-form) can be written as [3,14]:

$$\theta^p = \sum_{i_1 \ldots i_p} a_{i_1 \ldots i_p} dx^{i_1} \wedge dx^{i_2} \wedge \ldots \wedge dx^{i_p} \quad 0 \leq p \leq n \tag{1}$$

Here $a_{i_1 \ldots i_p}$ is the function of independent variables x^1, ..., x^n, n is the space dimension, and dx^i, $dx^i \wedge dx^j$, $dx^i \wedge dx^j \wedge dx^k$, $\ldots$ is the local basis subject to the condition of skew-symmetry:

$$\begin{aligned} & dx^i \wedge dx^i = 0 \\ & dx^i \wedge dx^j = -dx^j \wedge dx^i \quad i \neq j \end{aligned} \tag{2}$$

The exterior form differential θ^p is expressed by the formula

$$d\theta^p = \sum_{i_1 \ldots i_p} da_{i_1 \ldots i_p} \wedge dx^{i_1} \wedge dx^{i_2} \ldots \wedge dx^{i_p} \tag{3}$$

The importance of exterior form for description of physical fields and manifolds relates to the fact that the closed exterior form possess the invariant properties, correspond to the conservation law and describe the differential-geometric structures from which the pseudometric and metric manifolds can be made up.

The form called as a closed one if its differential equals to zero:

$$d\theta^p = 0 \tag{4}$$

From condition (4) one can see that the closed form is a conservative quantity. This means that such a form can correspond to the conservation law, i.e. a conservative quantity.

If the form be closed only on pseudostructure, the closure condition can be written as

$$d_\pi \theta^p = 0 \tag{5}$$

In this case the pseudostructure π obeys the condition

$$d_\pi {}^*\theta^p = 0 \tag{6}$$

here ${}^*\theta^p$ is the dual form.

From condition (5) one can see that the closed inexact exterior form is a conservative quantity, and condition (6) describes the pseudostructure on which the closed exterior form is defined. It appears that the closed inexact exterior and dual forms describe a differential-geometric structure, i.e. a structure with conservative quantity.

Such structures made up pseudometric and metric manifolds.

It is evident that for description of the process of manifold forming one has to understand how the closed inexact exterior forms, which describe relevant differential-geometric structures, are realized.

Below it will be shown that the closed inexact exterior forms are obtained from the skew-symmetric differential forms, which possess the evolutionary properties. These forms are obtained from the equations which describe any processes. The skew-symmetric forms, which are obtained from the conservation law equations for material system, are just such evolutionary forms.

Distinction of Evolutionary Forms from Exterior Forms

The skew-symmetric differential form defined on nonintegrable (deforming) manifold is an evolutionary form. The evolutionary form can be written in a manner similar for exterior differential form [7,15]. However, in distinction from the exterior form differential, an additional term will appear in the evolutionary form differential. This is due to the fact that the evolutionary form basis changes since such a form is defined on nonintegrable manifold.

The evolutionary form differential takes the form

$$d\theta^p = \sum_{i_1 \ldots 1_p} da_{i_1 \ldots 1_p} \wedge dx^{i_1} \wedge dx^{i_2} \ldots \wedge dx^{i_p} + \sum_{i_1 \ldots 1_p} a_{i_1 \ldots 1_p} d(dx^{i_1} \wedge dx^{i_2} \ldots \wedge dx^{i_p}) \tag{7}$$

where the second term is connected with the basis differential being is nonzero: $d(dx^{i_1} \wedge dx^{i_2} \wedge \ldots \wedge dx^{i_p}) \neq 0$. (For the exterior form defined on integrable manifold one has $d(dx^{i_1} \wedge dx^{i_2} \wedge \ldots \wedge dx^{i_p}) = 0$). The peculiarity of skew-symmetric forms defined on nonintegrable manifold can be demonstrated by the example of a skew-symmetric form of first-degree. Let us consider the first-degree form $\omega = a_\alpha dx^\alpha$. The differential of this form can be written as $d\omega = K_{\alpha\beta} dx^\alpha dx^\beta$, where $K_{\alpha\beta} = a_{\beta;\alpha} - a_{\alpha;\beta}$ are components of the commutator of the form ω, and $a_{\beta;\alpha}$, $a_{\alpha;\beta}$ are covariant derivatives. If we express the covariant derivatives in terms of connectedness (if it is possible), they can be written as $a_{\beta;\alpha} = \partial a_\beta/\partial x^\alpha + \Gamma^\sigma_{\beta\alpha}a_\sigma$, where the first term results from differentiating the form coefficients, and the second term results from differentiating the basis. If we substitute the expressions for covariant derivatives into the formula for commutator components, we obtain the following expression for commutator components of the form ω:

$$K_{\alpha\beta} = \left(\frac{\partial a_\beta}{\partial x^\alpha} - \frac{\partial a_\alpha}{\partial x^\beta}\right) + (\Gamma^\sigma_{\beta\alpha} - \Gamma^\sigma_{\alpha\beta})a_\sigma \tag{8}$$

Here the expressions $(\Gamma^\sigma_{\beta\alpha} - \Gamma^\sigma_{\alpha\beta})$ entered into the second term are just components of the commutator of the first-degree metric form that specifies the manifold deformation and hence is nonzero. (In the commutator of exterior form, which is defined on integrable manifold, the second term absents: the connectednesses are symmetric, that is, the expression $(\Gamma^\sigma_{\beta\alpha} - \Gamma^\sigma_{\alpha\beta})$ vanishes). [It is well-known that the metric form commutators of the first-, second- and third degrees specifies, respectively, torsion, rotation and curvature.]

Since the commutator, and hence the differential, of skew-symmetric form defined on nonintegrable manifold are nonzero, this means that such a form cannot be closed one. Such form is obtained from the equations for material system and possesses evolutionary properties.

The evolutionary form possesses an unique property, namely, this form can generate a closed inexact exterior form, which corresponds to the conservation law and describes the differential-geometric structure. The physical structures are just such structures.

Appendix 2

About Interactions

Below we present the table where physical fields and interactions are demonstrated in their dependence on the parameters p, k, n of the evolutionary and closed exterior forms. (Here p is the degree of evolutionary form in nonidentical relation, which is connected with the number of interacting balance conservation laws, k is the degree of closed form generated by nonidentical relation and n is the dimension of original inertial space.) [It should be emphasized the following. Here the concept of "interaction" is used in a twofold meaning: the interaction of balance conservation laws that relates to material systems, and the physical concept of "interaction" that relates to physical fields and reflects the interactions of physical structures, namely, it is connected with exact conservation laws].

The table corresponds to elementary particles.

In the Table the names of the particles created are given. Numbers placed near particle names correspond to the space dimension. Under the names of particles the sources of interactions are presented. In the next to the last row we present particles with mass (the elements of material system) formed by interactions (the exact forms of zero degree obtained by sequential integrating the evolutionary relations with evolutionary forms of degree p corresponding to these particles). In the bottom row the dimension of the *metric* structure created is presented.

Table 1.

interaction	$k\backslash p, n$	**0**	**1**	**2**	**3**
gravitation	**3**				**graviton** ⇑ electron proton neutron photon
electro-magnetic	**2**			**photon2** ⇑ electron proton neutrino	**photon3**
weak	**1**		**neutrino1** ⇑ electron quanta	**neutrino2**	**neutrino3**
strong	**0**	**quanta0** ⇑ quarks?	**quanta1**	**quanta2**	**quanta3**
particles material nucleons?	exact forms	**electron**	**proton**	**neutron**	**deuteron?**
N		1 time	2 time+ 1 coord.	3 time+ 2 coord.	4 time+ 3 coord.

From the Table one can see the correspondence between the degree k of closed the forms realized and the type of interactions. Thus, $k = 0$ corresponds to strong interaction, $k = 1$ corresponds to weak interaction, $k = 2$ corresponds to electromagnetic interaction, and $k = 3$ corresponds to gravitational interaction.

The degree k of closed forms realized and the number p connected with the number of interacting balance conservation laws determine the type of interactions and the type of particles created. The properties of particles are governed by the space dimension. The last property is connected with the fact that closed forms of the same degrees k, but obtained from the evolutionary relations acting in the spaces of different dimensions n, are distinctive because they are defined on pseudostructures of different dimensions (the dimension of pseudostructure $(n+1-k)$ depends on the dimension of initial space n). For this reason the realized physical structures with closed forms of degrees k are distinctive in their properties.

References

[1] Tolman R.C. *Relativity, Thermodynamics, and Cosmology*. Clarendon Press, Oxford, UK, 1969.

[2] Clark J.F., Machesney M., *The Dynamics of Real Gases*. Butterworths, London, 1964.

[3] Weinberg S., Gravitation and Cosmology. *Principles and applications of the general theory of relativity*. Wiley & Sons, Inc., N-Y, 1972.

[4] Petrova L.I. Physical meaning and a duality of concepts of wave functional, entropy, the Pointing vector, the Einstein tensor. *Journal of Mathematics Research*, Vol. 4, No. 3, 2012.

[5] Smirnov V. I., A course of higher mathematics. -Moscow, *Tech. Theor. Lit*. 1957, V. 4 (in Russian).

[6] Petrova L.I. Discreteness of the solutions to equations of mathematical physics. *Theoretical Mathematics and Applications*, Vol.3, No.3, 2013, pp. 31-47.

[7] Petrova L.I., *Exterior and evolutionary differential forms in mathematical physics: Theory and Applications*. -Lulu.com, 2008, 157pp.

[8] Petrova L.I., The noncommutativity of the conservation laws: Mechanism of origination of vorticity and turbulence. *International Journal of Theoretical and Mathematical Physics*, Vol.2, No.4, 2012.

[9] Petrova L.I., Integrability and the properties of solutions to Euler and Navier-Stokes equations. *Journal of Mathematics Research*, Vol. 4, No. 3, 2012, pp. 19-28.

[10] Petrova L.I. *Skew-symmetric differential forms in mathematics, mathematical physics and field theory*. -Moscow, URSS, 2013, p. 234.

[11] Tonnelat M.-A., *Les principles de la theorie electromagnetique et la relativite*. Masson, Paris, 1959.

[12] Petrova L.I., The quantum character of physical fields. Foundations of field theories, *Electronic Journal of Theoretical Physics*, Vol. 3, No. 10, 2006, pp. 89-107.

[13] Einstein A. *The Meaning of Relativity*. Princeton, 1953.

[14] Bott R., Tu L. W., *Differential Forms in Algebraic Topology*. Springer, NY, 1982.

[15] Petrova L.I., *Role of skew-symmetric differential forms in mathematics*, 2010 http://arxiv.org/abs/math.CM/1007.4757.

INDEX

A

B

C

D

E

F

G

N

O

P

Q

R

S

T

U

V

W

Y